AF669189

Astronomical Photometry and Spectroscopy

in the Optical and Near-Infrared Regions

Astronomical Photometry and Spectroscopy

in the Optical and Near-Infrared Regions

B.G. Anandarao

Formerly with Physical Research Laboratory, Ahmedabad

V. Venkata Raman

Space Physics Laboratory, VSSC, Thiruvananthapuram

BSP BS Publications™

An Imprint of **BSP Books Pvt. Ltd.**

4-4-309/316, Giriraj Lane,

Sultan Bazar, Hyderabad - 500 001

Astronomical Photometry and Spectroscopy in the Optical and Near-Infrared Regions

by **B.G. Anandarao and V. Venkata Raman**

© 2025, *by Publisher,* All rights reserved.

No part of this book or parts thereof may be reproduced, stored in a retrieval system or transmitted in any language or by any means, electronic, mechanical, photocopying, recording or otherwise without the prior written permission of the authors.

Disclaimer: The authors and publishers have taken due care to provide authentic, reliable and up-to-date information related to the subject. However, neither the authors nor the publisher shall be responsible for any liability for any damage caused because of the use of this book. The respective users must check the accuracy from other sources too.

Published by:

BSP **BS Publications**™

An Imprint of **BSP Books Pvt. Ltd.**

4-4-309/316, Giriraj Lane,
Sultan Bazar, Hyderabad - 500 001
Phone: 040 - 23445688
e-mail: info@bspbooks.net
www.bspbooks.net

ISBN: 978-93-49562-99-8 (Paperback)

Foreword

For thousands of years, light has been the main messenger bringing to humans some information on the vast universe. For a long time, naked eyes were the only but remarkable tool to receive this light, transmit the information to the brain and hence allow astronomy to emerge and develop in most of the cultures. Myths were elaborated, questions were raised, stars were named, strange phenomena such eclipses or comets were reported, rulers believed that the sky was guiding them, motions were measured and analyzed with the emergence of powerful mathematics. With Galileo in 1610 and until today, the landscape totally changed. Telescopes of growing size and photography revealed ignored objects, sharp images provided their exquisite details, spectroscopy analyzed the light and led to discovering the physical nature of celestial bodies. Physics itself, so essential for this understanding of the universe, underwent the revolution of relativity (1905-1916) and quantum physics (1900-1954), opening to the emergence of observational cosmology which considers the past and present Universe as a whole. After 1950, the full spectrum of light became measurable, with instruments located either on the Earth's surface or in space: these two giant steps changed the face of astronomy, and keeps changing it today.

In 2025, a student or researcher who wants to understand this incredibly rich landscape may easily get lost into it, when confronted to the immense wealth of publications and data bases which convey and share the knowledge elaborated by thousands of astronomers, astrophysicists, astrochemists and astrobiologists all over the world. By writing this book entitled *Astronomical Photometry and Spectroscopy in the Optical and Near-Infrared Regions*, B.G. Anandarao & V. Venkata Raman offer a thorough introduction to this immense domain, expanding from their own research experience over several decades at the 1.2-meter optical telescope at Mt. Abu in Rajasthan, with now the addition of a 2.5-m instrument. They first and briefly introduce the reader to this contemporary landscape, choosing to focus on light – already a huge task –, hence not considering the various cosmic particles and the gravitational waves which, since 2016 when they were first observed, represent an entirely new field for astronomy in the XXIst century and certainly beyond.

Observations at optical wavelengths of light – visible, infrared and since 1933 radio ones – have played, and still do, an essential role to understand the Sun, the stars, the planets, the galaxies. Subtle techniques and instruments have emerged to detect the most minute amount of light, analyze and decompose its spectrum, fight the distortions of images due to a thermally inhomogeneous Earth's atmosphere and achieve unprecedented sharpness of details. From the conception of an instrument to the end product of trustable data on a particular celestial source, the authors are familiar with all these, and provide the reader with a well-organized, up-to-date and precious analysis.

Beyond their general introduction and given their own, and rich experience as researchers, the authors choose to avoid discussing those parts of light which begin with high-energy gamma- and X-rays and, at the other end of the spectrum, extend to radio waves, from millimeter to kilometer wavelengths. For these, contributing so much to our today's knowledge of celestial objects, the techniques and the process of data acquisition are quite different and would require specific chapters, beyond the goal of the present book.

In 2023 the National Air and Space Administration (NASA) in United States launched the James Webb Space Telescope, devoted to visible and infrared light covered by this book, placed at the Lagrange L2 point 1.5 million km away from our Earth. The launch, with the fully robotic installation of a 6.4-m primary mirror in space, and the outstanding operation of this observatory since, is probably the most remarkable technological success, collectively achieved by humans since the beginning of history!

To observe the same wavelengths domain of light, the European astronomers, organized within the European Southern Observatory, will open in 2028 the largest optical telescope ever built, having a 29.6-m primary mirror and installed in the Atacama Desert in Chile. Another technological feat indeed!

These two achievements show clearly that tracking the light coming from the depths of the Universe still has a long way ahead and prepares discoveries, some of which and probably the most fascinating ones are difficult to predict. What a challenging future for young students, who have experienced the beauty of a night sky full of stars, or enjoyed learning that space is curved all around us, with triangles of light rays where the sum of angles departs from 180°! As having myself had in France the privilege to introduce one of the authors, Professor Anandarao, to these beauties nearly half a century ago, it is with deep thanks that, today, I am glad to provide to him and his co-author this Foreword for an excellent book aimed at the future generations of astronomers, in India and beyond.

Pierre Léna

Emeritus Professor, Observatoire de Paris & Université Paris-Cité
Member of the Académie des sciences, France

Medon, France
January 11, 2025

Preface

If you can't explain it simply, you don't understand it well enough

- Albert Einstein

Since time immemorial, using un-aided eye humans have been observing the skies to record the sidereal motion of stars and constellations and other prominently visible Celestial objects as well as their morphological and characteristic features. Thus, the optical observations of celestial objects are some of the most ancient scientific studies ever undertaken. Thanks to their diligent and monumental efforts we now know the sidereal motion of the Sun and the constellations in the Cosmos as well as some of the extended sources namely the Galactic Gaseous Nebulae and nearby external Galaxies (as opposed to a point source such as a star). With the advent of photography and later, of electronic detectors, observational astronomy has taken a giant leap forward. With the development, in the later part of the last century, of digital panoramic detectors of great sensitivity in optical/visible (used here synonymously) as well as in infrared wavelength regions, the advancement in our understanding of the celestial phenomena has been unprecedented.

The aim and objective of writing this monograph are to bring out in detail the two physical techniques of observational astronomy namely the photometry and spectroscopy. Towards this endeavour we present here a concise note on the variety of astronomical sources in the cosmos that employ these techniques to understand the underlying physical processes. This will be followed by a detailed account of the two major techniques used in astronomy. We provide several examples and describe some important analysis techniques for reduction of noise and enhancement of quality of the observed data. We deal with mainly the two important electromagnetic radiation wavelength regimes pertaining to the optical and the (near-) infrared radiation. The observational examples cited in various chapters (mainly in Chapters 4, 5 and 6) are taken mostly from published work of our own, and of our colleagues at Physical Research Laboratory, Ahmedabad. These are meant for illustrative purposes.

The book is divided into seven chapters: Chapter 1 gives a general introduction to the astronomical objects and general terminology; Chapter 2 gives an outline of radiation processes encountered in astronomical objects and atomic and molecular spectroscopy; Chapter 3 describes the telescopes and detectors pertaining to the two wavelength regimes, viz. optical and infrared; Chapter 4 gives a description of photometry and the relevant instrumentation; in Chapter 5 the principles of spectroscopy and the relevant instrumentation are described; Chapter 6 gives data processing and enhancing techniques; lastly in Chapter 7 a future perspective is given on the foregoing topics.

The book is intended to address the post-graduate students in the universities where Astronomy and Astrophysics is an elective subject and the young research workers trying to make their mark as observational astronomers. We hope the book serves this purpose well.

This monograph is the culmination of a journey that has been supported by many extraordinary individuals and institutions, to whom we express profound sense of gratitude. Particularly we wish to thank Prof. Pierre Lèna (Observatoire de Meudon, France), for many discussions and for agreeing to write the Foreword. We appreciate the collaboration of our PRL colleagues Jyotindra bhai Desai, Dipankar Banerjee, Seema Ahmed, Abhijit Chakraborty, Nanda Kumar, Muthumariappan and Lokesh Dewangan. We acknowledge the help of the technical staff of our group, the mechanics at PRL workshop and all the staff at the Mt Abu Observatory. We gratefully acknowledge the PRL directors officiated during the period 1980-2010, for their wholehearted support to the works at Mt Abu Observatory.

BGA affectionately thanks his family members Janaki, Prashant, Soumya, Abhinav and Prashanthi, for always standing by him. He expresses his gratitude to Dr. B.V. Krishna Murty for his unstinted support and for being such a great inspiration.

VV would like to express his heartfelt thanks to his beloved spouse Punitha and his wonderful kids Karunyan and Aathreyan for their unwavering support and joy throughout this journey thus far. He expresses his deep gratitude to his brother-in-law and friends, for being there to cheer him during every small victory, and for making him smile during challenging moments. He would like to acknowledge all his colleagues at the Physical Research Laboratory and Space Physics Laboratory for their support.

We are also grateful to the Department of Space & Indian Space Research Organisation for their generous support that made this work possible.

*- **Authors***

Contents

CHAPTER 1

Introduction

What emptiness do you gaze upon! – Rabindranath Tagore

The gamut of astronomical sources consists of the relatively near-earth solar system objects to the far-flung galaxies and the active galactic nuclei such as quasi-stellar radio objects or quasars. Planets in the solar system, stars in a galaxy, the interstellar medium, galaxies and clusters of galaxies are the building blocks that constitute the physical Universe. These astronomical sources represent various physical sizes, distances (from the earth) and other physical parameters such as temperature, density, fields, and timescales characterising the physical processes that account for their behaviour. In a way one may categorize the objects into two groups namely, the point sources such as stars that generally cannot be resolved by a telescope, and extended sources such as gaseous and diffuse nebulae in our Galaxy and nearby external galaxies. This categorization necessarily depends on the distance as well. Let us loosely categorize that any object of angular size less than a second of arc may be considered as a point source and any object that has an angular size more than a second of arc as an extended source. With that being said, we hasten to add here that using modern technological advances, sub-arcsecond observations have become possible with telescopes of large diameters. This categorization reflects basically the observer's inability to resolve a celestial object beyond a certain degree of spatial dimension. We shall elaborate this aspect in Chapter 2. Celestial objects in general reveal themselves by the emission of electromagnetic radiation. While it is conceivable that an object emits radiation, in principle, in the entire electromagnetic spectrum (like for instance a black body), the possibility of detection of the radiation depends on the wavelength range in which the emission occurs predominantly, as decided by various physical conditions of the source. This aspect will be elaborated in Chapter 3. As shown in Table 1.1, the wavelength (λ) or frequency (ν) of the radiation and hence its associated energy varies from milli-electron volts in the radio regime to higher than mega-electron volts in the most energetic radiation towards the gamma-ray end of the spectrum. The table also lists out observable features that include continuum and/or discrete radiation processes (see Chapter 3 for details). Clearly, therefore, the observational techniques involving telescopes and detectors would be quite different for different wavelength regimes. Accordingly, based on the wavelength regime and hence on the observational and detection technique, astronomical observations may be divided into various sub-branches, namely, Optical (visible), Infrared (IR), Ultra-Violet (UV), Radio, X-ray, and gamma-ray astronomies. Further, the presence of the earth's atmosphere intervening the line of sight between the observer and the celestial object, the celestial radiation gets impeded due to absorption and scattering by the constituents of the atmosphere. Fig 1.1 depicts the atmospheric effect on the radiation for the entire electromagnetic spectrum as it penetrates through the atmosphere. While the visible and radio regimes get through unhindered, other wavelength ranges such as ultraviolet, infrared, and sub-millimetre radiation undergo severe extinction (a term that accounts for both scattering and absorption). The main absorbers are water vapour, carbon dioxide and ozone in the lower atmosphere for infrared region, nitrogen and oxygen gases in the upper atmosphere and ozone in the stratosphere for energetic radiation. Radio waves of long wavelengths (greater than about 30 m or low frequencies less than about 10 MHz) are reflected out by the plasma of the Earth's topside (above the height of maximum plasma density) of Ionosphere. At frequencies below 1 MHz the interplanetary and interstellar plasma makes it impossible for observations (Longair 2011).

Table 1.1: Electromagnetic Radiation and observable features.

EM Radiation	E(eV)	λ	ν(Hz)	Observables Features
Gamma Rays	$1\text{-}1000\text{x}10^{6}$	$< 10^{-11}$ m	$2\text{x}10^{20}\text{-}2\text{x}10^{23}$	Nuclear
Hard X-rays	$2\text{-}10\text{x}10^{3}$	0.025–0.25 nm	$4\text{x}10^{17}\text{-}2\text{x}10^{20}$	Thermal Bremsstrahlung
Soft X-rays	$0.1\text{-}2\text{x}10^{3}$	0.25-10 nm	$3\text{x}10^{16}\text{-}4\text{x}10^{17}$	Atomic Inner Shell
EUV	13.6-100	10 – 91.2 nm	$3\text{x}10^{15}\text{-}3\text{x}10^{16}$	Atomic / Continuum
UV	4-13.6	91.2 - 300 nm	$1\text{x}10^{15}\text{-}3\text{x}10^{15}$	Atomic / Molecular
Optical	1.25-2.5	400 - 800 nm	$4\text{x}10^{14}\text{-}7\text{x}10^{14}$	Atomic / Molecular
Near-IR	0.17-1.25	0.8 - 6 μm	$4\text{x}10^{14}\text{-}7\text{x}10^{14}$	Vibrational
Mid-IR	0.03 -0.17	6 - 35 μm	$1\text{x}10^{13}\text{-}5\text{x}10^{13}$	Rotational / Dust
Far-IR	$3\text{x}10^{-3}$ -0.03	35 - 300 μm	$1\text{x}10^{12}\text{-}1\text{x}10^{13}$	Rotational / Dust
Sub-mm	$1\text{x}10^{-3}\text{-}3\text{x}10^{-3}$	300 - 1000 μm	$3\text{x}10^{11}\text{-}1\text{x}10^{12}$	Rotational
Radio/mm	$1\text{x}10^{-7}\text{-}1\text{x}10^{-3}$	0.1 cm – 10 m	$3\text{x}10^{7}\text{-}3\text{x}10^{11}$	Non-thermal

To observe the radiation in the "block out" frequency regimes it is necessary to circumvent the atmospheric effect by flying instruments onboard balloons, aircrafts and spacecrafts as the situation demands (see Fig 1.1). Based on this we can also divide the astronomical observations into two broad classes - ground-based (for optical, near-infrared and radio wavelengths) and space-based astronomies (for UV, far-infrared, X-rays). Very high energy radiation detection is made from ground-based secondary radiation detection.

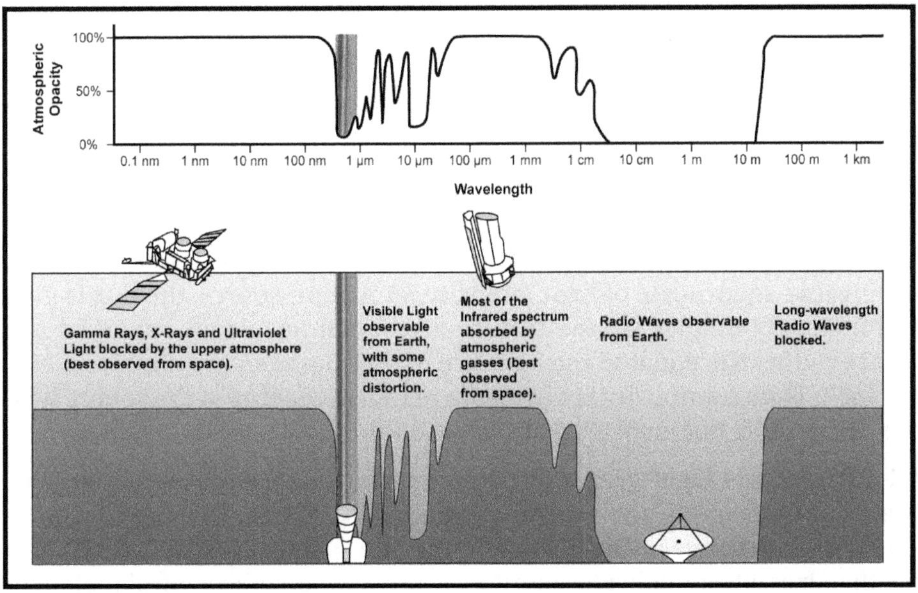

Figure 1.1: Atmospheric opacity for electromagnetic spectrum from high energy gamma rays (at left) to low energy radio waves (to the right). The top panel shows the atmospheric opacity as a function of wavelength. The bottom panel shows the observational means at different wavelength regimes. The narrow vertical dark band shows the visible wavelength range that gets through to the ground level along with some bands in near-infrared. Radio waves get through unhindered (from NASA website).

1.1 Definition of Important Terminology used in Astronomy

Astronomers use certain scientific terminology specific to celestial observations to quantify the measured radiation and other physical parameters. We define such terminology in the following sub-sections.

1.1.1 Intensity and Specific Intensity: Intensity is source specific – measured at the source of radiation. Intensity is the power emitted by unit area of a source into a unit solid angle. It is designated as I and is given in

Watts/steradian/m^2. Specific intensity I_λ is intensity measured in each narrow interval of spectral wavelength and is given in Watts/steradian/m^2/μm. In the units for specific intensity, /μm is usually used for infrared radiation; and /nm or /Å may be used of visible light.

1.1.2 Luminosity: Luminosity, like Intensity, is also source specific. It is the total flow of radiation emanating from a source of effective temperature T and radius R. It is measured in Watts (Joules/s). It is given by,

$$L = 4\pi R^2 \sigma T^4 \tag{1.1}$$

where σ is Stefan-Boltzmann constant. The luminosity in each narrow spectral bandwidth is called monochromatic luminosity (L_λ) given in W/μm.

1.1.3 Flux Density: While intensity and luminosity are defined specifically to characterise a source, flux density is what the observer measures on a detector. It is the power received from a source in unit frequency or wavelength measured in W/m^2/Hz or W/m^2/μm respectively. Following the inverse-square-law of light intensity for a point source, the flux is given by luminosity divided by the surface area of an imaginary spherical shell whose radius d is equal to the distance of the source to the observer: $F = L / (4\pi d^2)$. Thus, luminosity is a measure of radiant energy at the source while flux is the same but measured at the observer.

1.1.4 Parsec and Light-year: Astronomers use distance scales that are very specific. For example, the astronomical unit (AU) is the average distance ~ 1.49 x 10^{11} m between the Earth and the Sun. This unit is specifically used for the sizes of the debris around a star, or the distances of planets from their parent star. The most familiar term, known as the light-year, is the distance travelled by a photon in one year and is given by 9.46 x 10^{15} m. The unit of distance that is more popular among astronomers is parsec (short for "parallax second"), for distances larger than the size of the solar system. Parallax (see Fig. 1.2 for definition) is a measure of distance based on the apparent shift of the position of a celestial object as observed from the Earth at two diametrically opposite positions in Earth's orbit around the Sun. Parsec is defined as the distance at which the radius of the Earth's orbit around the Sun (= 1 AU) subtends an angle of parallax of one arcsecond. It is equal to 3.08 x 10^{16} m. A parsec is thus ~ 3.25 times the light year. In general, if the angle of parallax for a star is p arcsec, then its distance is $d = p''$ parsec or pc (see Fig. 1.2).

1.1.5 The Magnitude System: Traditionally astronomers (since Hipparchus in the 2nd century BCE) use a brightness unit, known as the magnitude system, to quantify the brightness of a star. Magnitude denoted as 'm' is a relative logarithmic scale giving the brightness of a star with respect to that of a standard star whose brightness or magnitude, m_0 is already known.

$$m - m_0 = 2.512\ log_{10}\left(\frac{F_0}{F}\right)$$

or (1.2)

$$F = F_0 10^{-0.4(m-m_0)}$$

with F and F_0 being the fluxes (Watts/m^2/Hz). According to this definition, a star of higher magnitude is fainter than a star of lower magnitude. A magnitude difference of 5 between two stars indicates 100 times in their brightness ratio ($2.512^5 = 100$). It is normal practice to consider a main sequence (dwarf) star of spectral type A0 (having effective temperature of about 10000 K; in comparison the Sun is also a dwarf star but of G3 spectral type having an effective temperature of 5875 K) for a standard star wherein its magnitude in all the spectral bands as considered as 0. The star α Vega of spectral type A0V fits the bill for flux-calibrating the optical and infrared observations (see Appendix A3 for Vega standard fluxes).

Using the above equation for flux and Vega as standard star having magnitude zero, we can also compute the number of photons emanating from a star of a given magnitude within a wavelength band, impinging upon a telescope,

$$n_p = \frac{3.64 \times 10^{-9}}{10^{0.4167m}}\ \Delta\lambda\ \frac{\pi D^2}{4}\ \frac{1}{h\nu}\ \frac{\pi\theta^2}{4} \tag{1.3}$$

with D the aperture or diameter of the telescope primary mirror (in cm), $\Delta\lambda$ bandwidth of incoming radiation of central wavelength λ (in Å units), θ the angular size of the star image formed on the detector and ($h\nu$) the photon energy at 0.55 µm and the numerical factor denotes the total energy flux (in ergs/s/cm^2/Å) from a standard star of 0th magnitude at 0.55 µm.

1.1.6 Apparent and Absolute Magnitudes and the Distance Modulus: The magnitude that we determine using the equation (1.2) gives apparent magnitude of a star relative to a standard star that has magnitude 0 at all wavelength bands. The apparent magnitude is related to the flux (luminosity divided by the square of the distance to the object) – the larger the distance the larger would be the magnitude for the same luminosity. It depicts the inverse square law for luminosity variation with distance to the source. So, the farther a star of a particular luminosity is situated the fainter it appears and hence the bigger would be its apparent magnitude (the bigger the magnitude the dimmer the object appears).

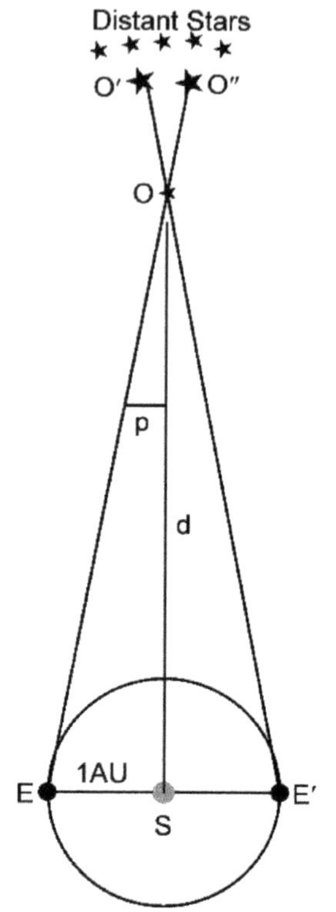

Figure 1.2: Definition of parallax. S and E denote the sun and the earth and the circle the orbit of the earth. A star O, at a distance *d*, will have an apparent shift in its position (denoted by O' and O" with respect to relatively distant stars when observed from two diametrically opposite positions E and E' in Earth's orbit around the Sun. The half-angle *p* (being very small) is called the parallax. The distance at which a parallax of 1 arcsec occurs is defined as 1 parsec.

For instance, the apparent magnitude of the Sun is -26.74 due to its proximity to the Earth, while the actual or Absolute Magnitude is +4.83.

The absolute magnitude scale is normalised with reference to a distance of 10 parsecs. If we know both these the distance may be calculated using the "distance modulus" equation given by

$$m_\lambda - M_\lambda = 5\,log_{10}\,D - 5 = -5\,log_{10}\,p - 5 \qquad (1.4)$$

Here D is the distance in parsec and p the annual parallax in arcsec.

We can define also the bolometric magnitude representing the total integrated flux in all wavelengths given by

$$M_{bol} = 4.75 - 2.5\log_{10}(L/L_\odot) \qquad (1.5)$$

where $L_{\odot}$ is the luminosity of the Sun and the value 4.75 is the bolometric magnitude of the Sun. The bolometric correction is defined as

$$BC = m_{bol} - m_{\lambda} = M_{bol} - M_{\lambda} \tag{1.6}$$

Objects appear small or point-like (of the order of arcsec or smaller) because they are single sources and far-off; they can appear big if they are bigger intrinsically (extended sources more than several arcsec to several degrees). At sufficiently large distances even the extended sources appear point-like. For **extended sources** (sizes more than a few arcsec in size) the magnitude scale usually refers to per square arcsec (the brightness of the object divided by its solid angle, (see Chapter 4).

1.1.7 Doppler Shift: Analogous to sound waves, light waves also undergo Doppler shift. For a celestial object moving in space with reference to an earth-based observer, the central wavelength/frequency (λ/ν) of a spectral line from the object shifts by an amount proportional to the velocity of the object (v). This is given by,

$$\frac{\Delta\lambda}{\lambda} = \frac{v}{c} = \frac{\Delta\nu}{\nu}$$

Here c is the velocity of light. If the object is receding from the observer, the frequency of the radiation decreases, or the wavelength increases, and such a shift is called a "red shift". Blue shift occurs when the object approaches observer.

1.2 A Glossary of Astronomical Objects

We present in this section a brief outline of characteristic features of various astronomical objects. This does not claim an exhaustive description; hence the reader may refer to more general books on astronomy for a comprehensive treatment. Over the past decades, there have been several excellent books written by eminent authors for general and specific topical reference on observational astronomy and theoretical astrophysics (e.g., Spitzer 1978; Lèna 1979; Dyson & Williams 1980; Shu 1982; Prialnik 2000; Whittet 2003; Smith 2004; Stahler & Palla 2004; Schulz 2005; Tielens 2005; Carroll & Ostlie 2007; Bradt 2008; Longair 2011; Sutton 2012; Lang 2013; Barbieri & Bertini 2020; Heidt 2022).

Figures 1.3 and 1.4 show images of the celestial sky mapped in the visible and infrared wavelength regions (see Longair 2011 for details). As may be made out from these images there appears a variety of objects in the sky:

stars, interstellar medium containing nebulae and molecular clouds and distant external galaxies. In the figures we can see bright regions of stars and emission nebulae as well as dark regions due to extinction by gas and dust. Such dark regions also reveal cold regions emitting in much longer wavelengths than visible and near-infrared wavelengths.

Astronomy and Astrophysics is the branch of Physical Science that strives to understand the origin, and evolution of various objects that constitute the celestial sky. Towards this goal, we need to understand these celestial objects in terms of the underlying physical processes. Therefore, Photometric and Spectroscopic investigations of various celestial objects become one of the most important aspects of observational Astrophysics. We give below short notes on celestial objects from solar system planets to external galaxies via stars and interstellar medium.

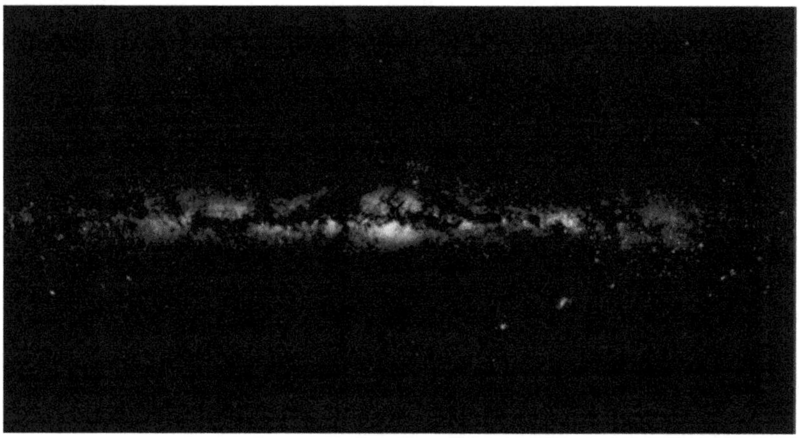

Figure 1.3: Celestial Sky seen in visible light (Longair 2011).

Figure 1.4: Celestial sky seen in near-infrared light (Longair 2011).

1.2.1 The Sun: The star of our solar system, the Sun, is a G2V type star. Due to its proximity to an Earth-bound observer, it is the only star that facilitates studies at an unprecedented spatial resolution (in terms of surface area covered per picture element) so that the regions of its photosphere regions were discerned into sub-regions such as chromosphere and corona. From this close-range, astronomers were able to identify unique phenomena such as spots, coronal emissions, coronal holes, coronal mass ejections, prominences, flares, plages, granules and super granules and spicules or filaments and so on. Therefore, the Solar studies have a bearing also on stars that are Sun-like. With the help of a great volume of multi-wavelength observations from both ground-based and space-based solar-telescopes, the physical phenomena responsible for the behaviour of the Sun have been primarily understood, although several issues remain unsolved. It is in its middle age in its main sequence phase. The star's age is estimated to be 4.57 x 10^9 years. The Sun has a radiative core and a convective envelope. Its magnetic field plays a major role in its evolution. Due to its magnetic field its tenuous atmosphere called corona is heated to a few million degrees K. The solar plasma also leads to a atomic and ionic excitation by collisions with electrons. The magnetic field also leads to generation of synchrotron radiation. For a comprehensive review of the Sun and Solar Physics we may refer to Aschwanden (2019).

Figure 1.5 illustrates the structure of the Sun in a schematic way. The central core where nuclear reactions occur is shown in dark with regions of radiative equilibrium and convective equilibrium are shown surrounding it respectively in grey. In the radiative region the energy transport to the upper layers is done predominantly by radiation. In the outermost convective region, the transport of energy is done by convection. In this region the solar magnetic field is also generated by the process called dynamo. The dynamo process itself is triggered by differential rotation in these layers. These regions are shown nearly in the relative proportion. The outer regions namely the photosphere, the Chromosphere and the Corona are shown in relative proportions but in comparison with the interior regions. Above the dark line (circle) where the optically thick region exists is the Sun's atmosphere. The effective temperature of the Sun refers to the equilibrium temperature of the photosphere, a very thin region of a few hundred kilometres. Above the photosphere occur the Chromosphere and a thin (of nearly a hundred kilometres thick) region called Transition region. Above the transition region is the Solar Corona (the outer boundary shown in the figure is only symbolic and not to scale in comparison with the lower layers of the Sun's atmosphere. The coronal plasma is called "the million-degree corona" energised purportedly by the dissipation of solar magnetic fields.

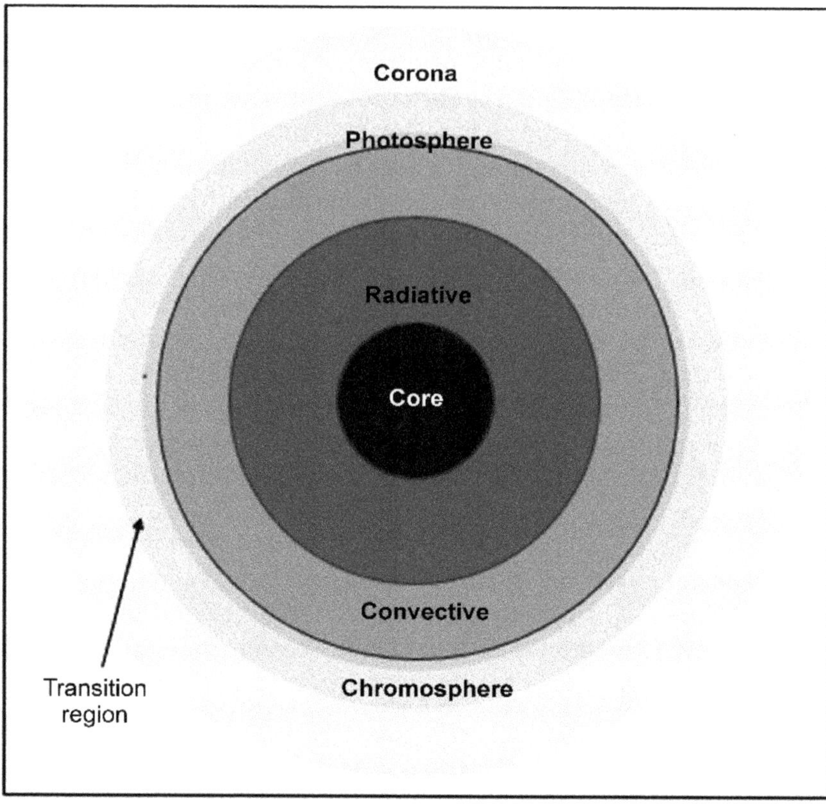

Figure 1.5: Solar structure shown schematically: the sizes from the core to photosphere are in relative proportion while those from photosphere outwards are separately in relative proportion.

1.2.2 Planets: Planets around stars, such as the Sun, are believed to be formed in the debris disk around protostars (or the proto-Sun in the case of the Solar system) by condensation of planetesimals or gravitational collapse before the disk is blown off by stellar winds and radiation pressure. A planetary body having a mass less than about 0.001 $M_\odot$ has only its potential energy converted to heat energy for its sustenance. It shines due to its albedo (reflected light) from the parent star. Sizes of planets vary from a fraction of Earth radius to a few times the size of the largest Solar system planet Jupiter. The distances from their parent star (the Sun for the Solar System) vary from a fraction of to several AU (astronomical unit). Table 1.2 gives some important parameters for Solar system planets. Without considering the effect of possible atmosphere of a planet, its effective dayside temperature due to heating by the parent star may be estimated from the radiation equilibrium equation (i.e., radiation absorbed equals radiation emitted) as follows:

$$T_P = T_S\,(R/2D)^{1/2}\,K \tag{1.7}$$

where T_S, and T_P are effective temperatures of the star of radius R, and its planet situated at a distance D. The above relation may be derived from a simple consideration that the absorbed radiation by the planet must be equal its emitted radiation. For the Solar System case this may be simply written as,

$$T_P = (1-A)^{1/4}\,\frac{278.3}{D(AU)^{0.5}}\,K \tag{1.8}$$

The solar effective temperature was taken as $T_\odot \sim 5770$ K and 1 AU = 1.4959 x 10^{11} m. The effect of the planetary (bond) albedo (A) is given by the factor $(1 - A)^{1/4}$. This will give a temperature of 255 K for Earth while the actual average surface temperature is 283-293 K. The difference may be attributed to the effects of the atmosphere of the earth.

Table 1.2 lists the effective temperatures of solar system planets (without Albedo factor) along with the albedo factors. For comparison, measured mean temperatures (T_M) are also listed. Also given in the table are, distances, masses (in terms of Earth mass, $M_\oplus$), average densities, orbital eccentricities and, inclinations of the eight planets. These data are from: http://nssdc.gsfc.nasa.gov/planetary.

Table 1.2: Solar System Planets – Physical Parameters.

Planet	Distance (AU)	Mass ($M_\oplus$)	Density ρ (gm/cc)	e	i deg	Albedo Factor $(1\text{-}A)^{1/4}$	T_P (K)	T_M (K)
Mercury	0.39	0.055	5.42	0.205	7.0	0.968	447	440
Venus	0.72	0.815	5.24	0.007	3.38	0.562	327	737
Earth	1.0	1	5.51	0.017	0	0.911	278	288
Mars	1.52	0.107	3.93	0.093	1.85	0.931	225	208
Jupiter	5.20	316.58	1.33	0.048	1.3	0.901	122	163
Saturn	9.57	95.14	0.69	0.054	2.48	0.901	90	133
Uranus	19.29	14.54	1.27	0.047	0.77	0.915	64	78
Neptune	30.24	17.08	1.64	0.009	1.77	0.918	51	73

1.2.2.1 Exoplanets: It is very reasonable to assume that stars other than the Sun may also host planets. The genesis for such observations came when the Infrared Astronomical Satellite (IRAS) with its discovery of the presence of large dust grains or planetesimals around main-sequence stars such as Vega, indicating that the debris (discs) around the stars formed during their formation stage may indeed become source of material to form planets (see e.g., Zuckerman 2001). Attempts were made to detect planets outside the solar system during the last quarter of the 20th century. First successful detection of an extrasolar planet or simply "exoplanet" was made by Mayor & Queloz (1995) in the nearby star 51 Pegasi, using high-resolution spectroscopy. From Newtonian mechanics, we understand that a pair of a host star and its planet, forming a binary system, revolve around the centre of mass of the system. While the planet in the system makes a much larger orbit, the star makes a much smaller one around the centre of mass. For an Earth-bound observer, the periodic change in the position and velocity of rotation and the time of arrival of light from the parent/host star form the observables for the detection of a planet. Further the periodic dimming of light from the star in question can also be detected when its planet passes in front while orbiting. Due to the very large difference in the brightness of the star and that of a planet (amounting to typically about 9 orders of magnitude for Sun/Jupiter), it is very difficult to observe the planet directly by imaging. By using specially designed masks planets separated by 1 arcsec from its host star may be detected. Several space-borne photometers (for Kepler, see Howell et al 2014, for Transiting Exoplanet Survey Satellite (TESS), see Ricker et al 2016 and, for JWST, see Carter et al 2023) were flown to detect planets around nearby stars using transit technique. In order to estimate reliable masses of planets however the transit observations need complimentary/supplementary ground-based radial velocity observations using high resolution and high precision spectrometers. Masses of planets may also be determined using the gravitational lensing technique; such observations are however few and far between. One of the important criteria of whether an extrasolar planet may possibly be habitable is its surface temperature to be conducive for the existence of water in liquid phase. Using the host/parent star parameters, equivalent blackbody temperature of the planet may be estimated using the eqn. 1.8. Lately, attempts are being made to detect the planetary atmospheric composition as well using transit spectroscopy method. Such observations will also be useful in determining the habitability of exoplanets. For a comprehensive review of the detection and science of exoplanets we may refer to Kaltenegger (2017).

1.2.3 Stars: Stars constitute the most important components of a galaxy. They account for more than 90% of the mass and luminosity of a galaxy. The rest being from gas and dust. The formation and evolution of stars account for the chemical composition and evolution of a galaxy. Star formation and evolution are thus one of the most important problems of astrophysics. Stars are believed to form by gravitational collapse of denser parts of a molecular cloud caused by a perturbation in the interstellar medium (ISM) of a galaxy. During its contraction phase under its own gravity, a protostar (star in its formation stage) of sufficient mass attains temperatures of $T_c \geq$

$10^7 K$ at its core. At such high temperatures nuclear reactions take place involving hydrogen nuclei to form helium nuclei with the release of nuclear fusion energy. Masses less than about 0.08 $M_\odot$ cannot attain such high temperatures and end up as what are called Brown Dwarfs. These objects are however capable of burning deuterium at their cores for a brief period in their initial phases. The mass limit for Brown Dwarfs is 0.001 $M_\odot$ below which begins the planetary regime. Stars of different masses have different surface temperatures (and hence different spectra) and sizes. The higher the mass the higher will be its effective temperature and size and shorter will be its lifetime.

1.2.3.1 Classification of Stars: Stars are classified into various types based on their surface or effective temperatures (T_{eff}) or colours – from the hottest O type stars (T_{eff} around 3000 K) to the coolest M type stars (T_{eff} around 3000 K). The spectral types are designated by the letters O, B, A, F, G, K, M in the order of decreasing effective temperatures with sub-types within each type designated by numerals from 1 to 9; 1 being the hottest and 9 the coolest of the subgroup. For example, an O1 star is hotter than an O2 star and so on. Stars of cooler surfaces than M types, discovered lately, are called very low mass stars (VLM stars) and their spectral types are called L and T types (Kirkpatrick 2005). It is to be noted that this temperature or spectral classification is based on the so-called dwarf or main sequence phase during which stars remain stable with core hydrogen-burning as the energy source. We can understand the significance of the spectral classification (based on the surface or effective temperature of a star) from Fig 1.6. The figure shows the line strength (in arbitrary units) of different ionization stages of various important and abundant elements as a function of stellar (effective) temperature (hence the spectral type). We may notice that the hottest stars of O type show in their spectra He II (singly ionized Helium) and HeI (neutral Helium) lines (in absorption). The stars of spectral type A0 having a surface temperature of about 10000 K show predominantly the Hydrogen recombination lines in absorption. The cool stars (such as K0) show Ca II (singly ionized Calcium) absorption lines while still cooler stars (M types and even cooler stars) show molecular lines (such as TiO, VO) in their spectra. We may note here that Saha's thermal ionization equation gives the theoretical (physical) foundation for the spectra exhibited by stars of different colours. Table 1.3 gives some important parameters (in Solar units) for stars of different masses (spectral types) in the main sequence phase (these are mean parameters for dwarfs compiled by Pecaut & Mamajek 2013). A second type of classification called the luminosity classification is based on the evolutionary stages of stars (discussed below).

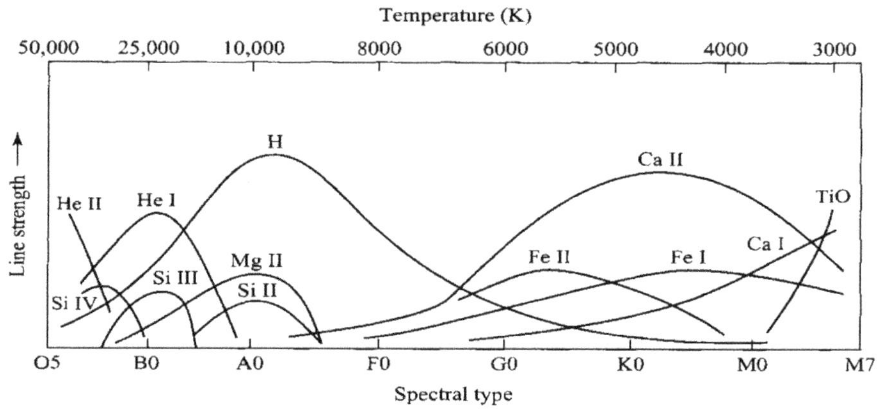

Figure 1.6: Ionization stages of abundant elements in stars of different effective temperatures or spectral classes (from Carroll & Ostlie 2007).

1.2.3.2 Main Sequence: The longest phase in the life of a star of any mass is known as the Main sequence. This phase is signified by nuclear reactions at the core of a star involving hydrogen nuclei (proton-proton or p-p chain) fusing to form Helium nuclei. Such stars also called dwarf stars in the luminosity classification. Stars of different masses make a sequential track in the luminosity-temperature space as shown in Fig 1.7. The time that a star spends in this phase depends up on its mass and luminosity, M/L. We notice that since L increases with M as $L \sim M^{3-4}$, the main sequence timescale decreases rapidly with increasing M: the smaller the mass the longer is the main sequence lifetime. The main sequence timescale may also be termed as nuclear timescale (see later sections of this chapter).

As mentioned above, stars spend most of their lifetimes in the hydrogen burning main sequence phase. After the exhaustion of hydrogen fuel at the core (which is now helium-rich), a star starts contracting under the inexorable pressure of gravity. This leads to the rise of the core temperature. The envelopes however are cooler and start to expand to maintain hydrostatic equilibrium thus making the star expand in size to become a Red Giant. The contracting core which is now predominantly helium-rich will eventually attain temperatures more than ~ 10^8 K. At this stage nuclear reactions will be rekindled fusing 3 or 4 helium nuclei to respectively synthesise carbon and oxygen nuclei. This stage lasts for much shorter timescales than the hydrogen burning main sequence stage. After the exhaustion of the helium the core becomes once again inactive, and the star sustains on the shell-burning – the helium shell surrounding the core and the outer hydrogen shell surrounding the latter. For most part of this stage, it is the hydrogen burning shell that supplies the fuel while helium shell accounts for about 10% of the time. During this alternate hydrogen-helium burning cycles the star's envelope once again expands and becomes what is called an asymptotic giant branch star (AGB star). The luminosity of the star rises asymptotically with temperature. Now depending upon the initial mass of the star further nuclear reactions may take place at the core. For a star of initial mass typically less than about 8 $M_\odot$ the core remains inactive, and the envelopes keep expanding. These stars, called intermediate mass stars, eject their envelopes into the ISM as planetary nebulae (PN) while their cores cool down and eventually end up as White Dwarfs (WD).

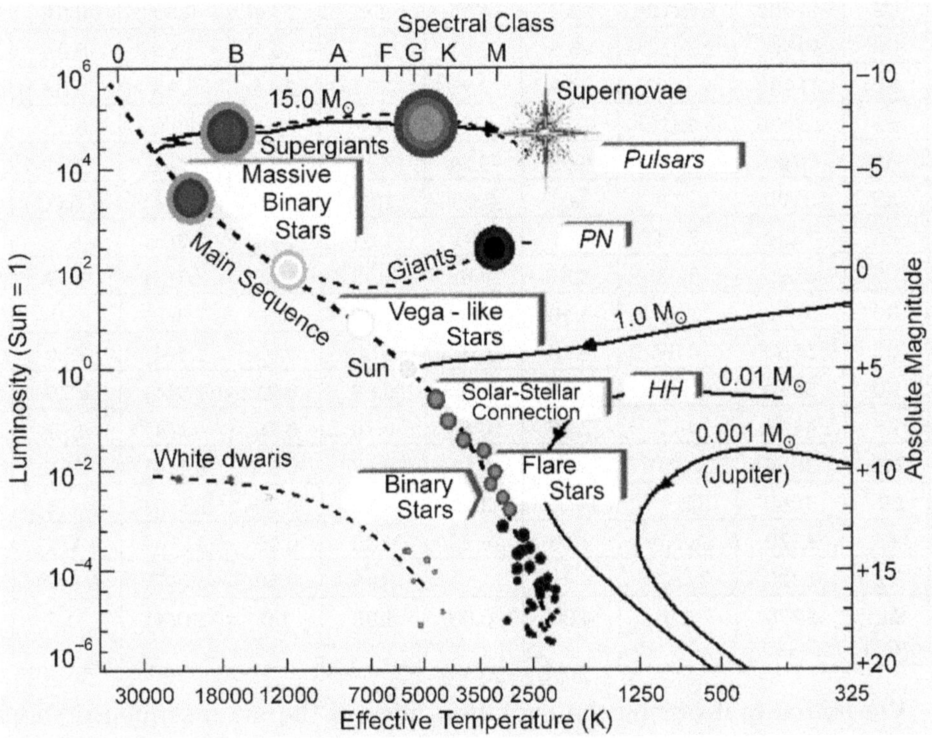

Figure 1.7: Herzsprung-Russell Diagram showing main sequence stars in the Effective Temperature- Luminosity plane. Also shown are various pre- and post-main sequence phases of stars (from G. Carenini - www.ca.ubc.ca).

Stars with masses larger than 8-10 $M_\odot$ carry forward the core nuclear activity. For stars having initial mass more than ~ 15 $M_\odot$, the core becomes predominantly an Iron core whence the nuclear fusion reactions cease to continue. At this stage, the star becomes unstable due to the onset of a series of fission reactions that are endothermic in nature (requiring energy). This leads to a catastrophic core collapse followed by energetic explosion. The resulting tenuous nebular object containing the ejected matter, rich of heavy elements, is called a Supernova. Depending upon the core mass left behind the less massive core becomes a Neutron Star (NS) by contraction, while a more massive core becomes a Black Hole (BH) for the gravity finally becomes inexorable.

Table 1.3: Physical Parameters of Main Sequence Stars (Luminosity Class = V). Numbers in the parentheses are in powers of 10.

Sp. Type	T_{eff} (K)	L ($L_\odot$)	M_V (mag)	BC (mag)	M ($M_\odot$)	R ($R_\odot$)	τ_{ms} (10^6 yr)	τ_{KH} (10^6 yr)
O3	44900	6.6×10^5	-5.8	-4.01	59	13.4	0.89	0.01
O6	39500	2.3×10^5	-5.1	-3.57	35	10.3	1.50	0.02
B0	31400	4.5×10^4	-3.9	-2.99	17.7	7.16	3.95	0.03
B5	15700	5.9×10^2	-0.85	-1.34	4.7	3.36	80.0	0.35
A0	9700	38	0.99	-0.21	2.18	2.19	570	1.81
A5	8100	12.3	2.01	0.00	1.88	1.79	1.5(3)	5.0
F0	7220	7.24	2.57	0.01	1.61	1.73	2.2(3)	6.6
F5	6550	3.63	3.37	-0.02	1.33	1.47	3.7(3)	10.5
G0	5930	1.35	4.48	-0.065	1.06	1.1	7.8(3)	4.0
G5	5660	0.89	4.98	-0.105	0.98	0.97	1.1(4)	35.0
K0	5270	0.46	5.78	-0.195	0.88	0.81	1.9(4)	65.9
K5	4440	0.17	7.28	-0.63	0.70	0.70	4.1(4)	130
M0	3850	0.07	8.80	-1.15	0.57	0.59	8.1(4)	254
M5	3060	3×10^{-3}	14.15	-3.11	0.16	0.20	5.3(5)	1.4(3)
L0	2270	2×10^{-4}	20.0	-6.25	0.08	0.10	4.2(6)	1.1(4)
T2	1220	2×10^{-5}	-	-	0.06	0.10	3.0(7)	6.3(4)
Sun (G2)	**5770**	**1.0**	**4.80**	**-0.08**	**1.00**	**1.0**	**1.0(4)**	**31.7**

We notice that during the evolution beyond the main sequence stage stars tend to increase their sizes in order to maintain hydrostatic equilibrium. This process leads to increasing of their initial (main sequence) luminosities (recalling that luminosity is proportional to R^2 and T^4). Based on their evolutionary stage, further classification of stars, called the luminosity classification is made. This classification serves complementary to the colour-classification discussed earlier. Accordingly, the Luminosity Class I stars are called Super Giants; Class II and III are called Giants; Class IV are Sub-Giants and Class V are known as Dwarfs that represent the main sequence phase; Class VI are Sub-Dwarfs. For example, a G3 dwarf is designated as G3V, an F4 giant as F4III and an O5 super giant is denoted by O5I and so on. Fig 1.7 shows a useful depiction of stars of different masses and luminosity stages in the Temperature-Luminosity plane and such a diagram is called Herzsprung-Russell diagram or H-R diagram. In this diagram stars in all stages of their lives may be depicted in the luminosity-temperature space – right from stars in making (protostars) to evolved giant stars through the hydrogen burning main sequence stars (see the Figure 1.7). The typical loci of luminosity classes of stars are shown clearly and separately in Fig 1.8 in the luminosity-temperature space – their evolution being from the dwarf (V class) stage towards right.

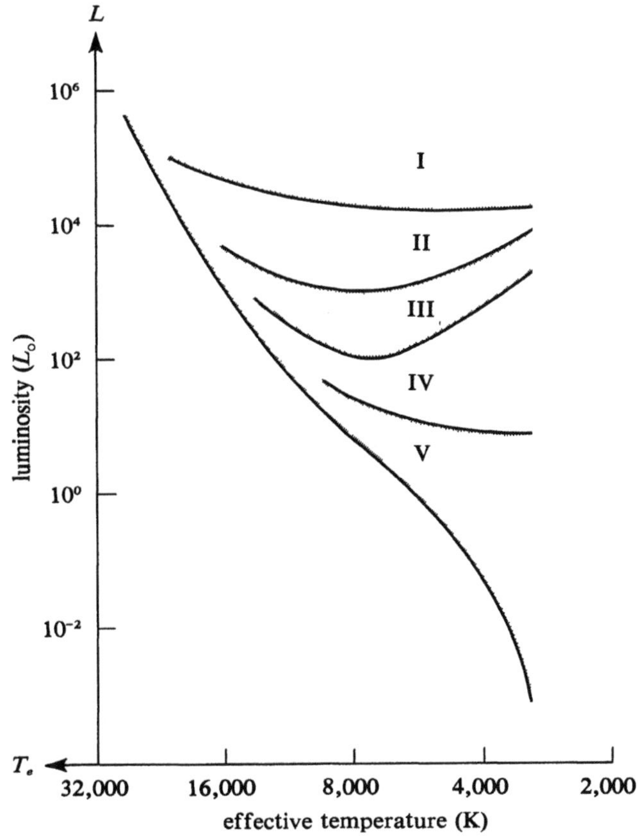

Figure 1.8: Luminosity Classes of Stars in the HR Diagram (Shu 1982).

1.2.3.3 Equations Governing the Stellar Interiors: The interiors of stars are basically governed by hydrostatic and thermal equilibria, the continuity (conservation) of mass and luminosity and finally, the energy release by nuclear reactions and the transport of this released energy to the surface layers either by convection or by radiation. The most important inputs to these equations are the initial mass and the elemental composition of the star, although magnetic field and rotation also may contribute significantly. Such a theoretical foundation allows us to understand the behaviour of the fundamental parameters pertaining to the stability and evolution of stars – the variation of temperature, pressure, density and luminosity of matter from the core towards the surface. Firstly, it is important to establish the equation of state in the stellar interior which defines the relationship between the gas pressure and the temperature and density (see, for a detailed discussion, Prialnik 2000). For an ideal gas we have, $P = (\rho kT/\mu_m m_H)$, where T is the gas temperature, ρ is the pressure, μ_m is the mean molecular weight (determined by the composition of the gas) and m_H is the mass of a hydrogen atom. Without going into the details of their derivation we give below the equations governing the stars for reference:

$$\frac{dM_r}{dr} = 4\pi\rho_r r^2 \ (conservation\ of\ mass) \tag{1.9}$$

$$\frac{dP_r}{dr} = -\frac{G\rho_r M_r}{r^2} \ (hydrostatic\ equilibrium) \tag{1.10}$$

$$\frac{dL_r}{dr} = 4\pi\rho_r r^2 \varepsilon \ (continuity\ of\ luminosity) \tag{1.11}$$

$$\frac{dT_r}{dr} = -\frac{3\kappa\rho_r}{16\sigma T_r^3}\frac{L_r}{4\pi r^2} \ (radiative\ transport) \tag{1.12}$$

$$\frac{dT_r}{dr} = \frac{(\gamma-1)T_r}{\gamma P_r}\frac{dP_r}{dr} \ (convective\ transport) \tag{1.13}$$

In the above equations the temperature T_r, pressure P_r, mass density ρ_r and luminosity L_r are all functions of the radial distance r from the core centre to the surface of a star; ε is the rate per unit mass at which energy is produced from thermonuclear fusion reactions (such as the proton-proton (p-p) chain to convert H into He nuclei); κ is the mean opacity of the stellar layers; σ is the Stefan-Boltzmann constant; and γ is the adiabatic constant (ratio of heat capacities at constant pressure and constant volume). These equations are for a static situation since the time dependency is not considered. Time dependency would be of importance especially in the case of rapid phenomena such as the process leading to supernova explosion in massive stars and the pulsations of intermediate mass stars. For details of this topic, we may refer to the excellent book by Prialnik (2000).

1.2.3.4 Stability Criteria and Timescales of Stars: It is instructive to give important stability criteria and timescales of stars based on some of the above equations. We already introduced the timescale for hydrogen-burning stage called the main sequence timescale.

Dynamical Equilibrium: As mentioned before the interiors of stars are in hydrostatic equilibrium (eqn. 1.10) – the pressure gradient balancing the gravitational force against dynamical instability or collapse. It is given by.

$$\frac{dP_r}{dr} = -\rho \frac{GM_r}{r^2} \tag{1.14}$$

where P_r is the pressure (kinetic and radiation) and M_r is the mass of the star at a radial distance r from the centre and ρ is the average mass density. Situations arise during the life of a star when the equilibrium may be upset.

Chandrasekhar Limit: For a degenerate core, in which the gravitational contraction is countered by the degeneracy pressure of electrons or neutrons (in case of White Dwarfs or Neutron stars respectively), the limit on its mass above which it collapses under its own gravity is given by Chandrasekhar limit.

$$M_{Ch} = 5.83\,\mu_m^{-2}\,M_\odot \tag{1.15}$$

where μ_m is the mean molecular weight; in case of He, C, O cores its value is 2 and the limit reduces to 1.44 $M_\odot$. For a Fe core the limiting mass is 1.26 $M_\odot$. Masses above these limits would undergo inexorable gravitational collapse (see Prialnik 2000).

Eddington Limit for Luminosity: Eddington examined the case where radiation pressure alone balancing the gravitational pressure. Under this radiative equilibrium condition, the limit on the luminosity above which the star becomes unstable (leading to disruption) due to radiation pressure is evaluated by Eddington as

$$L_{Ed} = \frac{4\pi cGM}{\kappa} = 3.2\,\times 10^4 \left(\frac{M}{M_\odot}\right)\left(\frac{\kappa_{es}}{\kappa}\right) L_\odot \tag{1.16}$$

where κ is the mean opacity which is a function of temperature and composition of the star, taken relative to the opacity due to electron scattering, κ_{es}. $L_\odot$ is the solar luminosity. The star loses its radiative equilibrium and disintegrates if its luminosity exceeds the Eddington limit (see Prialnik 2000).

Important Timescales: Based on the equations governing the interiors of a star we can define a few important timescales relevant to stellar physics.

Dynamical Timescale: As we have mentioned a star maintains a state of hydrostatic equilibrium by balancing of the inward pulling gravitational pressure with kinetic gas pressure. If there occurs at some of point of time an imbalance between the two competing forces, there would be a quick restoration of equilibrium within a short time. This is called the dynamical timescale (see Prialnik 2000). It may be derived approximately as the rate of change of size of a star as a particle on its surface tries either to escape (with an escape velocity v_{esc}) or undergoes a free-fall (with a free-fall velocity v_{ff}). Thus,

$$\tau_{dyn} \approx \frac{R_*}{v_{esc}} \approx \frac{R_*}{v_{ff}} \approx \frac{1}{\sqrt{G\rho}} \approx 1000 \left[\frac{R}{R_\odot}\right]^{3/2} \left[\frac{M_\odot}{M}\right]^{1/2} sec \qquad (1.17)$$

where G is the gravitational constant and ρ the average mass density of the star under consideration. For the case of the Sun the dynamical timescale is ~ 1000 sec, while for a O3V star it is about 6 days.

Thermal/Kelvin-Helmholtz Timescale: Another timescale of importance is that at which the star's internal energy arising due to gravitational contraction is expended by emission of radiation. This is given by.

$$\tau_{th} \approx \frac{U}{L} \approx \frac{GM^2}{R\,L} \approx 3.17 \times 10^7 \left[\frac{M}{M_\odot}\right]^2 \left[\frac{R_\odot}{R}\right] \left[\frac{L_\odot}{L}\right] yrs. \qquad (1.18)$$

For the Sun's case the timescale is approximately equal to 3x10^7 yrs. This is the same as the familiar Kelvin-Helmholtz timescale (τ_{KH}) on which the gravitational contraction raises the core temperature and density of a star to the values conducive for the onset of the hydrogen-burning nuclear reactions. This timescale also defines the time taken by a star (of mass M) to reach the onset of main sequence phase or the zero-age main sequence (ZAMS) phase (see Zinnecker & Yorke (2007), for τ_{KH} as a function of $M_\odot$, $R_\odot$, and $L_\odot$). Essentially, we understand that a star must maintain both hydrostatic and thermal equilibrium for its stability against phases of expansion (leading to cooling) and contraction (leading to heating). Table 1.3 lists the thermal timescales for stars of various masses (from models of Schaller et al 1992; Pecaut & Mamajek 2013).

Nuclear Timescale: The theory of stellar interiors predicts that the core of a star is hot enough to trigger nuclear fusion reaction in which four hydrogen nuclei (protons) fuse to convert into one helium nucleus. Nuclear energy source is thus necessary and sufficient for the star to sustain on timescales that are estimated in the case of the solar system age which are very much longer than the K-H timescales.

The timescale that a core of certain mass completely converts itself from hydrogen dominated one to helium dominated one defines the nuclear timescale or the main sequence timescale. Here in this case the energy source is the nuclear energy at the core and is given by the Einstein's well-known equation $E=Mc^2$ and if the luminosity pertaining to this is L_{nuc} or L, then the timescale is simply given by (with ε as a constant that depends on the binding energy of a nucleon divided by its rest mass energy),

$$\tau_{nuc} \approx \frac{\varepsilon\, Mc^2}{L} \approx \varepsilon\, 4.5 \times 10^{20} \left[\frac{M}{M_\odot}\right] \left[\frac{L_\odot}{L}\right] sec. \tag{1.19}$$

Since the luminosity is a function of mass of a star the timescale becomes a very strong function of mass (related inversely). Again, for comparison for the case of the Sun this timescale is about 9 billion years (nearly the age of the Universe!). For more massive stars the times come down rapidly.

Adopting a value of $\varepsilon \sim 10^{-3}$, we obtain the nuclear timescale (which is orders of magnitude longer for Hydrogen burning cores than Helium cores) or the main sequence lifetime of a star in terms of years as

$$\tau_{nuc} \sim 1 \times 10^{10} \left(\frac{M}{M_\odot}\right) \left(\frac{L_\odot}{L}\right) yrs. \tag{1.20}$$

Table 1.3 lists main sequence or nuclear timescales for stars of various masses (taken from the models of Schaller et al 1992; Pecaut & Mamajek 2013).

A comparison of the three timescales as defined above gives us the inequality, valid for stars of all mass ranges, as

$$\tau_{dyn} \ll \tau_{th} \ll \tau_{nuc} \tag{1.21}$$

Thus, stars adjust themselves to changes in temperature and pressure on dynamical timescales of the order of hours at the maximum much shorter than the thermal and nuclear timescales. The thermal timescale determines the time required to attain core temperatures needed for the onset of nuclear reactions while the nuclear timescales represent the lifetimes of stars.

1.2.3.5 Stellar Evolution: For the sake of completeness and continuity, we give here a short rudimentary account of stellar evolution. As we have seen, stars spend a major part of their lifetimes in the main sequence stage, signified by the core-hydrogen burning. The higher the mass of a star the shorter its lifetime will be. Following the exhaustion of hydrogen, due to the absence of any energy source, the core undergoes a contraction while the envelope expands to maintain the hydrostatic equilibrium. This is the state when the star enters the red giant branch. The ongoing contraction heats up the core leading to the attainment of higher temperatures (> 10^8 K) triggering the nuclear reactions involving helium nuclei. These reactions, called the triple alpha processes, synthesize carbon and oxygen nuclei in the core. We note here that stars of masses less than 0.8 $M_\odot$ end up with degenerate helium cores, but in nature we do not find these (except those evolved from binary stars), for the main sequence lifetime of a single star of such a mass is much longer than the time of universe or Hubble time (see Table 1.3). In stars of masses more than 0.8 $M_{\odot}$, after the exhaustion of core helium, once again, the contraction-expansion cycle starts. It is important to note that for stars of masses between 0.8 and 2.25 $M_{\odot}$ the cores become electron-degenerate before helium ignition starts leading to a runaway core-helium flash (Iben 1974, 1985; Prialnik 2000). The inactive core of the star has now hydrogen and helium shells burning alternatively providing energy for the star. For stars of masses less than 8 - 10 $M_{\odot}$, the core remains unignited, as the electron degeneracy pressure supports it from gravitational collapse. Their cores, initially very hot, will cool down to become white dwarfs, while their envelopes, ionized and excited by the central hot core, become planetary nebulae. For stars having masses more than 10 $M_{\odot}$, however, the ignition of successive cores will continue (beyond C-O) with heavier nuclei, till it reaches the stage of having a core of iron (Prialnik 2000; Langer 2012). At this stage, fusion reactions are no longer possible. The iron core will undergo a collapse following a series of endothermic nuclear fission reactions. This is followed immediately by a catastrophic highly energetic explosion called supernova (type II supernova). The degenerate cores of high mass stars will become neutron stars or black holes, depending upon their core masses. Thus, stars of all masses lose a major portion of their mass, and this mass loss enriches the surrounding medium called the interstellar medium and provides matter for the formation of next generation of stars.

1.2.3.6 Binary Stars: It is certainly possible and highly probable that stars mostly occur in binaries (for understanding origin of binaries, see Tohline 2002). It is estimated theoretically that more than 60% of the stars may have companions (see e.g., Iben 1985). Evolution of such a composite system of double stars is significantly different from that of a single star. This is due to the gravitational or tidal interaction between the two stars in the system affecting their dense extended atmospheres generated by mass loss during evolution beyond main sequence. We know that a star of higher mass evolves faster than the one that has a lower mass. Such a situation where the components of binary stars have different evolutionary timescales is highly dynamic. This leads to very interesting and often exotic possibilities arising in binary stellar systems.

Shown in Fig 1.9 is a schematic representation of a binary system in its composite gravitational field potential. The stars of masses M_1 and M_2 are shown as dark spots in the figure. The solid circles and dashed contours show the cross-sections of the iso-potential surfaces for the binary system:

the solid curves representing envelopes around individual components and the dashed curves the envelopes that are common for both the stars. The masses considered for the stars are nearly equal and hence the contours for each star look nearly the same. The two lobes (shown as solid contours) called the Roche lobes represent the two zero-gravitational potential surface regimes corresponding to the two stars. Within these lobes each star would govern the dynamics of a test particle. The point at which the two lobes meet is called the first Lagrangean point or L_1. Exactly at L_1 a test particle is floating in space freely being inside a potential well (or minimum). Four more such minima exist in the system as shown in the figure, called the Lagrangean points, L_2, L_3, L_4, and L_5. When the more massive star of the binary evolves (faster than its companion) it undergoes mass loss leading to filling of its own Roche lobe. The pressure at L_1 due to this filling ushers the matter into the Roche lobe corresponding to the second star of lower mass. The mass thus gained by the second star creates a hot spot on its surface as shown in Fig 1.10 and starts heating the surface matter. This situation, occurring in close-binary star systems, potentially changes the course of evolution of not only the receptor star but also the donor star (see e.g., Iben 1985; Marchant & Bodensteiner 2024). Their evolution would be very different from that of a single star of similar mass. Even intermediate mass stars in binary could end up as neutron stars or even black holes. Further, for instance in the general class of cataclysmic variable stars (see Warner 1995; Smith 2006), this type of mass transfer takes place in a binary system in which one of the stars (being more massive to start with) evolved into a white dwarf having lost the envelopes to the inter-binary or interstellar medium, the subsequent mass loss from the evolving secondary star (less massive to begin with) would lead to an explosive situation of rekindled nuclear reactions by CNO cycle on the surface of the white dwarf. This triggers a runaway of radiation transport. Such catastrophic situations lead to sudden brightening of the system known as Classical Nova (new star) phenomenon (see e.g., Chomiuk et al 2021). The situation may also lead (depending upon the masses of the stars) to what is known as a Type I supernova phenomenon. More massive stars in binary systems evolve into more exotic objects such as Low Mass X-ray Binaries and High Mass X-ray Binaries (LMXB and HMXB) in which the accreting components are believed to be neutron stars or black holes (for details on evolution of massive binary stars, see Marchant & Bodensteiner 2024). We may refer to excellent articles by Iben (1974, 1985) on the evolution of a single star and that of a binary.

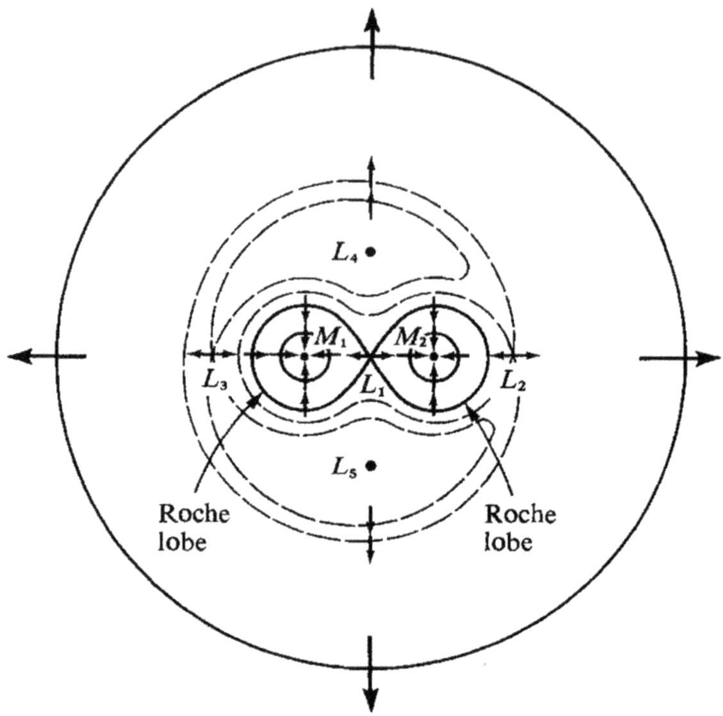

Figure 1.9: Gravitational iso-potential curves for a binary system of stars. The mass transfer takes place through the Lagrangean point L_1. The Roche lobes indicate the individual gravitational regimes of the two stars (from Shu 1982).

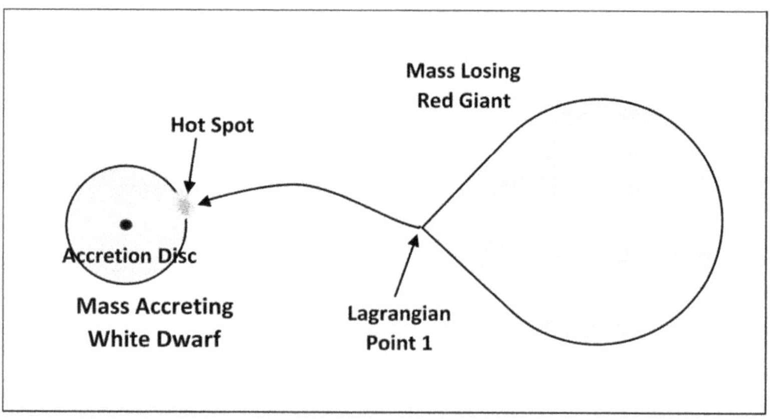

Figure 1.10: Mass transfer through L1 in a binary star – consisting of a compact object evolved from an initially more massive star and the currently evolving lower mass companion. The presence of the compact object may be inferred from energetic radiation emanating from the accretion disk since the cooler star cannot account for it (adapted from Shu 1982).

1.2.3.7 Mass Loss from Stars: Spectra of stars in main sequence phase show predominantly absorption lines from various constituent elements (such as H, He, Mg, Ca, Fe) in their envelopes against the source of hot inner regions (see Fig 1.6 and the right top panel in Fig 1.11). Cooler stars in the main sequence, in contrast, exhibit molecular absorption lines (such as TiO, VO). Spectral lines arising from the atmospheres of stars not only reveal the physical parameters but also the physical conditions of the star: whether the star is in the main sequence phase or has evolved into the giant phase and so on. Spectra can also identify the pre-main sequence phase of protostars as opposed to evolved stars.

Stars lose mass at different stages of their lives, particularly and predominantly after the main sequence phase. While in the main sequence phase stars lose mass at a very low rate (typically $< 10^{-12}\ M_{\odot}/yr$) by way of highly tenuous stellar wind, the mass loss rate increases dramatically by several orders of magnitude following the onset of the giant phase in the case of low and intermediate mass stars (see van Winkel 2003; Herwig 2005). When stars evolve into the giant stages copious mass loss occurs causing extended tenuous atmospheres normally called circumstellar regions. The line profiles from such stars appear a bit more complex than those from main sequence stars. Further, stars which are in the process of formation also possess circumstellar matter; the major difference in their case is that the matter is being accreted rather than ejected as in the case of evolved stars. The question of whether matter is ejected (in evolved stars) or accreted (in protostars) may be answered by spectroscopic and photometric observations.

Mass loss rate at a radial distance r from the star of mass M where the average density is $\rho(r)$ and the wind has reached its terminal velocity v_∞, may be given by,

$$\frac{dM}{dt} = 4\pi r^2 \rho(r) v_\infty \tag{1.22}$$

During the red-giant stage (following the exhaustion of core hydrogen nuclear fuel), stars lose mass at a rate of $10^{-9}\ M_{\odot}/yr$ while during their asymptotic giant branch stage (second red-giant stage after helium exhaustion at the core) the mass loss rate is as high as $10^{-6}\ M_{\odot}/yr$. Such a heavy mass loss leads to obscuration of visible light from the stellar photosphere due to large extinction. Stars are observable only in the longer wavelengths during these giant stages especially for cooler stars.

Massive stars of mass more than $\sim$ 8-10 $M_{\odot}$ also lose most of their mass during the onset of supernova phase leaving behind cores that evolve to become neutron stars or black holes (see Vink 2022). In the case of binary stars, the common envelopes contribute to mass loss (see e.g., Taam & Sandquist 2000). In the case of nova phenomenon, the surface matter of the accreting component contributes to the mass loss. Spectral lines arise from the tenuous gaseous matter thus ejected from these scenarios and typically help the observer unravel the underlying physical processes.

P Cygni Line Profiles: Several hot stars exhibit peculiar line profiles indicative of mass loss or wind - a combination of both absorption and emission lines known as P Cygni profiles. These are a result of blue-shifted absorption from out-flowing wind.

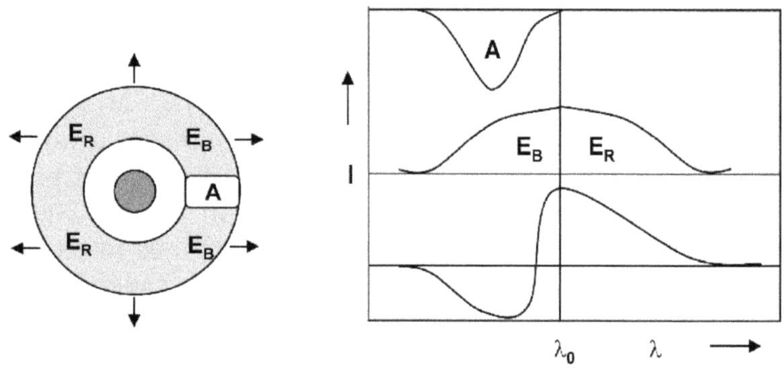

Figure 1.11: P Cygnii Profile formation. The patch A in front of the star accounts for the absorption profile (shown on top panel of the schematic graph on the right; the emission blue and red components are depicted in the middle panel and the resultant profile called P Cygnii profile is shown in the bottom panel.

As illustrated in Fig. 1.11 the star at the centre (which is either in a pre-main sequence or post-main sequence stage) is surrounded by a presumably spherical shell of matter colder than the stellar surface or the photosphere.

Under these conditions an observer would record an absorption profile of a spectral line from a given atom or molecule as shown in the right-side part of the figure marked A. This is the result of matter directly in front of the star (shown in grey in the left part of the figure) absorbing the radiation emanating from the hotter stellar photosphere; blue-shifted for an observer in front of the absorbing matter. The atoms/molecules present in the other regions of the circumstellar matter would also emit due to the excitation by either collisions or by radiation from the central star. These emitting regions are marked E_B and E_R for blue and redshifted regions respectively. The emission profile thus formed is shown in the middle panel of the right-side figure. The resultant of the two profiles, viz. the blue-shifted absorption profile and the emission profile symmetric about the rest wavelength as shown in the bottom panel of the right-side figure, is known as a "P Cygni profile". Such a phenomenon was first observed in the case of the star P Cygni, a pre-main sequence star and hence the name. As mentioned already the P Cygni line profile has a typically shallow absorption line followed by a steep rising emission part that slowly tapers off towards the redshifted side of the spectrum as shown in the schematic figure. A star that accretes matter (such as a proto-stellar object) shows the absorption part in the redshifted part and such a profile is called an "inverse-P-Cygni" profile. For a detailed treatment of P Cygni profile formation, we may refer to Lamers & Cassinelli (1999).

1.2.3.8 Variable Stars: Stars exhibit variability in their overall light output and even their spectral line strengths in various stages of their evolution as well as during their formation stages. These are termed as intrinsic variable stars. Stars may also vary due to extraneous factors for instance due to occultation by another astronomical source such as a binary component or a planet. Such stars are called extrinsic variable stars. Variability may be periodic as well as aperiodic or quasi-periodic in both single and binary stars. For a general description of variable stars, we may refer to Percy (2007).

Periodic variability occurs in single stars mainly during their evolution into giant stages wherein they undergo periodic oscillations whose period depends upon the average density. In this category are the RR Lyrae variables having variability periods of a few hours, Cepheid variables with periods of a few days to a few tens of days and Mira variables having periods of a hundred to a few hundred days. RR Lyrae variables exhibit change in light output of 1 magnitude while the Cepheid variables show a linear relationship between their period and the magnitude of variability. These are examples of stars showing variability due to oscillations of their interior regions. Single stars also exhibit aperiodic variability during mass loss episodes.

It may be noted here that the discovery of the period-luminosity (or absolute magnitude) relationship discovered in Cepheid variables (by Henrietta Leavitt in the beginning of the last century) established a distance candle in the form of these stars for measuring distances to galaxies (we can see farther into the universe since these stars are bright enough to take us as far as the Virgo cluster galaxies).

Binary stars exhibit periodic variability due to the mutual obscuration of the binary components. Periodic changes occur in their spectral lines as well – single-line wherein only one of the two components exhibits spectral lines and double-line wherein both the binary components exhibit. Double stars also show variability due to mass loss or mass exchange between the two components. This includes the phenomena of Novae (and Supernovae) which exhibit aperiodic variability in both their photometric output as well as their spectral characteristics over a period of a few weeks or longer (see recent review of Chomiuk et al 2021).

Monitoring of variable stars in both multi-band photometry and spectroscopy gives valuable inputs on the processes leading to such behaviour.

1.2.4 Interstellar Medium (ISM): The medium or matter between the stars in a galaxy is in general termed as interstellar medium or interstellar matter (ISM; for details on the topic, see Tielens 2005). The matter between stars is generally very tenuous but there exist regions of cold matter of high density. The temperature and density in ISM vary largely from one region to the other. The ISM comprises of gaseous and diffuse nebulae that are remnants from evolved stars including planetary nebulae, ejecta from intermediate mass stars having masses 1-8 $M_{\odot}$, supernova remnants, ejecta from Supernova explosion from massive stars with masses > 15 $M_{\odot}$ and giant molecular clouds, containing gas and dust grains, that are sites of new star formation. Depending on the regional physical parameters, in particular the density and temperature, the ISM is divided into 5 main phases – (i) hot tenuous ionized regions; (ii) warm ionized regions; (iii) warm neutral regions; (iv) cold neutral regions and (v) cold molecular cloud regions. Each of these regions of ISM is distinct and distinguishable by its physical parameters such as the kinetic temperature and the density. Table 1.4 (data from Smith 2004) lists these phases along with their typical physical parameters.

Owing to the constituent gas and dust grains, the ISM absorbs/scatters the light from bright sources such as stars and makes them appear fainter for an observer. Dust grain formation takes place in ISM due to coagulations of matter in the cold environments. The gas molecules and dust particles scatter and absorb radiation from the hot sources contributing to the extinction of light along the line of sight for an observer.

The cold and visually dark regions of giant molecular clouds are the sites of star formation. The hotter regions such as the supernova remnants and planetary nebulae on the other hand indicate the presence of evolved stars that are losing or have lost mass back into the ISM. Thus, there exists such a cyclic phenomenon that is relentlessly recurring within the ISM. Since the stars are cauldrons of nuclear reactions, with every generation of stars the metallic content (elements heavier than Hydrogen and Helium) increases. This in turn changes the chemical composition of ISM itself and hence that of its host galaxy.

Fig 1.12 (adapted from Whittet, 2003) shows the phases of ISM in gas temperature – density space. We can see the large range of objects that constitute ISM from hot tenuous inter-cloud medium to cold dense molecular clouds. The lines that bound the marked regions show linear trend that maintains the pressure equilibrium between various phases (see Whittet (2003) for more details). Essentially the ISM gas pressure is given by $p = nkT$ with n and T denoting the number density and temperature and k the Boltzmann constant. If pressure were to be in equilibrium, nT (approximately equal to 3×10^{9} m^{-3} K: from Mathis 2000) must be invariant or n should be inversely proportional to T. This is what the figure indicates. The equilibrium temperature is determined by the balance between heating and cooling processes (Spitzer 1978). The two main constituents of ISM namely the gas and the dust are heated by absorption of energetic photons from the interstellar radiation field or collisions with cosmic rays. The dust grains thus heated would cool by emitting photons at longer wavelengths (by the so-called reprocessing). The gas atoms and molecules not only absorb photons but also get heated up by collisions with photoelectrons from small dust grains; and the cooling mechanism is by emission of spectral lines.

Table 1.4: Phases of Interstellar Medium.

Phases of ISM	T (K)	n (m^{-3})	M ($10^{9} M_{\odot}$)	Size (pc)	Object
Hot Ionized	$3\text{-}20 \times 10^{5}$	3×10^{3}	0.003	10^{3}	SNR
Warm Ionized	10^{4}	3×10^{5}	0.05	10^{3}	PNe
Warm Neutral	8000	4×10^{5}	0.20	100	PDR
Cold Neutral	40-100	6×10^{7}	3.0	10	GMC
Molecular	3-20	3×10^{8}	3.0	1	GMC

We note here that the region marked 'HII region' representing the ionized Hydrogen region does not fall in this equilibrium regime. Other factors especially, the interstellar magnetic field also plays a role depending upon the ratio of kinetic pressure to magnetic pressure p_B ($\beta = p/p_B$ with $p_B \sim B^2$, where B is the Magnetic flux density). Magnetic field effects become significant in ionized or partially ionized regions. Turbulence also helps in keeping the regions dynamic and hence the equilibrium situation (as the figure seems to suggest) is rather an ideal one (Elmegreen 1997).

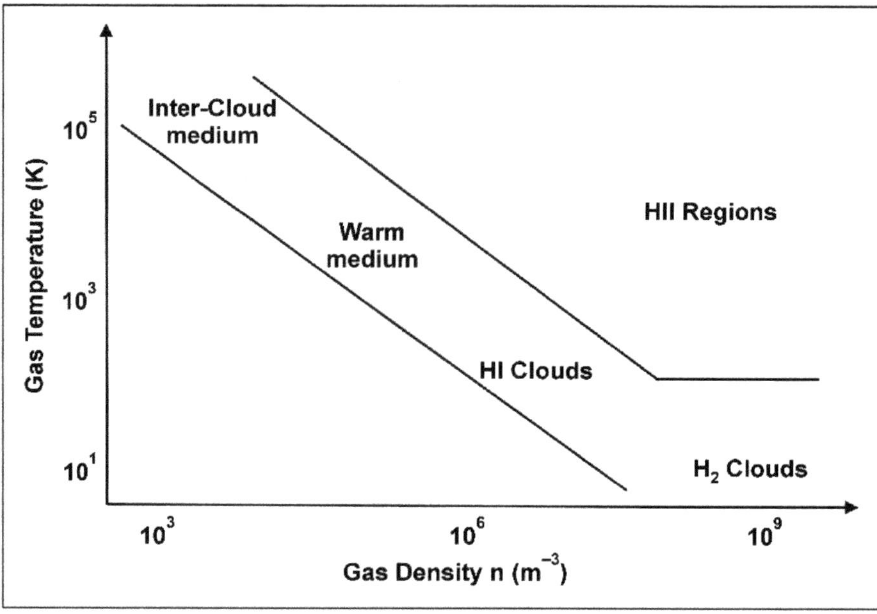

Figure 1.12: Interstellar Medium - temperature versus density for different phases (from Whittet 2003).

1.2.4.1 Star Forming Regions: Stars are believed to form from the small molecular clouds and the giant molecular clouds in the ISM. When a molecular cloud is perturbed by a disturbance such as a passage of supernova wind or AGB star wind, gravitational instability would set in breaking up the cloud into several smaller entities of higher density compared to their surrounding medium. In general, as a consequence of the Virial theorem under static equilibrium conditions (see e.g., Carroll & Ostlie 2007), we expect the following relation between various competing forces, namely the gravitational force acting towards the centre countered by the force due to kinetic pressure and that due to magnetic field,

$$2\{(KE)_p - (KE)_s\} + B + W = 0 \tag{1.23}$$

where the first term in the equation gives the kinetic energy of the cloud in excess of that of background cloud with $(KE)_p$ and $(KE)_s$ representing kinetic energy of cloud and in the medium surrounding it respectively. The second term B represents the magnetic energy and the last term W the gravitational potential energy. If we drop the magnetic force term in the above equation, the static Virial theorem suggests that the total kinetic energy of the cloud is equal to half of the potential energy due to its gravitational force. Overcoming the gravitational force to establish a hydrostatic equilibrium calls for certain limits on the temperature and mass density of the cloud. Now under certain special cases let us look at some important timescales.

Free-fall timescale: If gravity is the only force relevant, then the particles in a cloud experience a free-fall. The timescale of the free-fall can be obtained from the balance equation (1.23) as

$$\tau_{ff} = \left\{\frac{3\pi}{32G\rho_0}\right\}^{1/2} \tag{1.24}$$

where ρ_0 is the initial mass density of the collapsing object under its own gravity and G is the Newton's gravitation constant. Note that the free-fall timescale depends only on density irrespective of the initial mass of the collapsing cloud. For a typical average density in giant molecular clouds (GMC) of $\sim$ 100 cm^{-3} we get the free-fall timescale of $\sim 10^7$ yrs.

Kelvin-Helmholtz timescale: As discussed earlier, this timescale indicates the timescale over which the gravitational energy alone can sustain the luminosity of the star. During this time, the central core temperature attains values more than 10^7 K at which stage nuclear fusion reactions involving Hydrogen nuclei are triggered. We can therefore interpret this timescale as the time taken by an object of mass M to reach the hydrogen burning stage or zero-age main sequence (ZAMS) stage.

Massive stars (with masses much more than the mass of the Sun) have the KH timescales (about 10^4 yrs for a collapsing cloud of mass enough to make an O type star) much shorter than their free-fall timescales and reach ZAMS stage even as the accretion is still on-going. This is in stark contrast with the formation of Sun-like or intermediate-mass stars. Let us recall that for the Sun, for instance, the K-H timescale is 30 million years. This difference in the K-H timescale is a major factor in deciding the processes leading to the formation of massive stars as opposed to the formation of low-mass or the Sun-like stars.

If only the kinetic energy term is considered to counter the gravitational potential energy, then we can derive a limiting mass above which the cloud becomes unstable against gravity and collapses. This exercise was first done by Sir James Jeans nearly a century ago when he considered the stability in a molecular cloud region subjected to a small perturbation and derived a limiting (perturbation) wavelength called Jeans length above which gravity prevails over kinetic pressure and the region collapses under its own gravity separating itself from its surroundings. The limiting wavelength is given by (see Smith 2004, for details)

$$\lambda_J = \left(\frac{\pi c_s^2}{G\rho}\right)^{1/2} \tag{1.25}$$

In the above equation c_s is the local sound speed and ρ the average mass density of the region under consideration. We can now use this length to evaluate the mass of the region under perturbation as (with temperature T in degrees Kelvin and particle number density n in m^{-3}),

$$M_J = \left(\frac{4\pi}{3}\right)\left(\frac{\lambda}{2}\right)^3 \rho = 1.18 \times \left(\frac{T}{10}\right)^{3/2} \left(\frac{n}{10^{11}}\right)^{-1/2} M_\odot \tag{1.26}$$

This limiting mass above which the cloud becomes unstable to perturbation is known as the Jeans mass.

When magnetic field is considered separately in the balance equation given earlier, we may derive the corresponding magnetic critical mass for a spherical cloud as (see Smith 2004)

$$M_\phi = 0.17 \frac{\phi}{G^{1/2}} \tag{1.27}$$

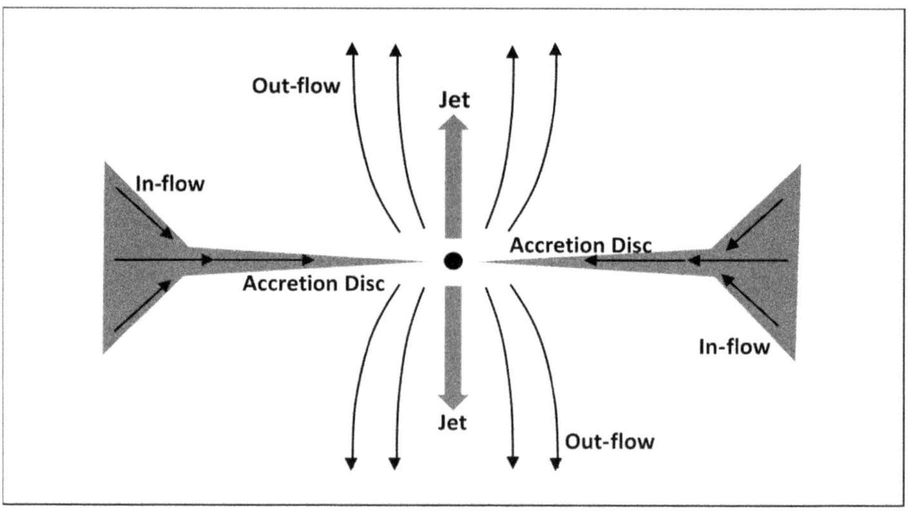

Figure 1.13: Structure of a protostar showing an accretion disk, outflows, and jets (adapted from Zinnecker & Yorke 2007).

where φ is the magnetic flux ($= \pi R^2 B$ with B as the magnetic field and R the size of the cloud). We may write the ratio of cloud mass to the magnetic critical mass as follows:

$$\frac{M}{M_{\Phi}} = 4.6 \times 10^4 \left(\frac{N}{10^{22}\,cm^{-2}}\right)^3 \left(\frac{B}{10^{-6} Gauss}\right)^{-3} \quad (1.28)$$

In the above equation N is the column density of Hydrogen nucleons (see Smith 2004). It gives an extremely high critical mass for magnetic support at large scales of cloud size.

We can now write down the modified Jeans critical length (considering both kinetic and magnetic forces separately and adding) as

$$\lambda_{JM} = \frac{3GM}{\pi\sigma^2}\left\{1 - \left(\frac{M_{\Phi}}{M}\right)^2\right\} \quad (1.29)$$

where σ is the velocity dispersion or line width. This defines the critical size of the cloud that would collapse when compressed. Depending upon the strength of magnetic field we can have two cases – sub-critical case where magnetic field is strong in which the cloud would never collapse and super-critical case where magnetic field is weak in which the cloud would collapse. This latter case is believed to be the process by which high mass stars are formed: at larger sizes the molecular clouds are super-critical since the weak magnetic field fails to provide support the cloud against gravitational collapse. At smaller sizes the clouds may be sub-critical with strong magnetic field supporting the cloud against gravity; unless the effect of strong magnetic field is alleviated by the process called "ambipolar diffusion" (ion-drag due to strong neutral-ion bonding by collisions). It is not clear observationally as to the formation of star formation in sub-critical case. We may refer to Smith (2004) for an excellent treatment of the subject. The observable manifestations of a protostar, namely jets and accretion disk, are schematically shown in Fig 1.13 (adapted from Zinnecker & York 2007). Fig 1.14 shows these structures in a young stellar object (YSO), observed in various diagnostic spectral lines.

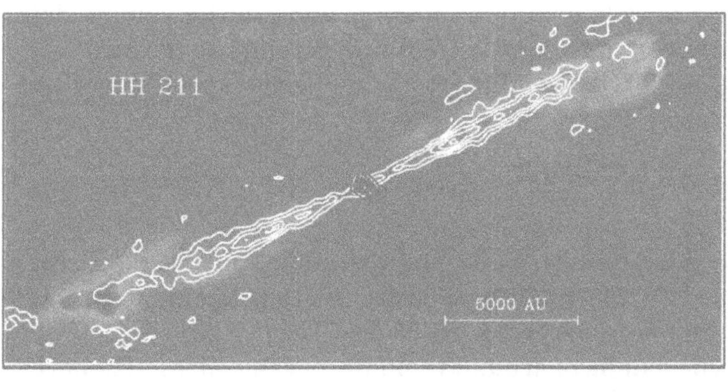

Figure 1.14: Jets and disk in a young stellar object (from IRAM website).

Figure 1.15: Orion Nebula - nearest star formation region (from Hubble gallery).

1.2.4.2 HII Regions: These are regions of ionized hydrogen (HII) usually associated with and present in the giant molecular clouds. Fig 1.15 shows the most extensively observed nearby HII region in Orion (bright regions) created by massive stars. The figure also shows the on-going star formation in nearby regions (darker regions). When massive young stars are formed embedded in a dense molecular cloud, those at the edges of the cloud where the density is not so much as it is in the inner regions, the Extreme-UV radiation flux (Lyman continuum) from the massive protostars ionizes the surrounding matter rich in Hydrogen. The ionization front is impeded at a radial distance R_S from the source (massive young star) when the number of ionizing photons is equal to the number of events of recombination (giving rise to Balmer α line radiation) within the volume defined by the radial distance. Under this assumption Stromgren derived (see e.g., Smith 2004) the size of a HII region for a star of given spectral type as,

$$R_S = 0.032 \left(\frac{N_{UV}}{10^{49}}\right)^{1/3} \left(\frac{n_e}{10^5}\right)^{-2/3} \quad parsec \tag{1.30}$$

where N_{UV} is the number of UV photons emanating from the star per second (normalised to the numerical factor in the denominator) and n_e is the electron number density normalised to 10^5 per cm^3. Clearly only hot stars of early type can have around them observable sizes of spheres of ionized hydrogen. Therefore, the presence of an HII region is indicative of the presence of a massive star. Depending upon the angular sizes (based upon the distance of the young star as well as the density of the surrounding matter) the HII regions are divided into categories of normal HII regions (e.g., those in Orion molecular cloud and Trifid nebula); Ultra-Compact HII (UCHII) Regions and Hyper-compact HII (HCHII) regions (see e.g., Churchwell 2002; Zinnecker & York 2007).

As mentioned earlier, massive protostars have K-H timescales much shorter than the free-fall (condensation) timescales (depending only on density) in contrast with the low-mass protostars. For this reason, massive stars attain the zero-age-main sequence phase while still accreting the mass from surrounding parent cloud (see Zinnecker & Yorke 2007). The environment of a massive star can be schematically understood as shown in Fig 1.16. Shown in the figure are layers of regions around a massive star from ionized regions to molecular regions farther from the star where the photon energy decreases below ionizing energy level. The photo-dissociation or photon-dominated regions (PDRs) represent essentially regions where photons having energies around 4.5 eV or larger can dissociate hydrogen molecules (being the most abundant molecule).

During their proto-stellar stages stars of all masses may exhibit features such as jets or outflows and accretion disks. In the case of the low-mass protostars the timescales associated with these manifestations are much longer than for their massive counterparts. For this reason, these features are more widely studied in low-mass stars compared to the massive protostars.

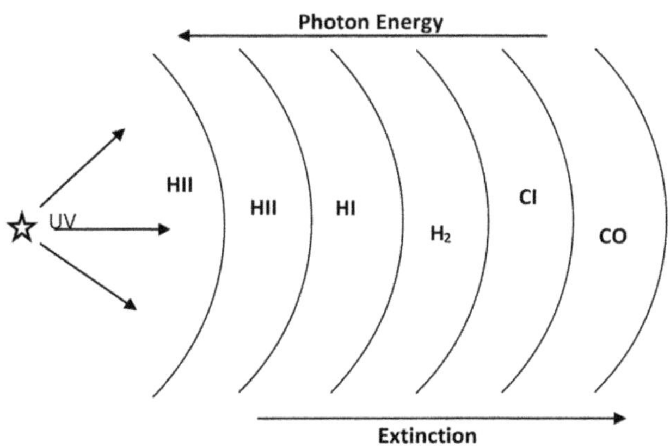

Figure 1.16: Layered structure of a photo-dissociation region (PDR) powered by the UV flux from a massive or hot central star. The energy of photons from near the HII regions (marked HII) towards the edge of neutral atomic region (CI) varies from a few tens of eV to 1 eV. The visual extinction varies from 1 to 10 from the star outwards.

1.2.4.3 Planetary Nebulae: Intermediate stars below about 8 $M_{\odot}$ fail to ignite the nuclear reactions (involving C and O) after the core becomes predominantly made of Carbon and Oxygen. The resulting gravitational collapse is halted only by the electron degeneracy pressure in the core that eventually becomes a White Dwarf. Planetary Nebulae are the result of an episode of heavy mass-loss via a super-wind mechanism following a steady mass-loss during both RGB and AGB stages. The mass loss rates at various stages are:

$$\begin{aligned} \frac{dM}{dt} &= 10^{-9}\, M_{\odot} \quad RGB\ wind\ Mass - loss \\ &= 10^{-7} M_{\odot} \quad AGB\ wind\ Mass - loss \\ &= 10^{-4}\, M_{\odot} \quad Superwind\ mass - loss \end{aligned} \tag{1.31}$$

where $M_{\odot}$is the solar mass.

The remnant hot core having temperatures more than 50000 K ionizes and excites the surrounding ejected matter. As a result, the tenuous surrounding nebula gives rise to emission in optical and UV region from various atoms of various ionization stages. Some of the most intense line emissions in the visible region come from Hα (Balmer series) at 656.3 nm and doubly ionized oxygen [OIII] at 500.7 nm. More than 60% of the PNe display axially symmetric shapes in the plane of the sky. While it is possible that a large fraction of the galactic intermediate-mass stars could be binaries it is yet to be understood if the tidal force shapes the initial mass-loss during RGB or AGB stage (see Balick & Frank 2002). One of the most intriguing questions is in what way the observed equatorially enhanced density gets formed. The study of the PNe will throw light on the ejection and subsequent shaping mechanisms of the nebular matter and hence the palaeontology of mass loss. Another issue with PNe is to account for the mass that is lost to the ISM – the central object being less than about 1$M_{\odot}$ the remaining mass of about a few solar masses needs to be accounted for. The visible nebula accounts for only about a fraction of solar mass; so, it is expected that a large amount of mass is in the form of molecules and possibly dust as well as some portion diffusing away into ISM. Planetary Nebulae enrich the ISM by slow neutron-capture elements by mass loss. For details on planetary nebulae, we may refer to excellent monographs by Pottasch (1983) and Kwok (2000).

In addition to the conventional processes of photoionization and photoexcitation, shock ionization and excitation in ISM also leads to emission from several atoms and their ions.

$$T_s = 1.38 \times 10^5 \left[\frac{v_s}{100} km\, s^{-1}\right]^2 K \tag{1.32}$$

Here the shock velocity is given by v_s and T_s the temperature attained by the gas in the wake of the shock. While the presence of shock-like projections may be seen in the high-resolution sharp images of several planetary nebulae (see e.g., Fig 1.17), Reipurth & Bally (2001) showed (see Fig 1.18) that several abundant species in the gaseous layers of PNe can be ionized and excited by the shock fronts. High velocities up to 100 km/s are not uncommon in the PNe as several kinematic observations indicate (see e.g., Muthu et al 2001).

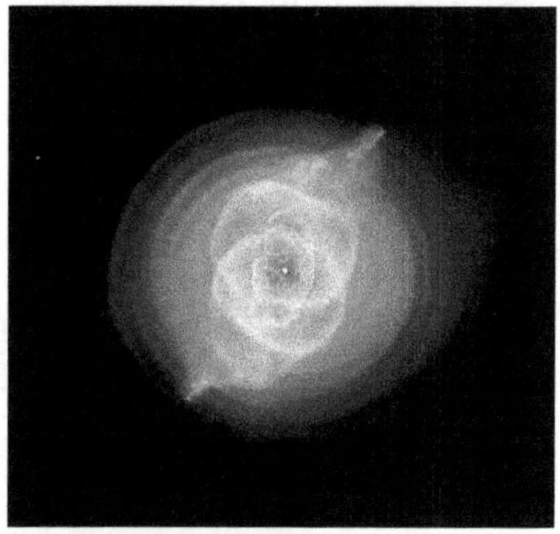

Figure 1.17: Planetary Nebulae - late stages of intermediate stars. The Cat's eye (From Hubble Gallery).

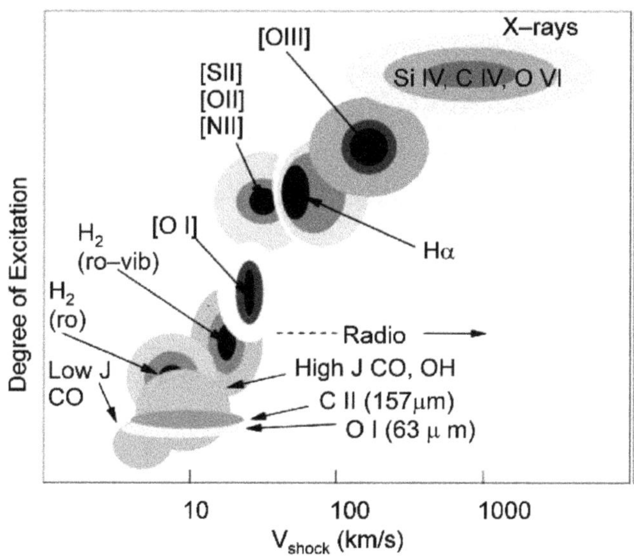

Figure 1.18: Shock Excitation in Nebulae. From Reipurth & Bally 2001.

1.2.4.4 Supernova Remnants: These are the diffuse gaseous matter ejected by a supernova explosion at the end of the journey of evolution of a massive star having mass higher than about 15 $M_\odot$. The gas is rich with heavy elements. Supernova remnants (SNR) are irregular in their morphology unlike the PNe. Fig 1.19 shows the well-known SNR called the Crab Nebula formed from the Supernova SN 1054 first identified in the year 1054 CE. The diffuse gas may be observed in emission lines from several ionized and neutral species. The energetics of SNR may be derived from observations of such emission lines. Their central objects are either neutron stars (pulsars) or black holes depending upon the initial core mass of their progenitors, required to be more than 3 $M_{\odot}$. SN/SNR enrich the ISM by rapid neutron-capture elements. For details on SN/SNR we may refer to detailed review articles by Weiler & Sramek (1988), Smartt (2009), Janka (2012), Marchant & Bodensteiner (2024).

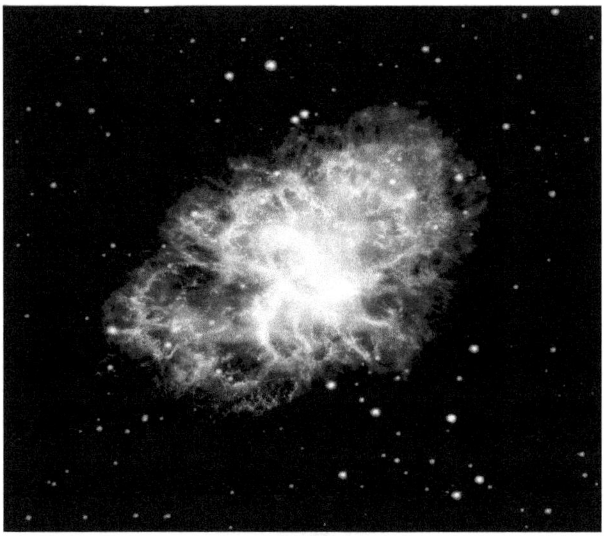

Figure 1.19: Supernova Remnant - debris from supernova explosion in massive stars - picture shows the famous Crab nebula (from Hubble gallery).

1.2.5 Galaxies: Galaxies such as our Milky Way are large (several kpc in extents) conglomerates of stars and gas (we may refer to the excellent book on the subject by Binney & Tremaine 2008). The number of stars in a galaxy is about a few billions and the total mass that it contains in the form of stars (and gas and dust) is more than 10 billion solar masses. Galaxies are classified into mainly four categories based on their morphology: elliptical (denoted by E) having round shape and smooth distribution of light, spiral (S or SA) having a central bulge and an outer flattened disk, barred spiral (SB) with a central bulge and a bar as well as a flattened disk and irregular galaxies (Irr) having irregular geometrical shape. The elliptical galaxies are further sub-classified depending upon the ratio of their major-to-minor axes (from E0, E1 ....). Similarly, the spirals and barred spirals are sub-classified (into Sa, Sb and so on; and SBa, SBb and so on respectively) by the "tightness" with which the two spirals are wound. Figure 1.20 shows the classification (known as "Hubble Classification" after the pioneering work of Edwin Hubble) schematically. Fig 1.21 shows the spiral galaxy Messier 81 (M 81) observed by the Hubble Space Telescope (HST).

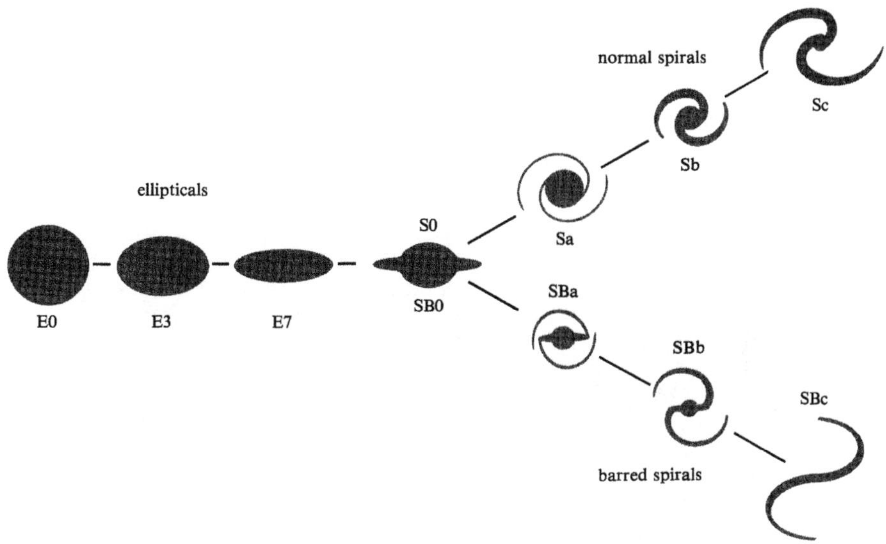

Figure 1.20: Hubble Classification of Galaxies (taken from Shu 1982).

The elliptical galaxies have mostly stellar content with very little gas and dust. Therefore, we usually do not find evidence for recent star formation leading to the presence of young stars in ellipticals. Their component velocities are predominantly random compared to rotational velocities. In spirals the rotational velocities are more predominant. This difference could account for their shapes – roundness in elliptical galaxies and flattened disks in spirals. The elliptical galaxies are also termed as early type while spirals are called late type. Star formation occurs in the spiral arms of both spirals and barred spirals.

The rotational velocities in spirals are expected to be following a rigid body law – increasing with the radial distance from centre – indicating high density of stellar content. Following the bulge however we expect a Keplerian law for rotational velocities - like the planets in Solar System – decreasing with radial distance as $R^{-1/2}$. But the spectroscopic observations of spirals show flattened rotational curves as schematically shown in Fig 1.22. This is indicative of some form of unseen or undetected and yet unknown matter called "dark matter" surrounding the galaxies.

Hubble's law or Hubble-Lemaitre Law: From the redshifts of galaxies Edwin Hubble (Hubble 1929) found out that the receding velocity of a galaxy is proportional to its distance. This shows that the farther the galaxy is the faster it is receding from the local group. This is called the Hubble's law given by

$$V_G = H_0 \, D_G \tag{1.33}$$

Here V_G and D_G denote the velocity of the galaxy and its distance. H_0 is the proportionality constant famously known as Hubble constant having a value of 50-75 km/s/Mpc. The most inevitable explanation for this path-breaking observational finding is that the Universe is expanding; and hence the Hubble "constant" plays the most crucial role in reconciling with the existing Cosmological theories.

Figure 1.21: The great spiral galaxy M81 (from Hubble gallery).

Apart from the classical elliptical and spiral and irregular galaxies, the extragalactic cosmos presents also an exotic type of sources called Active Galactic Nuclei (AGN) which typically show extraordinarily bright central (nucleus) source putatively embedded in a host galaxy. The central source is thought to be powered by a supermassive Black Hole. Of the AGNs, the Seyfert type of galaxies are relatively nearby as compared to the other types called Quasars and Blazars. For a recent detailed exposition, we may refer to Netzer (2015). A typical spectrum of such sources presenting a bevy of emission lines is shown in Fig 1.23: from Circinus Galaxy (a Seyfert type AGN).

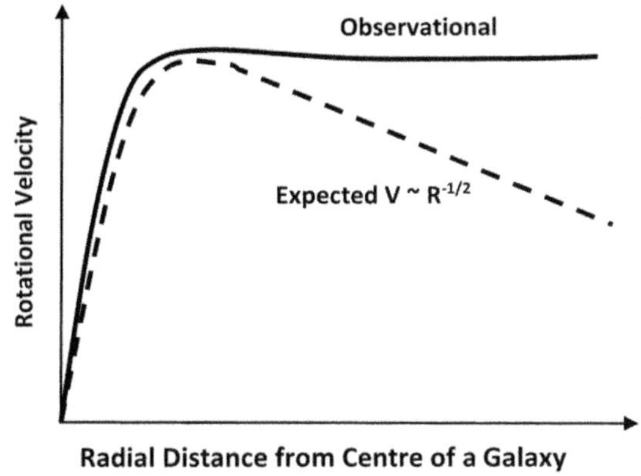

Figure 1.22: Rotational Velocity Curves for Spiral Galaxies showing flatness (from Rubin et al 1985).

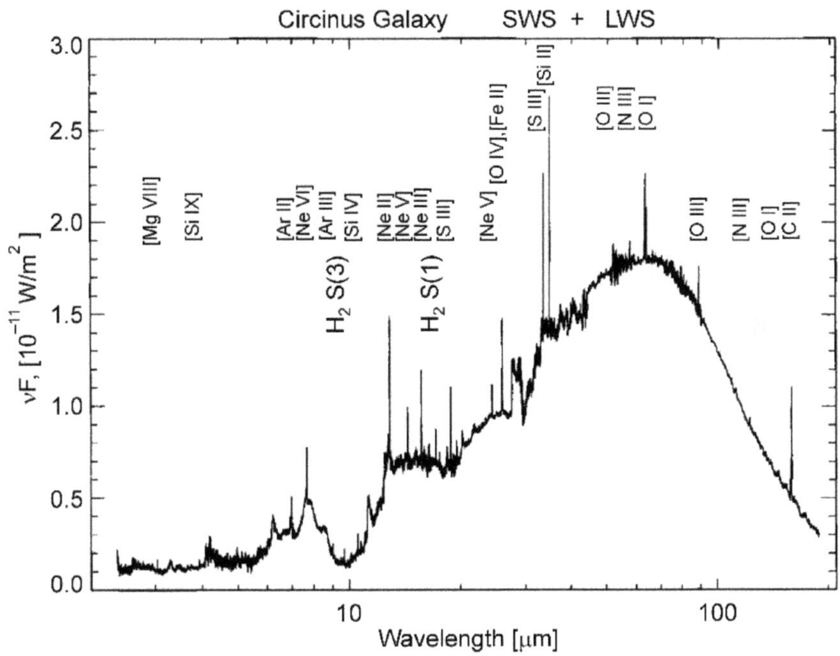

Figure 1.23: ISO infrared spectrum of Circinus Galaxy (from Glass 1999).

1.2.6 Transient Sources: These sources are time-varying on scales of the order of 1 sec and longer. Time resolved and time tagged observations are very important to track the dynamic behaviour of the source. Some well-known examples of such sources are Novae, Supernovae, Outbursts from Young Stellar Objects (YSOs) and Gamma Ray Bursts (GRBs, see e.g., van

Paradijs 2000; Berger 2014). These are objects that display aperiodic or random variability. Objects that vary systematically with highly precise periods such as variable stars and Pulsars do not fall into this category.

1.3 The Forthcoming Chapters

As discussed in the foregoing sections, the celestial sphere offers a variety of sources (called astronomical sources) that are not only very far-off but also very faint in comparative terms, because of their large distances.

We need special instruments to make observations that could unravel the behaviour of these sources. Naked-eye observations, the most ancient and serious pastime for mankind, fascinating even today as it were, had given qualitative account of some of the brightest sources in the sky. However, such unaided observations cannot give deeper (fainter) insight and reliable, non-subjective (objective) and quantitative information. Observations of celestial objects involve measuring the number of photons per unit time, in unit energy (frequency or wavelength) interval and in unit view angle (steradian). These physical quantities are defined in Section 1.1.

Firstly, we need a device, called the telescope, to magnify and collect sufficiently large number of photons. Telescope (usually a parabolic mirror) itself however does not record the photons as they bounce off from its highly reflective surface. We need a suitable optical arrangement to collect the light from the telescope and focus onto a detector to detect and make a record of the detection quantitatively. We discuss these two important tools of observations viz. telescopes and detectors (for visible and near-infrared wavelength regions of electromagnetic spectrum) in Chapter 2 entitled "Telescopes and Detectors".

In Chapter 3 titled "Radiation Processes and Atomic and Molecular Spectroscopy" we discuss the most common or predominant radiation mechanisms in celestial objects which depend on the prevalent physical processes.

Chapter 4 ("Photometric Techniques") discusses astronomical photometry. We discuss instrumentation and techniques used to derive physical parameters concerning celestial objects.

In Chapter 5 titled "Spectroscopic Techniques" we present astronomical spectroscopic instrumentation and techniques to obtain important physical parameters.

Data quality enhancement techniques are discussed in Chapter 6. We conclude in an "Epilogue" by mentioning future directions in Chapter 7.

CHAPTER 2

Telescopes and Detectors

By night the Glass of Galileo … observes, Imagin'd land and regions in the Moon – John Milton

In observational astronomy, the Telescope is the most fundamental front-end instrument which gathers electromagnetic radiation reaching terrestrial environment from a celestial object. From the earlier discussion on flux density/photon number density (eqn. 1.3), clearly the larger the size of the telescope the more would be the number of photons it will collect. Since the light from cosmic sources is so faint that it requires light gathering capability for us to detect the source above the noise from extraneous and internal (detector) sources. A telescope of a given aperture or size would collect fewer photons from a far-off object than from the one of same luminosity but closer to it. The light gathered thus by the telescope requires to be detected and recorded by a back-end instrument called detector either by analog or digital means. The principal light gathering optics of a Telescope can be made either using lenses (refractive optics) or mirrors (reflective optics). The refractive telescopes however become heavy and cumbersome and unwieldy especially when larger aperture is required. Here in this chapter, we discuss only the reflective telescopes and the corresponding types of foci.

Further, we discuss the types of detectors used for visible and infrared wavelength ranges and the process of matching the detector with the telescope for obtaining optimal signal. Firstly, we describe the celestial coordinate system using which coordinates of any object in the sky may be uniquely determined. Such a system of coordinates facilitates pointing and tracking of a celestial source for by the telescope.

2.1 Celestial Sphere and Equatorial Coordinate Systems

Celestial sphere may be defined as a sphere concentric with the Earth with its equator coplanar with that of the Earth. The North and South poles of the Earth are extended to meet the North Celestial and South Celestial poles respectively (NCP and SCP), as shown schematically in Figure 2.1. All cosmic sources can then be imagined to be situated on the surface of the celestial sphere. As shown in the figure, we can draw a great meridian circle passing through the NCP, any given celestial object S and SCP. Such a circle intersects the celestial equator at a point that makes an angle of δ with the line joining the centre of the celestial sphere and the star S. This angle called declination is analogous to latitude in the geographical coordinate system. The declination varies from 0 to +90 degrees (positive) in the northern hemisphere, and 0 to -90 degrees (negative) in the southern hemisphere. When the Sun traverses in the celestial sphere in the ecliptic plane (making an angle of $23^0\,27'$ with the celestial equator) in an apparent great circle during the course of a year, it meets the celestial equator at a point called Vernal Equinox (χ in the figure) while going from south towards north.

North Celestial Pole

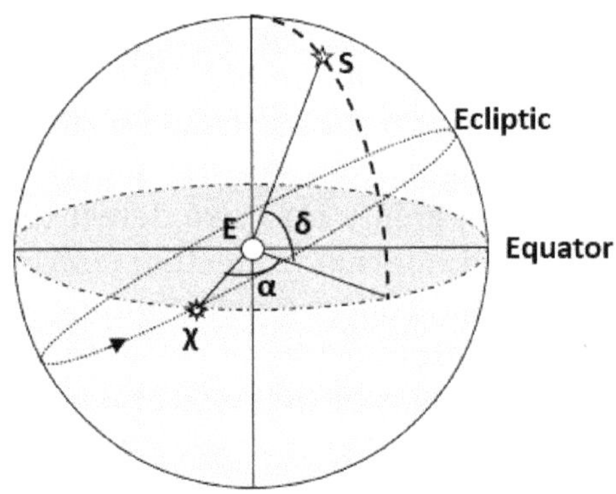

South Celestial Pole

Figure 2.1: The Celestial Sphere (shown by the full circle) in a geocentric configuration. The Earth is marked E being the tiny circle at the centre and its equator is in the same plane as the celestial equator (the dash-dotted curve). The star is marked S. NCP is the north celestial pole and at the nadir is the south celestial pole (SCP – not marked). The dashed curve is the great circle from NCP to SCP passing through the star S. The position of S is uniquely defined by the equatorial coordinates, right ascension, and declination (α, δ). The point where the Sun cuts the celestial equator while traversing the ecliptic circle from south towards north (marked by an arrow) is called the vernal equinox (χ).

Another coordinate for the object may then be designated as the angle that the line joining the point on the equator makes with the line joining the Vernal Equinox on the equator. This coordinate called Right Ascension (RA or α) along with the declination uniquely defines the position of a celestial object in the sky. RA is analogous to the longitude in the geographical coordinates. RA varies from 0 to 360 degrees in the anticlockwise direction along the celestial equator. The sidereal motion of the object is defined by the coordinate called Hour Angle (HA or Θ)

which is the difference between the Local Sidereal Time (LST) and the RA of the object. Due to the relative motion of stars in the background the RA changes very slowly. But HA changes with local sidereal time. This is further illustrated in the ensuing paragraphs.

2.1.1 Local Horizon Coordinate System: Another way of uniquely defining the position of an object in the sky is to define a local coordinate system. At any location we can define the horizon as the great circle intersecting the celestial sphere and the plane perpendicular to the observer's zenith at the position of the observer, as shown in Figure 2.2. The plane that is formed by the horizon is the horizontal plane. The local meridian is the great circle perpendicular to the horizontal plane passing through the zenith and NCP. We can define the local North and East to the right of the North on this plane. Any object in the sky can then be projected on to the horizontal plane. Then the altitude or elevation (*E*) and azimuth (*A*) as shown in the figure uniquely define the position of an object in the sky at any given time. Azimuth is always measured in the horizontal plane from north towards east. Elevation is always positive from 0^0 to 90^0. Zenith angle is defined as (90^0 – *E*) measured from the zenith towards the horizontal plane.

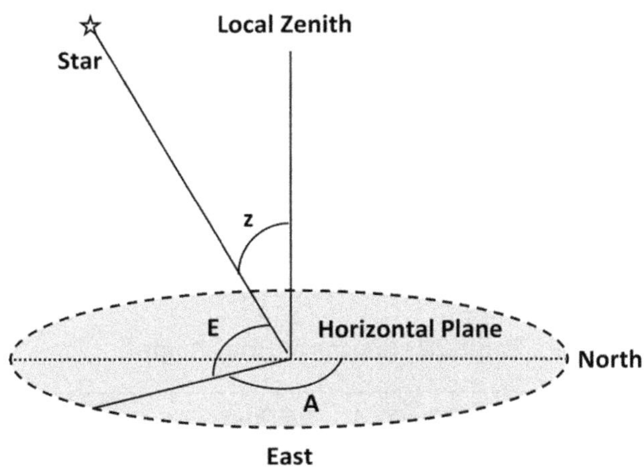

Figure 2.2: Altitude (Elevation) – Azimuth (E-A) coordinate system. The shaded disk shows the local horizontal plane drawn tangential to a position on the surface of the earth. The vertical from that point perpendicular to the horizontal plane is the local zenith. The star position then is uniquely given by the elevation (E) and the azimuth (A) measured from local north towards east. Elevation is sometimes replaced by the zenith angle z (= 90 - E).

The two coordinate systems, as defined above, are related by the following formulae that may be derived from spherical trigonometry (see e.g., Hann 1866). For a given site of latitude ϕ, the transformation equations for converting from equatorial coordinates (δ, Θ) to alt-az coordinates (E, A) are given by,

$$\begin{aligned} \cos E \sin A &= -\cos\delta \sin\Theta \\ \cos E \cos A &= \sin\delta \cos\varphi - \cos\delta \cos\Theta \sin\varphi \\ \sin E &= \sin\delta \sin\varphi + \cos\delta \cos\Theta \cos\varphi \end{aligned} \tag{2.1}$$

And for reverse calculation from (E, A) to (δ, Θ) the equations are,

$$\begin{aligned} \cos\delta \cos\Theta &= \sin E \cos\varphi - \cos E \cos A \sin\varphi \\ \sin\delta &= \sin E \sin\varphi + \cos E \cos A \cos\varphi \end{aligned} \tag{2.2}$$

with, $$\Theta = LST - \alpha\,. \tag{2.3}$$

2.1.2 Transit and Sidereal Time: A transit corresponds to the position of highest elevation of an object in the sky on the meridian – the great circle through the two celestial poles and local zenith. Objects transiting will gain ~ 4m every day due to the difference in the sidereal and solar days – transit occurs earlier each day passing. As mentioned earlier, Hour Angle (Θ) is defined as the difference between Local Sidereal Time (LST) and the Right Ascension (RA or α) of an object. The following table 2.1 gives the 4 epochs of Hour Angle in a solar year.

The time between two consecutive noontimes constitutes a solar day, while the time between two consecutive transits of a star is termed as a sidereal day. A sidereal day is 3m 56s (~ 4m) shorter than a solar day.

Table 2.1: Epochs of Hour Angle.

Date	Θ (hrs) at Noon	Θ (hrs) at Mid-Night	Epoch
March 21	00	12	Vernal Equinox
June 21	06	18	Summer Solstice
September 23	12	24/0	Autumn Equinox
December 21	18	06	Winter Solstice

The Sun is at 0h *HA* when at the vernal equinox while crossing the celestial equator from south towards north on Mar 21 (see Figure 2.3).

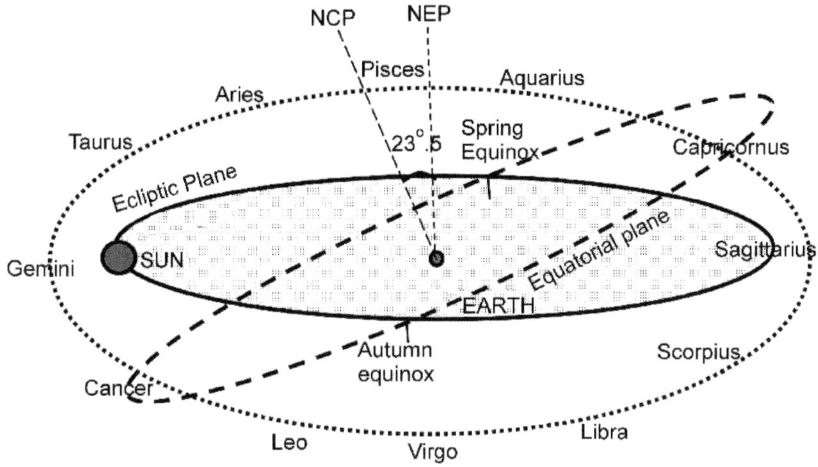

Figure 2.3: The Sun's apparent path seen by an earth-based observer. The zodiacal constellations and the seasonal epochs may be seen (adapted from Shu 1982). NEP is North Ecliptic Pole.

The local sidereal time (*LST*) may be calculated for a place of longitude Λ by using the relation by

$$LST = GMST + \Lambda/15 \tag{2.4}$$

with *GMST* as Greenwich mean sidereal time.

Julian Day: For observations on time varying phenomena, we generally use all observations referenced in one timescale, and this is called the Julian Day (*JD*). The *JD* may be found out for a particular day (*D*) in a month (*M*) of a year (*Y*) by using the relation given by,

$$JD = D - 32075 + \frac{1461\left(Y+4800+\frac{M-14}{12}\right)}{4} + 367\ (M - 2 - (\frac{M-14}{144})/12 \tag{2.5}$$

Now, with the advent of a new millennium the epoch being shifted to the year 2000.0, we use the Modified JD or *MJD* given by,

$$MJD = JD - 2400000.5$$

Julian centuries (since 1 January 2000) are given by

$$T = (JD - 2451545.0)/36525.$$

Now *GMST* is given by,

$$GMST = 6^h 41^m 50^s.54841 + 8640184^s.812866\, T + 0^s.093104\, T^2 - 0^s.0000062\, T^3 \quad (2.6)$$

The Zodiacal Belt: Zodiacal belt is an imaginary 18^0 wide circular strip centred on the ecliptic great circle defining the path of the Sun as depicted in Celestial Sphere (see Fig. 2.3). The Constellations that the Sun passes "through" (over) are called the zodiacal constellations. Early astronomers (and astrologers) believed that the Sun passes over 12 constellations taking 1 month to cross each constellation. However, this picture has changed since Hipparchus' time due to the precession of Earth's spin axis moving westwards at a rate of 1^0 of arc every 72 years. Accordingly, the first point of Aries or the Vernal Equinox now occurs at a different time of the year. Further, modern observational astronomy has established that the Sun crosses 13 constellations and not 12 and, in each constellation, the "resident" time varies and not 1 month in every case. The constellation Ophiuchus is now in the Zodiacal belt showing in between Scorpius and Sagittarius in the chain starting at the Vernal Equinox in Aries through Taurus, Gemini, Cancer, Leo, Virgo, Libra, Scorpius, Sagittarius, Capricornus, Aquarius, and Pisces. Obviously, therefore, the picture of the Zodiacal belt is changing with time, though relatively slowly comparable to human's life span. The orbits of some (outer) planets are also not confined to only these constellations and but extend to those in the neighbourhood - these constellations are called the "para-zodiacal". We must note here that, even today, the *astrological* predictions are made based on the earlier picture of 12 zodiacal signs with the Sun, solar planets and Moon crossing each sign in a month's time.

Apparent Positional Changes of Cosmic Sources due to Precession of the Earth's Rotational Axis: The celestial objects change their apparent positional coordinates due to the precession of the axis of rotation of the earth with a period of 25,770 years. These changes are given by (see Lang 1974),

$$\Delta\alpha = M + N \sin\alpha \tan\delta; \ \Delta\delta = N \cos\alpha$$

With M = 46.12436 arcsec/year and N = 20.04311 arcsec/year, for the Julian epoch J2000. These values further change with time but slowly: M changes by + 0.00186 arcsec/year and N by – 0.0085 arcsec/year per century. Changes occur also due to other factors such as nutation and, aberration (for details, see Lang 1974) but, for very distant sources such as stars, these are very small compared to the precession correction.

Proper Motion of Celestial Objects: All celestial objects are in motion in space: planets around the Sun, the Sun other stars around the centre of our galaxy called Milky Way, and galaxies around the centre of a cluster of galaxies and so on. Therefore, as seen by an earth-based observer, the positions of the astronomical objects do not remain constant but change albeit very slowly with time. When we observe a star, for example, using a spectrograph, we may detect its motion in our line-of-sight say V_r, called radial velocity, as shown in Figure 2.4. If the actual velocity of the star known as the space velocity V_s, makes an angle Θ with the radial velocity vector, we can obtain the proper motion, defined as the rate of change of this angle with time, from multiple observations at different times. The time interval Δt during which the star would have moved in the sky or transverse plane by a distance of Δr, determines the transverse velocity component V_Θ. The proper motion is then given by (see Carroll &Ostlie 2007, for details),

$$\frac{\Delta\theta}{\Delta t} = \frac{V_t}{d}$$

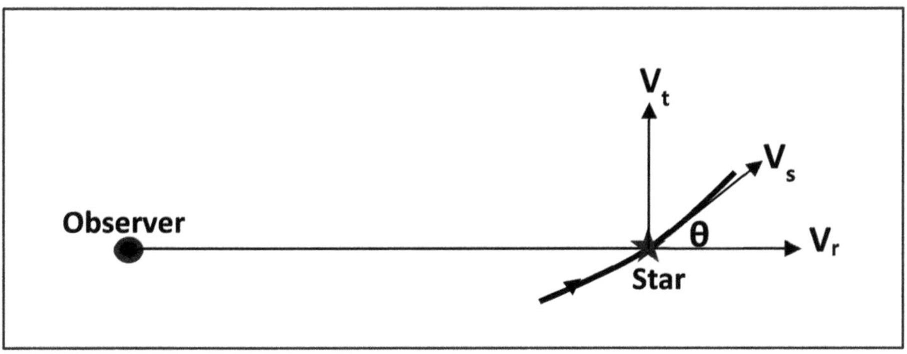

Figure 2.4: Definition of proper motion of a celestial object.

The changes in the equatorial coordinates (α, δ) of the star, due its proper motion, are given by,

$$\Delta\alpha = \Delta\theta \frac{\sin\varphi}{\cos\delta}\ ;\ \Delta\delta = \Delta\theta\cos\varphi$$

where φ is called the position angle of the star's motion on the celestial sphere and δ is the declination of the star.

In general, the changes due to precession are larger than those due to the proper motion.

2.2 Telescope Mirrors

A celestial object may be brought to focus by using a large enough convex lens (refractive) or by a concave mirror (reflective). However, as already mentioned, in the case of a lens the telescope optics becomes too lengthy and clumsy to handle and hence all the telescopes that are built or being built since the middle of the last century, employ large parabolic mirrors. Telescopes with spherical mirrors suffer from imaging defects such as spherical and chromatic aberrations and parabolic mirrors do not suffer from these basic aberrations and hence are extensively used.

The basic components of a reflecting telescope are the primary mirror that collects light from a distant celestial object and reflects the collected light into a detector. The primary mirror is therefore the heart of the telescope. In order to decrease the thermal effects (expansion) the telescope mirror blanks are made of glass materials that have a very low thermal expansion coefficient. These glasses are commercially brand-named as Cervit, Zerodur and ULE, having expansion coefficients of $\sim 10^{-6\text{-}7}$ per degree Kelvin. To eliminate the spherical aberration from peripheral rays the mirror surface is made into a parabolic surface, as mentioned earlier. There are a few other aberrations that need to be addressed in obtaining better image quality. When off-axis rays are brought to focus the image of a point source like a star appears elongated in one direction, called coma, like the coma of a comet. Another optical imaging defect, called astigmatism, occurs when different parts of the mirror produce image at different parts in the image plane. Curvature of field produces the same effect as astigmatism, but the images are produced on a curve. Field distortion is the defect that makes the off-axis field to have a different plate scale from near axis. For details on the astronomical optics and the associated aberrations, we may refer to Schroeder (2000).

After polishing the mirror to the desired precision surfacing, the mirror is coated with a thin film of aluminium or silver for good reflectivity in the visible and infrared wavelengths.

2.2.1 Telescope Mounts: As shown earlier there are principally two coordinate systems based on which an object in the sky may be uniquely identified. The telescopes are mounted in accordance with any of the two systems: Equatorial and Altitude-Azimuth or Alt-Az mounts. In the equatorial mount the telescope works in equatorial coordinate system or right ascension – declination (α-δ) coordinates. The mounting is done in such a way that the declination axis is aligned perpendicular to the polar axis of the earth and follows the sidereal motion. This mounting has the advantage of directly following the equatorial coordinates namely right ascension and declination. The alt-az mount is based on the local horizontal plane coordinate system. In general, telescopes larger and hence more massive than ~ 2 m aperture are mounted in Alt-Az configuration for better stability. In such large telescopes, a dedicated computer converts in real-time the equatorial coordinates of celestial objects to the Alt-Az coordinates. The conversion formulae of coordinate systems are given in the earlier section.

2.2.2 Light Collecting Capacity and Diffraction Limit: The aperture of the telescope, D, decides both. The light collecting capacity depends upon the area of the telescope aperture (D^2). The aperture also decides the angular resolution. The telescope mirror forms a diffraction pattern at the image (focal) plane when light is passed through a narrow slit or a circular aperture. The central bright spot in diffraction pattern is called Airy disc. Based on this theory, Rayleigh determined the angular resolution of the primary limited by the Airy disc as

$$\theta_{DL} = 1.22\,\frac{\lambda}{D} \tag{2.7}$$

where the subscript *DL* denoted Diffraction-limited resolution and λ the wavelength of light ray forming the image. The larger the telescope size, the greater is the angular resolution in addition to greater light gathering capability. Table 2.2 lists the diffraction limited angular resolution at various wavelengths for a 10 m telescope. For a 1m telescope the limits are 10 times lower.

Table 2.2: Diffraction limited angular resolution for 10 m aperture at different wavelengths.

λ	θ_{DL}
0.5 μm	0″.013
1.0 μm	0″.025
2.0 μm	0″.050
10.0 μm	0″.25
100 μm	2″.50
1 mm	25″.10
10 mm	4′.20
10 cm	0°.70
100 cm	7°.0

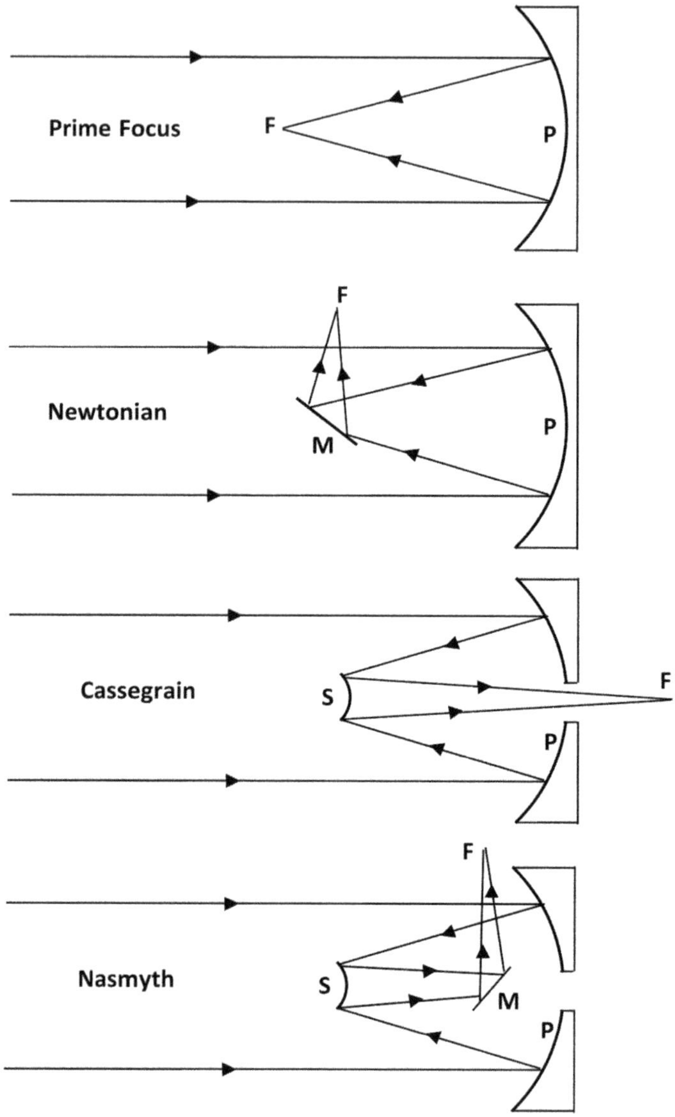

Figure 2.5: Telescope Optics: in the top panel is shown the classical prime focus where there is only the primary mirror (P); the second panel shows the Newtonian focus in which the primary focus is brought out of telescope cage by using a plane mirror (M); the third panel shows the Cassegrain focus in which the primary focus is brought out of the telescope cage though a small circular hole in the primary mirror by using a hyperbolic secondary mirror (S). In the fourth panel at the bottom is shown the Nasmyth optics in which the focus is brought out by using a hyperbolic secondary mirror and a plane tertiary mirror. The Nasmyth is akin to the coude focus except that in the latter case, the effective focal length is very long. In all panels the focus is marked 'F'.

2.2.3 Types of Telescope Optics: The focus that the rays reflected from the telescope primary mirror are brought to is called the **prime focus**. For most applications, a secondary mirror and sometimes a tertiary mirror is used to bring the focus of the primary to a convenient position for mounting instrumentation. Fig 2.5 schematically illustrates various arrangements of telescope optics. As can be seen from the figure (top panel), it would not only be extremely cumbersome to mount an instrument at the prime focus, but such an arrangement would also block part of the primary mirror and hence blocking the light from celestial objects. Prime focal arrangement is best suited for imaging large fields of view requiring a smaller focal ratio. Another way to bring the primary focus out of the telescope cage is to use a plane mirror in the path of the primary focus as shown in the figure. This arrangement is known as Newtonian optics. Alternatively, there are several mounts that have been in use which employ a secondary hyperbolic mirror, as illustrated schematically in the figure. The Cassegrain focal optics has a central hole on the primary through which the focus is brought out of the telescope cage by using a hyperbolic secondary mirror. Usually, the central hole of the primary is a little smaller in diameter than that of the secondary and the area of the primary traded off in the process is negligibly small compared to its total area. This arrangement accommodates large instrumentation at the focus. The Coudè optics (not shown), used for equatorial-mount telescopes, in which the focus is brought by secondary and tertiary mirrors, is for very massive and large instruments that require large focal distance. There is a variant to this called Nasmyth focus (see the bottom panel of the figure), which is generally used for large instruments on rotating platforms on large telescopes (having primary mirrors of size larger than 2 m) with Alt-Az mounted telescopes. By far the most versatile and convenient mount is the Cassegrain for 1-2 m class telescope primary mirrors. For large fields, a spherical primary mirror is used with correcting optics for aberrations and this arrangement is called a Schmidt Cassegrain telescope.

The combination of parabolic primary and hyperbolic secondary is very good for central rays or small sky angles. At large angles from the principal axis (called off-axis rays), aberrations such as coma and astigmatism will creep in distorting the image. A solution for this problem was found out by Ritchey and Chretien who used hyperbolic mirrors for both primary and secondary so that the off-axis rays are also nearly aberration-free. However, this arrangement suffers from an increase in astigmatism. For details we may refer to Schroeder (2000).

2.2.4 Surface Quality of the Primary: To achieve diffraction-limited accuracies in spatial resolution and image quality the surface of the primary mirror should have a good surface optical quality. The optical surface of the primary is made smooth at more than $\lambda/20$. The surface quality is also specified by the undulations that it may have – the peak-to-valley distance. This is usually made to less than 100 μm for best quality images. The quality of the mirror may also be specified by the encircled energy (*EE*) defined as the amount of light collected in the central maximum of the diffraction pattern called Airy disc specified in arcsec: for example, an *EE* of 90% in a circle of 0.3 arcsec at the image plane. In practice however the diffraction-limited resolution cannot be realised due to the distortion of light rays by the lower atmospheric turbulent eddies. This aspect will be discussed in detail in the ensuing subsections.

2.2.5 Focal Ratio and Plate Scale: The Focal-ratio (*f*-ratio or *f*-number) is defined as the ratio of the focal length of the optical system of the telescope that forms the focus to the aperture size of the primary mirror element. For purely primary mirror forming the optics, the ratio is $N_P = f_P/D$, where f_P is the focal length of the primary mirror and D its diameter (or aperture). For example, the Keck telescope has a diameter of 10 m and a prime focal ratio of 1.75. When we use one of the focal optics shown in Fig 2.5, the focal length of the overall optics needs to be considered for finding the focal ratio. The focal ratio is then the ratio (N_f) of the telescope aperture to its effective focal length as given below.

$$N_f = \frac{f}{D} \tag{2.8}$$

The focal ratio determines the size of the celestial image on the detector surface and hence the spatial resolution. It also determines the intensity of the light received by unit area at the focal plane. These factors need to be considered while manufacturing the primary mirror.

Plate scale refers to matching of the sky field to the effective focal plane for a given telescope configuration. It simply defines the amount of sky angle imaged per unit linear scale in the image plane. In Figure 2.6, the telescope mirror is replaced by a lens for illustrating the point in a better manner. As shown in the figure, let us assume that for an angle θ in the sky plane, the linear size in the image plane is x. If the effective focal distance is f, then the plate scale is given by $(x/f) = \tan\theta$. Considering $x = 1$ mm, we obtain, for small angles, the plate scale in arcsec per mm on the image plane. For example, for the Mt Abu Gurushikhar 1.2 m telescope at its *f*/13 Cassegrain focus, the plate scale is about 13 arcsec per mm at the image plane. The plate scale in arcsec/mm is given by the formula

$$Plate\ Scale = \frac{206265}{f}\ \frac{arcsec}{mm}. \tag{2.9}$$

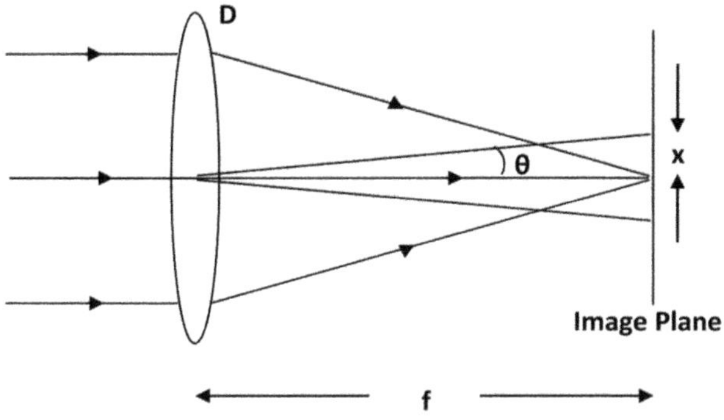

Figure 2.6: Plate scale: f is the effective focal length of the telescope and x is a measure of plate scale.

To satisfy the Nyquist criterion for adequate sampling, a sky angle equivalent to the PSF (e.g., 1 arcsec) should be imaged on two basic imaging elements or pixels of a modern digital detector array.

Matching the Telescope: Using a lens analogy for illustration, Figure 2.7 shows schematically the object and image planes and the solid angles (see also Figure 2.6) that the two subtend with respect to the centre of the telescope.

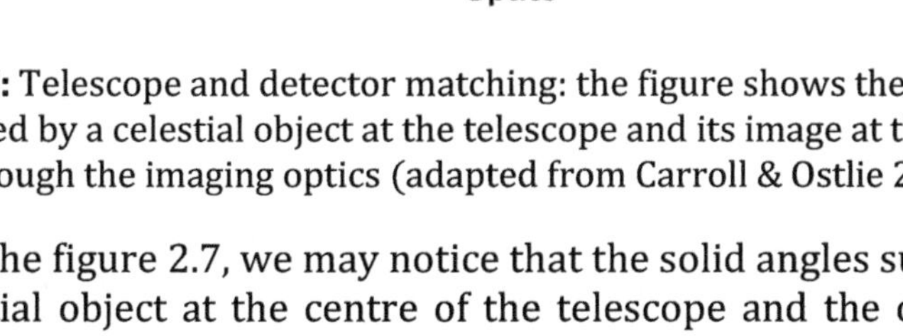

Figure 2.7: Telescope and detector matching: the figure shows the solid angles subtended by a celestial object at the telescope and its image at the detector through the imaging optics (adapted from Carroll & Ostlie 2007).

From the figure 2.7, we may notice that the solid angles subtended by the celestial object at the centre of the telescope and the one that the telescope pupil subtends at the image plane are equal, i.e., $d\Omega_o$ corresponds to $d\Omega_i$ (see Fig 2.7). Therefore,

$$\left(\frac{dA_o}{f_T^2}\right) = \left(\frac{dA_i}{f_i^2}\right) \tag{2.10}$$

where f_T and f_i are the effective focal lengths of the telescope and of the imaging optics, respectively. A_o and A_i are the areas of the object in the sky and its image on the detector, respectively. This is how we need to match the telescope with the instrument for a required image size on the detector by using suitable collimating and imaging lenses. This is done by matching the detector size (or pixel size) with the plate scale (or f-number) appropriately.

2.2.6 Point Spread Function: The 2-dimensional function that describes the diffraction pattern of a telescope image of a point source through a circular aperture is called the Point Spread Function (PSF). Figure 2.8 shows a typical image pattern for a point source in the sky. As briefly mentioned earlier, the Airy function describes the diffraction pattern for a point source.

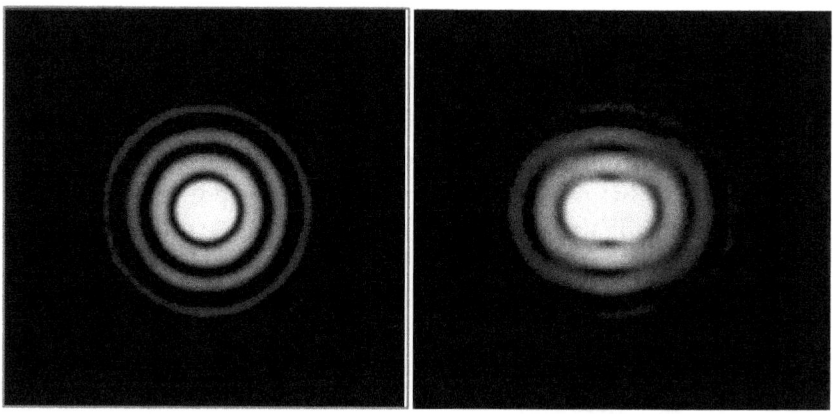

Figure 2.8: Diffraction pattern of a circular aperture. On the right is a resolved binary star as per Rayleigh criterion. (from Pedrotti et al 2017).

The central bright spot called Airy disc is enclosed within the first minimum. It contains 86% of the total light in the image plane. The first three minima occur at 1.22, 2.23, 3.24 λ/D as shown in the figure. The central bright spot has a full width at half-maximum (FWHM) of $1.028\lambda/D$. The angular size of the pattern is important for resolving two close-by sources as per Rayleigh criterion (illustrated in Chapter 5).

The Airy function representing the image (see Fig 2.8) in one-dimension, can be written as

$$I(x) = \frac{1}{(1-\epsilon^2)^2}\left[\frac{2J_1(x)}{x} - \epsilon^2\frac{2J_1(\epsilon x)}{\epsilon x}\right]^2 \tag{2.11}$$

with the dimensionless variable $x = (\pi/\lambda)\,D\theta$.

Here J_1 is the Bessel function of the order 1 of first kind. D is the telescope aperture and λ is the wavelength at which image is being recorded and ε is the fractional radius of the central obscuration of the primary mirror assumed circular. The dimensionless variable x is the distance from the optic axis in the focal plane and θ is the corresponding angular distance as measured from the aperture.

Figure 2.9 shows the Airy pattern in one-dimension, where the surface brightness is plotted against θ as given by the above equation.

Here it is assumed that the parameter ε is zero (i.e., no central obscuration of the primary mirror). We may note that the function $J_1(x)/x$ tends to 1 as x tends to zero. The successive bright rings show a cyclical pattern with the central bright maximum showing a narrowing trend for an obstructed (non-zero ε) primary mirror as described by Schroeder (2000).

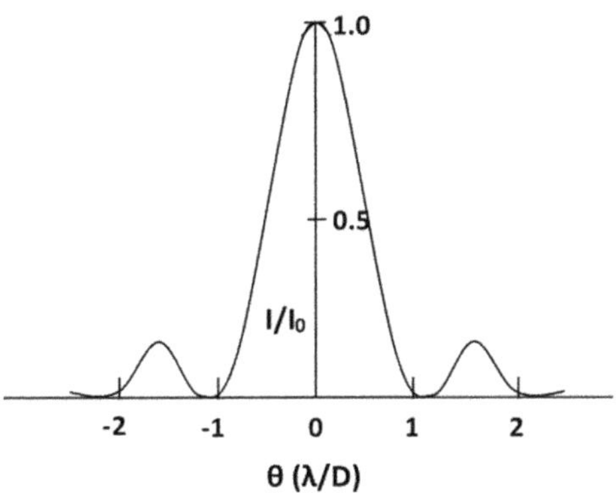

Figure 2.9: Airy function (surface brightness plotted against θ) showing the secondary and higher order maxima. The central brightest maximum is partially shown to show the smaller peaks more prominently. Assumed ε=0.

2.2.7 Encircled Energy: It is also important from the photometry point of view to know the fraction of energy embedded in the first maximum (i.e., how much of the starlight it contains). This parameter called "Encircled energy" (*EE*) is defined as the fraction of the integrated flux of the image within a specified distance x in the image plane and is given by,

$$EE(x) = \frac{(1-\varepsilon^2)}{2\,I_0} \int_0^{x_0} 2\pi\, I(x)\, x\, dx \tag{2.12}$$

With the total integrated energy or flux I_0 given by,

$$I_0 = \int_0^\infty 2\pi\, I(x)\, x\, dx \tag{2.12a}$$

The encircled energy *EE* (for a detailed diffraction theoretical treatment, see Born & Wolf 1999; Schroeder 2000) is the flux encircled within a radius x_0 as a fraction of the total integrated flux (I_0) in the image plane, centred on the Airy central maximum or PSF. The parameter ε is the fractional obstruction of the aperture of the primary mirror due to its central hole. Ideally, it should be as close to 1 as possible. Figure 2.10 shows the *EE* plotted against θ (in units of λ/D). The undulations on the top of the curve are the extrema. We notice that the first maximum (at 1.22 units) contains 86% of the energy.

The curve shown in the figure is for an un-obstructed primary mirror ($\varepsilon = 0$). For a mirror with central obstruction the EE decreases. As the obstruction (ε) increases a significant amount of energy is transferred from the Airy disc to the first maxima (see Schroeder 2000).

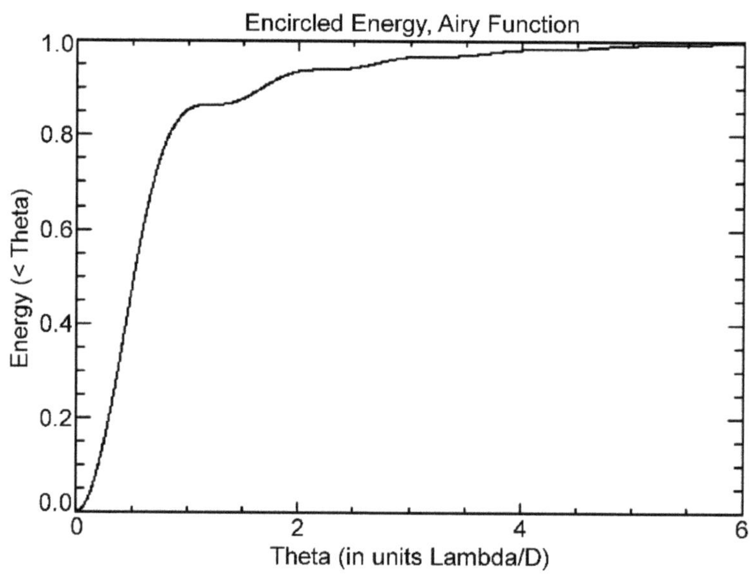

Figure 2.10: Encircled energy (*EE*) shown as a function of θ (in units of λ/D) for an un-obstructed primary (Schroeder 2000).

2.2.8 Strehl Ratio: In the presence of distortions in the incoming wave front the image quality suffers in comparison with an ideal wave front. Both the PSF and *EE* are impacted by a distorted wave front. In such a predicament, a parameter that characterises the point source image quality is Strehl ratio introduced by Karl Strehl of Germany at the end of

19th century. It is defined as the ratio of Airy disc intensity of a distorted wave front to the one produced by ideal wave front. Strehl ratio (R_S) decides the level of permissible tolerance for a distorted wave front.

$$R_S = \frac{I_A}{I_I} \tag{2.13}$$

where I_A is actual, and I_I is ideal intensities of PSF.

A telescope (including both primary and secondary is said to be diffraction-limited if $R_S \geq 0.8$. This demands a wave-front error (caused by aberrations in the optical system) of better than about $\lambda/14$ (see Schroeder 2000 for details).

2.2.9 Stops and Pupils: Stops and pupils are elements used in an optical arrangement using a telescope to form image of an object of interest.

Aperture Stop defines the amount of light gathered from an astronomical object and fed by such an arrangement for forming an image at the detector plane. It can be the rim of a mirror or a lens or a circular aperture. The size of it also determines the degree of optical aberrations and distortions. For a two-element telescope, for example, in a Cassegrain mount, it is the telescope primary mirror itself. For an infrared telescope it is the secondary mirror because it is smaller than the primary central hole to minimise the thermal emission from the rim of the hole.

Field stop is used to select relevant or required portion of a celestial object to be imaged by the telescope optical arrangement. It simply defines the field-of-view for an observation. If no diaphragm is used to select the field, then the edges of the detector would define the field stop.

Entrance Pupil is defined in general as the element that precedes the aperture stop in an optical arrangement. Since there is no element that precedes the aperture entrance pupil in the case of a two-element telescope, it is the primary mirror itself, or secondary mirror in the case of an infrared telescope.

Exit Pupil is defined as the image of the aperture stop formed by the optical system following the latter. In a two-mirror telescope the secondary is therefore the exit pupil. But in the case of an infrared telescope, the secondary is the aperture stop as well as the exit pupil. For all incident angles at the telescope primary (or secondary in the infrared case) the light beams from its rims approach the image plane as if emanating from the rim of the exit pupil.

A field lens that reimages the exit pupil on the surface of the detector is called the **Fabry lens** making the image stable and wandering-free.

Lyot Stop: In the case of near-infrared observations the telescope primary and secondary rims emit thermal radiation contaminating signal and increasing the background noise. This is particularly severe in K band and other longer wavelength bands. In the infrared bands even the spider that holds the secondary mirror contributes to the background noise. First conceived and devised by the French astronomer Bernard Lyot, the eponymous stop blocks the extraneous radiation either emitted directly or diffracted by the rims of the inner Cassegrain hole and the aperture of the primary mirror and the spider holding the secondary mirror. The stop devised precisely is placed at the plane where the primary is reimaged – after the field stop and collimator beam. The use of such a stop improves S/N and hence the limiting magnitude that may be reached especially in the K band and beyond.

2.3 Environmental Effects on Quality of Observational Data

For precision photometry and spectroscopy environmental effects need to be understood and taken into account when necessary. The environmental effects include the variation of temperature and pressure within vicinity of the observatory especially in the dusk period. These effects result in the "boiling-like" appearance of a point source such as a star. Before the start of observations, we need to allow sufficient time for attaining equilibrium between the observatory surroundings and the telescope and instrument chambers.

The atmosphere contributes to the deterioration of image quality in more than one way. The physical conditions that affect ground-based observations involve mainly the dependency of refractive index on temperature in lower troposphere, the upper troposphere as well as the tropopause region, and the subsequent turbulent mixing of several layers of different temperatures. The inversion layer and boundary layer would also contribute to the distortion. The wind effects near the mountain and nearby sea are some examples of local factors.

Boundary layer is that part of the lower atmosphere (nearly at the surface of the earth) where turbulent and other transport processes affect the physical properties of the layer. Inversion layer is situated in the tropopause where the temperature inversion occurs from decreasing to increasing trend with altitude. Establishing telescope at a higher altitude would greatly reduce the effects of these dynamic atmospheric layers on the image quality.

This is the reason why most observatories are built on mountain tops on the sides where wind blows usually into. Such high altitudes (typically a few km from the ground level) also ensure the site to be situated above the inversion layer. High-altitude observatories also have lower dust and water vapour content – the later particularly affects the infrared observations. The turbulence in the tropopause region which is about 10 km from the ground (for tropical regions) is the most important cause for image deterioration. The combined effect by which the image is deteriorated is called "seeing". This term usually includes the mid-troposphere effect, called scintillation, caused by a larger variation of the refractive index, and the "image motion" which makes the PSF peak to move around (or wander) randomly.

2.3.1 Atmospheric Turbulence and Seeing: Near the boundary between troposphere and stratosphere there exists a region of turbulent flow in the lower atmosphere at about 10 km. As shown schematically in Figure 2.11, the turbulent eddies distort the incoming wave fronts from cosmic sources causing corresponding distortions in the image quality. Even the wind at that level also can move the turbulent eddies across the field of view of the observed cosmic source causing wave front distortion. The turbulence is characterised by temperature, density, and velocity variations both spatially as well as temporally. The temperature differences between eddies lead to density differences making the refractive index of the medium inhomogeneous. This will lead to wave front distortions as it passes through the turbulent medium. The size of turbulent medium over which the phase variation is 1 radian is called Fried parameter r_0, an important length scale in seeing theory (see Lèna 1979 for more details). If we make a long enough exposure (say, longer than 1 second) of a celestial object in the sky with seeing characterised by r_0, then we obtain an angular resolution equal to the diffraction limit of a telescope of aperture equal to r_0. The value of the Fried parameter depends upon the integrated turbulence profile of the local atmosphere (which is time varying), the zenith angle of the object and the operating wavelength. It is given by,

$$r_0 \propto \lambda^{6/5}\, \chi^{-3/5} \left[\int_0^H C_n^2(z)dz\right]^{-3/5} \tag{2.14}$$

Here, χ is called the "air mass" and is equal to the secant of zenith angle z of the object under observation and H is the height of atmosphere considered relevant for turbulent effects. This equation shows that the Fried parameter decreases with the increase of turbulence (determined by the function C_n); it decreases by the increase of zenith angle and increases with wavelength. The seeing coherent timescale, defined as the time within which the turbulence may be considered coherent, is given by the following relation.

$$\tau_0 = 0.31\frac{r_0}{V} \tag{2.15}$$

where V is the macroscopic wind velocity that changes the turbulence characteristics. Assuming a wind speed of 10 m/s and the Fried parameter to be 20 cm in the visible region each turbulent cell or eddy would cross the field of view of the observer within a short time of 20 milliseconds. From the seeing point of view, it is this timescale that is important rather than the actual turbulent timescale which is expected to be much longer than the wind-induced timescale. The angle that the typical length scale subtends at the observer (see Figure 2.11) is given as follows,

$$\theta_0 = r_0/H \tag{2.16}$$

which is known as isoplanatic patch within which the turbulence remains coherent. Here "coherence" means that the distortions within the isoplanatic angle are correlated.

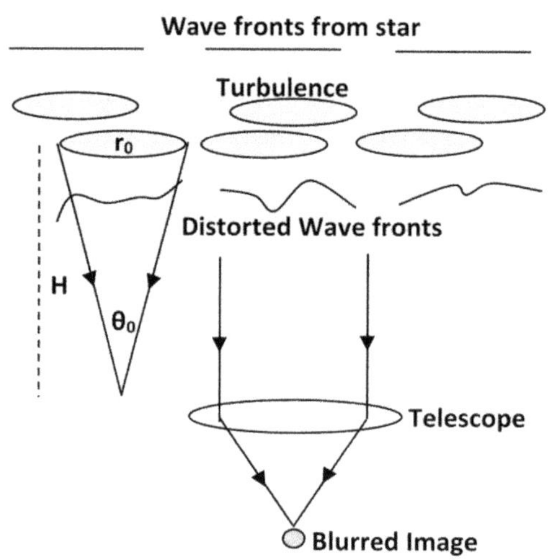

Figure 2.11: Atmospheric turbulence distortion of wave fronts. The turbulent scale size or Fried parameter is r_0 and the isoplanatic patch is θ_0.

If a telescope has an aperture D of the size of or less than the Fried parameter, usually $D \sim$ 0.5-3 r_0, then the image formed would retain the diffraction-limited quality of the image for exposures less than τ_0. However, the image motion would still be present which would deteriorate the image quality. For $D > 10\ r_0$ the image would be adversely affected by both image motion as well as wave front distortion. For large apertures much more than the Fried parameter, with long exposure times of significantly longer than 100 milliseconds, the angular resolution of a telescope is limited typically to about 1 arcsec independent of the size of the telescope aperture when used in normal uncorrected mode.

In Figure 2.11, the turbulent eddy size (Fried parameter) is denoted by r_0 and the angle that it makes at the observer, called the isoplanatic patch, is θ_0 considering the height of turbulent layer is H (taken as 10 km). The telescope aperture D is much larger than the eddy size. The turbulence distorts the incoming un-distorted plane wave fronts as shown in the figure. The distortion contains both tilt and undulations. Such distorted wave fronts when brought to focus would form a speckled image that retains the phase information, if exposure time is less the turbulent timescale. Extraction of diffraction-limited images from the speckled images (that retain both phase and amplitude information) was suggested by Labeyrie (1970). As already mentioned, if the exposure time is longer than the turbulent timescale then the speckles get smeared, and the image appears blurred. Hence the diffraction-limited angular resolution cannot be realised. The angular resolution is thus controlled by "seeing" that depends upon the atmospheric turbulence. The seeing-limited image size that is independent of the size of the telescope aperture is given as follows,

$$\theta_{SL} = \lambda / r_0 \tag{2.17}$$

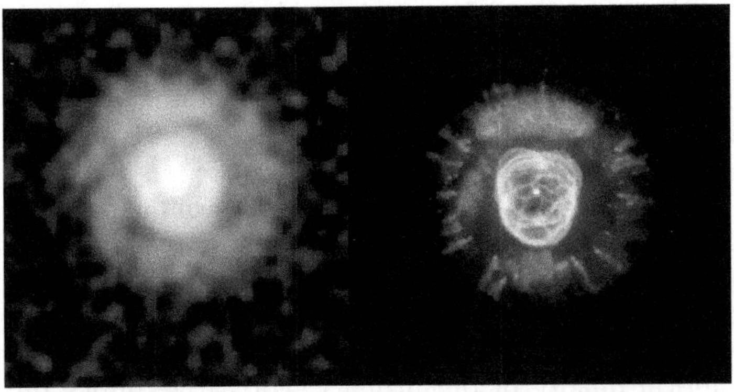

Figure 2.12: Effect of atmosphere. On right is HST image of Eskimo nebula (a Planetary Nebula) and on the left is the object in 2MASS imaging on the ground (from 2MASS and HST archives). Both are in the near-infrared).

Seeing may be estimated for a site by using a Differential Image Motion Monitor that uses two small mirrors (of sizes comparable to r_0) separated by one eddy size to take a series of short exposures (less than or equal to 10-20 milliseconds) of a star. Knowledge of wind speed at the site will then be used to estimate the seeing.

Figure 2.12 shows near-infrared images of the planetary nebula, NGC 2392 (the Eskimo) taken from a ground-based telescope (from 2MASS survey) and the Hubble Space Telescope (HST). The sharpness of the HST image is in stark contrast with the hazy ground-based image.

Table 2.3 gives the atmospheric seeing/turbulence parameters and the seeing-limited and diffraction-limited angular resolutions for a telescope aperture size of 10 m for various wavelength bands used in astronomical photometry. The isoplanatic patch θ_0 indicates the angular size in the sky within which the PSF of a stellar image remains steady or the turbulence may be considered "frozen". It is wavelength dependent, like the Fried parameter, being larger for longer wavelengths (see eqns. 2.14 and 2.16). It is assumed that the Fried parameter for reasonably "good" site is 20 cm at R band (0.70 μm). With respect to this value the parameters at other wavelengths are calculated and given in the table.

2.3.2 Active and Adaptive Optics: To illustrate the effect of the terrestrial atmosphere on the sharpness, two images of the "Eskimo" planetary nebula NGC are shown in Figure 2.12 – on the left is a ground-based observation under the 2MASS survey and on the right is the image taken by Hubble Space Telescope. The HST image is not only very sharp, but it reveals filamentary features that are smeared out in the ground-based image. The blurred image from the ground illustrates the need for correcting the seeing distortion. Several observatories where telescopes of apertures less than about 2m are situated tend to work at seeing-limited resolution that well suits for many scientific investigations. This way of using telescope without correcting for seeing is called "passive" optics.

The distortions caused by the telescope environment such as thermal and mechanical or orientation disturbance (due to gravity acting asymmetrically on the mirror), have timescales of the order of a second (or frequency ~ 1 Hz). These are corrected by active optics technique in which the telescope is adjusted by applying pressure at suitable points in the support system of the primary aperture to compensate for the distortions. Essentially the active optics would strive to maintain the original optical quality of the primary. For active optics the primary mirror must be thin. The ratio of diameter to the thickness of the mirror decides this. A thickness of less than ~ $D/12$ is considered thin. Usually, large telescopes are made very thin. As an example, we may recall that for the Very Large Telescope the mirror diameter is 8.2m while its thickness is 18cm!

The atmospheric seeing effects that have timescales of 10 Hz – 1 kHz need to be corrected by an involved process known as the adaptive optics (AO) technique (see e.g., Beckers 1993; Labeyrie 1995; Roddier 1999). The distortions in PSF in telescopes of apertures in the range D ~ 0.5-3 r_0 can be corrected with the technique of tip-tilt which is the simplest form of adaptive optics in which the compensation is made (to correct for the tilts in the incoming wavefronts) in real time by suitably adjusting one of the mirrors (usually the secondary) by a hexapod. AO systems deliver 10-100 times sharper images than un-corrected ones.

For larger mirrors (> 2 m) we need a full-fledged adaptive optics to achieve diffraction-limited angular resolution. In principle, the telescope senses and monitors a star within the isoplanatic angle of the programme object under study. The required wave front correction or compensation is estimated in real time and fed to a deformable small (~ 20 cm) mirror mounted behind the focus of the primary which is then focussed on to a detector. For this purpose, the deformable mirror should be able to respond, in real time, to an adequate number of actuators mounted underneath. For details of these techniques, we may refer to Lèna (1979). The isoplanatic angle for visible wavelengths is very small (see Table 2.3) and hence it is very difficult to find a guide star near the programme object. This shortcoming may be taken care of by using a laser guide (artificial) star monitored by an auxiliary telescope. Extreme Adaptive Optics (ExAO) was proposed for achieving unprecedented accuracy in wavefront correction using bright "natural" stars to reach diffraction-limited performance of large telescopes (see Guyon 2018).

Table 2.3: Atmospheric parameters and angular resolution limits at different wavelengths for a telescope aperture of $D = 10$ m.

Band	λ (μm)	r_0 (cm)	τ_0 (sec)	θ_{SL} (arcsec)	θ_{DL} (arcsec)
U	0.365	9.2	0.009	0.82	0.009
B	0.440	11.4	0.011	0.80	0.011
V	0.550	15	0.015	0.76	0.014
R	0.700	20	0.020	0.72	0.018
I	0.900	27	0.027	0.69	0.023
J	1.25	40	0.040	0.64	0.031
H	1.62	55	0.055	0.61	0.041
K	2.2	79	0.79	0.57	0.055
L	3.4	133	0.133	0.53	0.085
M	5.0	212	0.212	0.49	0.126
N	10.0	486	0.486	0.42	0.252

2.4 Detectors

In this section we discuss detector types used for detection of optical and near-infrared radiation. Photon detectors are of two different types: (i) incoherent detectors, in which the phase information is lost and only amplitude of the light wave is measured, are used in optical, UV, Visible and Infrared radiation, and (ii) coherent detectors in which both phase and amplitude are measured and are applied for longer wavelengths such as sub-millimetre radiation. The incoherent detectors are of two types – thermal detectors in which the change in temperature or pressure is measured as a function of wavelength of radiation; and direct photon or quantum detectors in which the radiation interacts with the electrons of the detector material. Table 2.4 lists various types of detectors and the main physical parameter measured in each detector. Since our scope is confined to visible and infrared wavelengths, we discuss here only the incoherent detector types. We may refer Kitchin (2003) and McLean (2008) for details on all classic and modern detectors in all wavelength bands. For particularly on infrared detectors, we may refer Rieke (2007).

Table 2.4: Detector types (adopted from Kitchin 2003).

Detector	Type	Measured Parameter
Human Eye	Quantum	Vision
Photo-emulsion	Quantum	Electron Excitation
Photomultiplier	Quantum	Electron Emission
Image Intensifier	Quantum	Electron Emission
Charge Coupled Device	Quantum	Charge
Photo-voltaic Cell	Quantum	Voltage
Photo-conductive Cell	Quantum	Conductivity
Phototransistor	Quantum	Resistance
Bolometer	Thermal	Resistance
Thermocouple	Thermal	Voltage
Golay Cell	Pneumatic	Voltage

2.4.1 Golay Cell for Infrared: For infrared radiation detection the Golay Cell had been used in laboratory during the middle of the last century. Invented by Marcel Golay in around 1947, this eponymous detector Golay cell belongs to thermal (pneumatic) category. A Golay cell consists of a chamber containing an infrared absorbing window adjoining a gas-filled chamber. Infrared radiation absorbed by the window results in increasing the temperature of a gas in the adjacent chamber. The other end of the chamber is sealed by a thin diaphragm shone by a light source externally. The heated gas produces pressure on the diaphragm which in turn produces a change in light output detected by a photodiode. This detector is quite sensitive and has linear response over a range of infrared wavelengths with response time of about a millisecond. However, the background noise from terrestrial environment is substantial rendering the applicability of Golay cell for cosmic observations very limited.

2.4.2 Bolometers: These are thermal detectors. When light falls on a detector made of a certain material the heat that it generates changes the resistivity of the material which is recorded suitably using associated electronics. This type of detectors is used in far-infrared wavelengths. Examples of such detectors include Ge:Ga.

2.4.3 Photographic Emulsion: One of the most (if not the most) extensively used quantum detectors in the last century and earlier had been the photographic emulsion for visible radiation. But the revolutionary developments that occurred during the second half of last century on semiconductor physics changed the outlook of observational astronomy and the photographic emulsion, being comparatively low sensitivity and slow in response, took a beating but did not fade out. It still survives for its sheer human involvement from loading the film for observations with right exposure to developing, analysing and archiving. For a detailed description of the technique reader is referred to Latham (1974) and Kitchin (2003). Typical photographic emulsion contains a thin layer of silver halide (mostly silver bromide with a little of silver iodide) crystals mixed with a supporting transparent medium of gelatine: this mixture is then deposited on a cellulose film (photographic film) or a glass substrate (photographic plate). The principal mechanism of image formation in photographic emulsion is the formation of electron-hole pairs when exposed to light: the electrons trapped in impurities in conduction band and holes removed chemically by gelatine itself. The electrons in the trap then neutralise silver ions to form pure silver atoms in the halide crystal. For silver bromide emulsion the minimum energy required to release electrons into conduction band corresponds to blue part of the spectrum. However, it is possible to pre-sensitise the emulsion to shift the peak response from blue to other wavelengths in the visible region. The grain size or the physical size of silver halide crystals of a few micrometres defines the minimum size of the sensing element or picture element (pixel).

The image thus recorded on emulsion is called "latent" image. To make it actually visible, further processing, called developing, is required to be done. In the development process the emulsion gets reduced – the silver atoms more rapidly than the silver halide hence those portions of silver halide that are exposed to light would trace out the image of the object being photographed. This is why not only does the exposure requires to be timed accurately but the development process also needs a fixed time (followed by washing of the excess developing solution using a chemical solution called fixer which in turn is followed by cleaning with distilled water), lest the entire emulsion gets reduced as time passes by. The image (negative) appears dark wherever light photons are absorbed, and silver grains are formed. If needed, a positive image may then be printed on a photographic paper by passing white light through the negative.

The optical or photographic density is defined as

$$D = \log_{10}\left(\frac{F_i}{F_t}\right) \tag{2.18}$$

In the above equation F_i and F_t are incident and transmitted flux from source. The transmission is through the emulsion given by

$$T_r = \frac{F_t}{F_i} = 10^{-D} \tag{2.19}$$

And the opacity of the emulsion, equals inverse of T_r, is given by

$$A = \frac{F_i}{F_t} = 10^{D} \tag{2.20}$$

Figure 2.13 shows a typical characteristic curve of an emulsion which is a plot of photographic response in terms of density D versus logarithm of exposure E. Here E is defined as the exposure time t multiplied by the intensity I of incident light. This plot is called Hurter-Driffield curve (see Born and Wolf 1999). There are essentially three regions in the curve shown in the figure – the ideal linear portion: the toe region at shorter exposures and the shoulder region or saturation region for longer exposures. The contrast of the emulsion is measured by the slope of the linear portion of the D-curve. Sensitometry or densitometry is the study dedicated to the D-curve. It may be noted that higher speeds of the emulsion and higher sensitivities lead to coarser grain sizes.

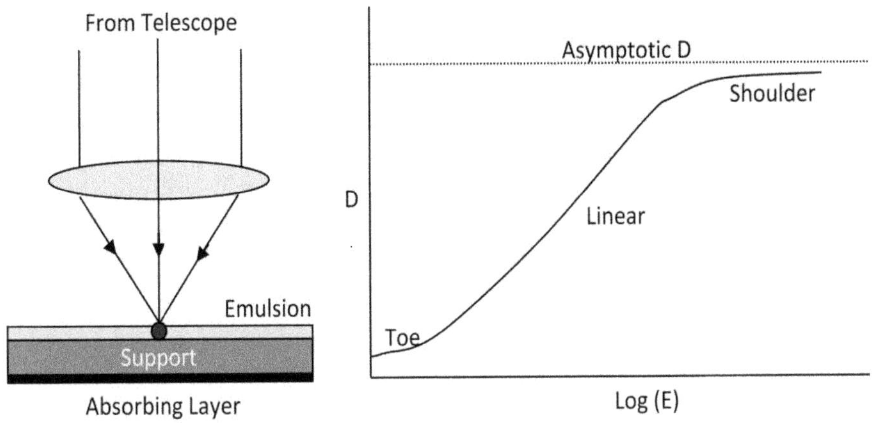

Figure 2.13: Structure of a photographic emulsion (left). On the right is shown a plot of optical density as a function of exposure for a typical emulsion (adopted from Kitchin 2003).

The resolution of the emulsion is decided by its structure and scattering that can happen in the supporting material leading to clouding around the image and this process is called halation. Halation is reduced by using an absorptive coating on the back of the film/plate. Photographic emulsion suffers from slow response times; and very low (about 1%) quantum efficiency. Even so the contributions that the photographic emulsions had made in advancing astrophysical quest are highly significant and remarkable.

2.4.4 Photo-Electric Effect: Following Albert Einstein's Nobel prize-winning discovery of photo-electric effect, technological advances have taken rapid and giant strides to produce detectors based on the effect, made of metals and semi-conductors. Since the middle of the last century, such detectors have started becoming available for faint light detection both as photo-voltaic and photo-conductive types. Later, detector arrays have come in to be used changing dramatically the scope of astrophysical observations in both sensitivity and ease. Here we discuss only the detectors that work on the principle of photo-electric effect. One may refer to Kitchin (2003) for a comprehensive treatment of various types of detectors.

When radiation of suitable energy falls on a conductor it releases electrons in the valence band to rise into the conduction band. The resulting current can be measured in terms of intensity of radiation. In metallic conductors, the photoelectrons collide with free electrons and hence some of them are lost and their valence bands are partially filled, since the gap between the valence and conduction bands is negligibly small, often overlapping. Also, for metals reflectivity is high and hence the efficiency of production of photoelectrons is very poor. Unlike metallic conductors, in semi-conductors however, the valence bands are filled, and the conduction band is well separated with small discernible energy gap as shown in the Figure 2.14. For completeness, we add here that insulators that are poor conductors have large energy gaps between their valence and conduction bands and hence make poor detectors of radiation. The energy balance equation for photoelectric effect may be written as follows,

$$E_m = h\nu - W = \frac{hc}{\lambda} - W \tag{2.21}$$

Here W is called the work function (see Figure 2.14), a characteristic energy of the material denoting the energy with which the electrons are bound. The excess energy (photon energy more than the work function) is imparted to the released electron and determines its kinetic energy. When the atoms are cold (at 0 K), the electron energy states are degenerate, electrons filling all the states up to the level of Fermi energy E_F, as shown in Figure 2.15. The continuum levels above Fermi energy up to the vacuum energy level E_V are all unoccupied at 0 K. The gap between E_V and E_F is called the work function, W. Since the metal is not at absolute zero temperature, electrons would have enough energy to cross over the energy barrier set by the work function.

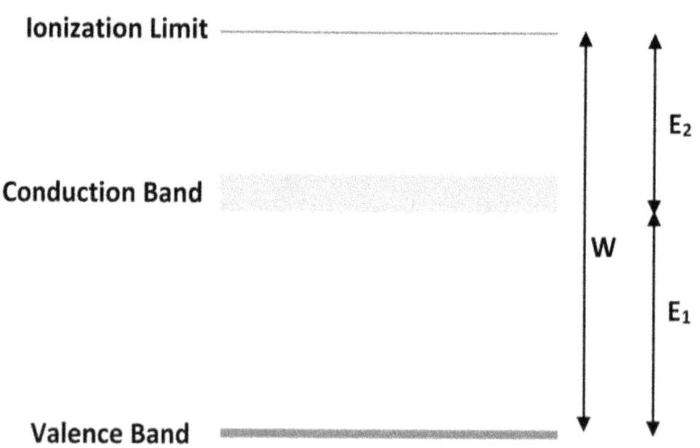

Figure 2.14: A schematic showing energy levels for a semiconductor. W is the work function or energy required to ionize the semiconductor. E_1 is the energy required to bring an electron from valence band to conduction band.

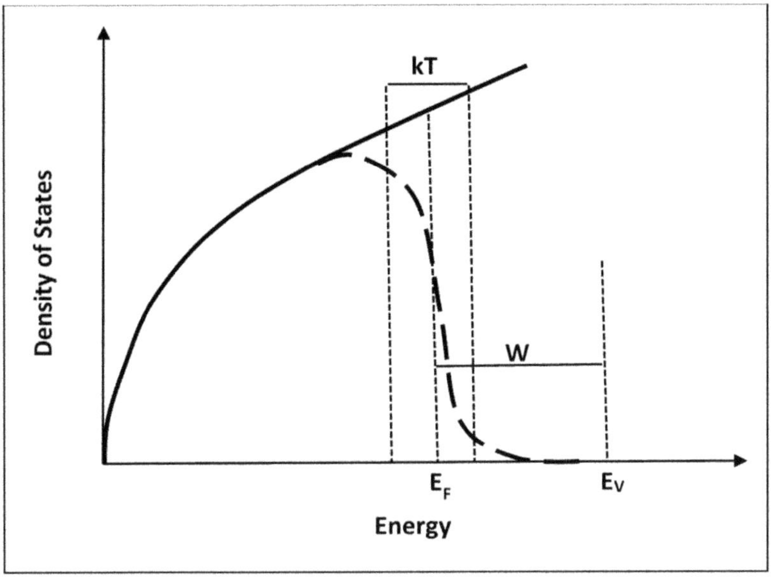

Figure 2.15: Density distribution of electron energy states and the energy of the highest occupied state, the Fermi energy. The dashed curve represents the distribution of density of filled states at a temperature T. The region to the left of the Fermi energy (E_F) line represents the distribution of filled states at 0 K. Work function W is defined as the difference between Fermi energy E_F and vacuum energy E_V (from Young 1974).

2.4.5 Parameters Defining the Performance of a Detector: While selecting a detector for a specific observation we need to consider the following characteristics.

Quantum Efficiency: Denoted by QE, it is the ratio (given in percentage) of the number of photons detected to the number incident on the detector. If QE is 90%, then we understand that the detector is capable of recording 90 photons out of 100. Since there are so many extraneous noise factors that work against the ideal figure defined above, we often come across what is called the detective quantum efficiency (DQE). It is the ratio of signal-to-noise ratio (SNR) at the output end to that at the input. The factors that contribute to the deterioration of SNR are too many – noise in the detector, its electronics, background noise from within the instrument as well as environment. It is thus the DQE that becomes the factor to reckon with.

Dark Current: This is the noise specific to the detector itself. When no light is incident on the detector – that is in totally dark condition - it still shows a "signal". It is due to the electron current owing to the background temperature. Specified in electrons per second (e^-/s) dark current can be reduced to negligible values by cooling the detector sufficiently, mounting it in a cold chamber with solid carbon dioxide or in a dewar filled with liquid nitrogen (77 K) or liquid helium (4 K). Sometimes when the detector requires cooling a few tens of degrees below 0 C, cooling by Peltier elements is also employed. Choosing a suitable coolant or cooling method depends upon the type of detector, desired SNR, and the wavelength region of interest.

Response: For a unit of incident light energy given in watts the detector and its associated electronics produce a measurable signal in terms of voltage or current defining the response or responsivity of the detector. The larger it is the better or more efficient is the detector.

Time Constant: The rapidity with which a detector responds to a change in intensity of the incident light defines the time constant.

Dynamic Range: It is the range of intensity that a detector is capable of detecting – from a maximum input intensity without getting saturated to a minimum that it barely detects. The minimum is taken as dark current level.

Spectral Response: It is the response of the detector with wavelength. The QE is evidently a function of wavelength. Associated with this are the two other parameters – the peak wavelength at which the response or QE is maximum and the cut-off wavelengths one on each side of the peak at which the efficiency drops to unacceptably low values. The range between the two cut-offs defines the wavelength response of a detector.

Noise Equivalent Power (NEP): It is the input radiation flux that gives a SNR of unity. It is usually a parameter defined for longer (far-infrared) wavelengths. But we can adapt it to the shorter wavelengths such as the near-infrared or visible regions. We can define NEP, for example, as the flux or magnitude that gives a SNR of 1 for an integration or exposure time of 1 second.

In the following sections we describe some specific detectors that are very extensively and popularly being used by astronomers.

2.4.6 Single Element Detectors: Photomultiplier

Basically, a photomultiplier tube (PMT) consists of a cathode that produces photoelectrons upon receiving photons from a source. These photoelectrons are then directed to several positively charged metallic "mirrors" called dynodes that multiply the electrons through production of secondary electrons. These are eventually collected by an anode at the end of a hermetically sealed glass vacuum tube. With respect to the negative cathode, a potential difference of about 2000 V is applied progressively from each dynode to the anode, by dividing the voltage using resistance dividers. An output current pulse is thus generated. Figure 2.16 shows a schematic of a photomultiplier tube with 5 dynodes. In principle, for every photoelectron received, the total number of secondary electrons produced by n stages of dynodes is m^n where m is the average number of secondary electrons generated at each dynode for every incident primary electron. Thus, the number of dynodes decides the multiplication factor or gain of the PMT, usually in the range of 10-100 million for 12-14 stage (each stage being a dynode) PMT. The anode current is then processed to obtain intensity of incident photons as a function of current strength. This way of analysing the current is called the analog mode. In many applications especially those involving faint sources, the PMT's quantum efficiency parameter is used to count individual photons by deciding a threshold for pulse height which either gives 1 or 0 for the photons counted. This is called the digital or photon-counting mode in which the QE is defined as the ratio of average number of photoelectrons emitted per second to the average number of photons incident on the photocathode. We need to also consider the collection efficiency by each dynode in deciding the effective QE. The cathode material decides the wavelengths response and range. The QE of a PMT is typically less than 40%. The principal source of noise is thermionic noise due to thermally (at room temperature) excited electrons contributing to the anode current. Clearly it increases as the area of the effective cathode is increased. A proper choice of the cathode size is thus important. For the detection of point sources, we may employ a cathode size of as small as a few millimetres. Further, for faint sources, the PMTs need to be cooled to reduce the thermal noise. When used in photon counting mode the distribution of photon arrival is given by Poisson distribution. Therefore, the uncertainty in signal counts N is given by its square root (i.e., $N^{1/2}$). PMT had been a very popular quantum detector for photometric studies for many decades in the last century. It had also been used in spectroscopy, for instance, in the case of axial fringe scanning Fabry-Perot Spectrometer.

As such the PMTs are useful only in the case of detection of light from a point source such as stars. Extended sources such as nebulae or galaxies can be observed by pointing the telescope to several regions on the object to ensure proper coverage. However, in addition to poor angular coverage, this way of sampling extended sources also leads to longer observation times required for using this single-element detector. For a general discussion on PMTs we may refer to Kitchin (2003) and for details to Young (1974).

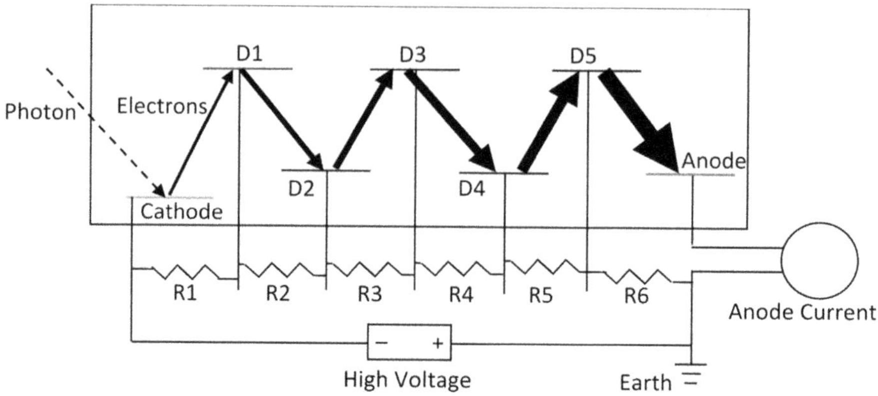

Figure 2.16: Sketch of a photomultiplier tube – showing photocathode in blue line and anode in red line. The dynodes are denoted by D1 to D5. The voltage dividing resistors are shown as R1 to R6 which progressively increase the voltage from D1 to D5 and the Anode (as shown by the increasing size of arrows).

We may mention here that the PMTs are very much in use as high energy astronomical detectors – for instance, arrays of PMTs are used for detection of Cherenkov radiation for high energy gamma-ray astronomy (Abe et al, 2023).

2.4.7 Array Detectors for Optical Region

2.4.7.1 Image Intensifier and Micro-channel Plate: An image intensifier consists of three principal parts namely, a photocathode, a microchannel plate, and a phosphor screen or other suitable imaging (position-sensing) detectors. The photocathode receives the incoming photons whereby photoelectrons are generated. These photoelectrons are intensified by the microchannel plate. The Micro-channel plate (MCP) is essentially a conglomeration of a large number of miniature PMTs of size of less than about 50 μm, working as independent channels for election multiplication. Modern MCPs are available with channels of a few micrometres. The thickness of the MCP is defined as the ratio of the length of the channel to its diameter (typically 40-80). The gain of the MCP is also dependant on its thickness. Typically, more than about 300 such micro-channels make one micro-channel plate as shown in Figure 2.17. At the anode-end, the MCP is fixed with a phosphor screen to capture the image visually. We may then use, for instance, a photographic film/plate to record the image produced by the phosphor screen. To increase the gain, often a stack of two (arranged in Chevron configuration) or three (in Z-stack configuration) MCPs are used trading with the effective pixel size. The effective "pixel" size for 3-stage MCP would be around a few tens of micrometres – the large size owing to the spreading of photoelectrons between the stacks of MCP. An extended object or a star cluster can be simultaneously imaged on the MCP. The MCPs are used not only in imaging photometry but also for spectroscopy. Since their advent a few decades ago, the Charge Coupled Devices (CCDs) replaced photographic film/plate for digitally recording the phosphor sensor image. Such CCD coupled intensifiers are known as Intensified-CCDs. For details on image intensifiers, we may refer to Kitchin (2003) and Hamamatsu manuals (www.hamamatsu.com). These detectors are being used not only for visible region but also for UV region, employing suitable photocathode materials. The QE is reasonably high for the MCPs. MCPs are also used for high energy physics and astronomy (see Wiza 1979).

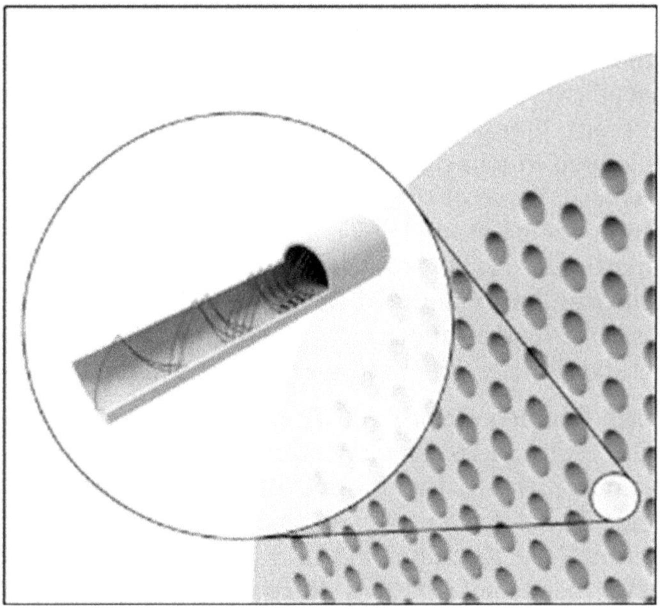

Figure 2.17: Micro-channel Plate (enlarged) from Chen & Wang 2017.

2.4.7.2 Imaging Photon-counting Detector (IPCD): This contains an MCP or a stack of MCPs coupled with a position-sensing resistive anode instead of a phosphor screen. The Image Photon Counting System (IPCS) originally conceived and developed by Boksenberg & Boggess (1973) especially for the purpose of imaging spectroscopy (e.g., Atherton et al 1982), used a 4-stage image intensifier coupled with a TV camera to record individual photoelectron events digitally (Jorden & Fordham 1986). In the original IPCS and the later versions, magnetic focussing (to bring down the pixel size) was employed to alleviate the photoelectron spread between stacks of MCPs (Jorden & Fordham, 1986). For comprehensive performance details on IPCS, we may refer to Jenkins (1987). The Imaging Photon-counting Detector (IPD) developed later is a worthy variant of the famous IPCS in which the position sensing is done by a resistive layer equipped with four anodes at each corner. The current generated by a photoelectron event measured at each of these anodes is a function of the spatial position of the event on the resistive layer (for details we may refer to Photek manuals at www.photek.com). The image could then be recorded digitally for further processing. There have been improvements in the resistive anode sensitivity and resolution (see Photek manuals). For IPDs, the effective "pixel" size (after the resistive anode sensing) is typically about a few to a few tens of micrometres. It may be noted here that there is no physical pixel in IPD, but each event may be measured with a spatial resolution as mentioned. This was used as a detector for Imaging Fabry-Perot spectrometer (IFPS) by Seema et al (1992) in the visible region. These detectors are also being used for UV imaging as well (for instance in ASTROSAT UV telescope, see Kumar et al 2012). In recent times, innovative improvements made on the MCP technology helped in using these systems for high energy astronomical observations also. Some of the images taken using the IPD-IFPS system may be seen in Seema et al (1992), and in Chapter 5 (Figures 5.39 and 5.43). As mentioned earlier, to achieve better sensitivity and position-sensing accuracy Charge Coupled Devices (CCD) are also used as back-end sensors with image intensifiers. These are called Intensifier-CCDs. The CCDs as stand-alone detectors are described in the following section.

2.4.7.3 Charge Coupled Device (CCD): Developed in Bell Labs by W.S. Boyle and G.E. Smith in 1970 who won the Nobel prize in 2009 for their invention (along with C.K. Kao for optical fibres), charge coupled device (CCD) works basically on the principle of photoelectric effect in semiconductors (see Kitchin 2003 and Lesser 2015 for details). The semiconductor used in CCDs is a silicon chip. An incident photon of appropriate energy releases a photoelectron and a hole in the p-type semiconductor. CCD contains a square or rectangular array of such units of semiconductors or pixels. The photoelectrons thus produced from each unit are accumulated in a potential well so that the charge can be read out by a single electrode at the end of a series of such semiconductor units (row by row) by a process known as charge coupling. It involves transport in one direction of accumulated charge by suitably varying the voltage given to each electrode.

The basic unit of a CCD consists of a metal-oxide transistor as shown in Figure 2.18. A small positive voltage is applied to this unit by an electrode. Thus, the holes are removed away from the photoelectrons and the latter are accumulated in a thin layer below the unit is shown in the figure. The individual units are insulated from each other by a very thin layer of intensified p-type semiconductor material. The accumulated charge is transported by progressively giving a time-variable sequence of higher positive potential to the electrodes such that it can be read out at the end of each row. In one direction each pixel is defined by the electronic gate on the silicon surface while in the second orthogonal direction by the electrical voltage applied to the electrode implant. The charge read-out is proportional to the intensity of radiation incident on the CCD; it is also proportional to the exposure time. Since Si has a band gap energy of 1.14 eV a CCD can be used for detection of photons corresponding to wavelengths from 300 to 1000 nm (near UV to near-infrared fringe). One of the important parameters in CCDs is their charge transfer efficiency (i.e., efficiency of charge transfer from one unit to another) which in the latest detectors is very close to 100%!

The CCDs have in general high quantum efficiency compared to many area detectors both analogue and digital types. CCDs are made in two configurations – the front illuminated and back-illuminated or back-thinned as shown in Figure 2.19. In the latter version the light falls directly on the silicon block (in the backside of the detector) thus getting rid of losses that the front illumination variety suffers. A potential difference applied between the front and backsides drives photoelectrons to the frontside where they are readout using gates in a way similar to that in the frontside type. With back-thinned CCDs, QE of nearly 90-95% is achieved.

Figure 2.20 shows QE as a function of wavelength for back-thinned or backside CCDs – different curves are for different anti-reflective coatings and silicon wafer thicknesses of the device – the QE increases by decreasing the reflectance and by increasing the thickness (see Lesser 2015). Pixel sizes of as small as a few micrometres are available while arrays sizes are available as large as 4kx4k. Dark noise is typically 0.0001 e^-/s; and read-out noise is 1 e^-. Full well capacity is a few times 10^5 electrons (see Figure 2.21).

To reduce the dark current noise cooling is employed (for operating temperature of $\leq$ - 100^0 C) either by mounting the CCD in liquid Nitrogen Dewar or by Peltier cooling. CCDs are extensively used in both photometry (including polarimetry) and spectroscopy in astronomy.

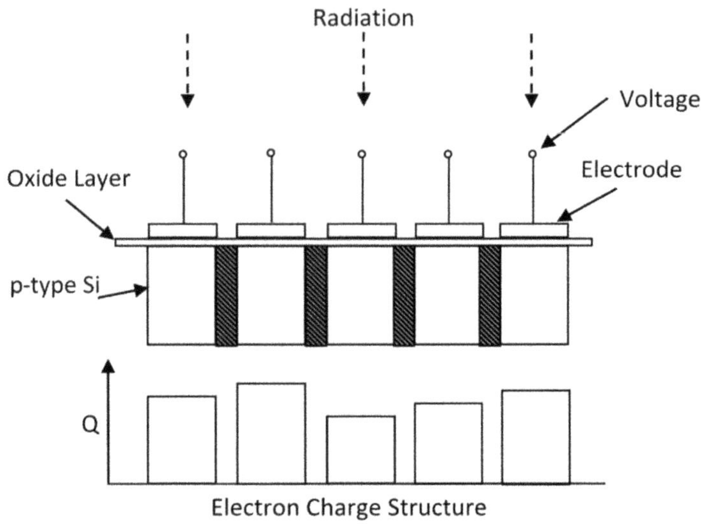

Figure 2.18: Sketch of the basic Structure of a CCD (adapted from Kitchin 2003).

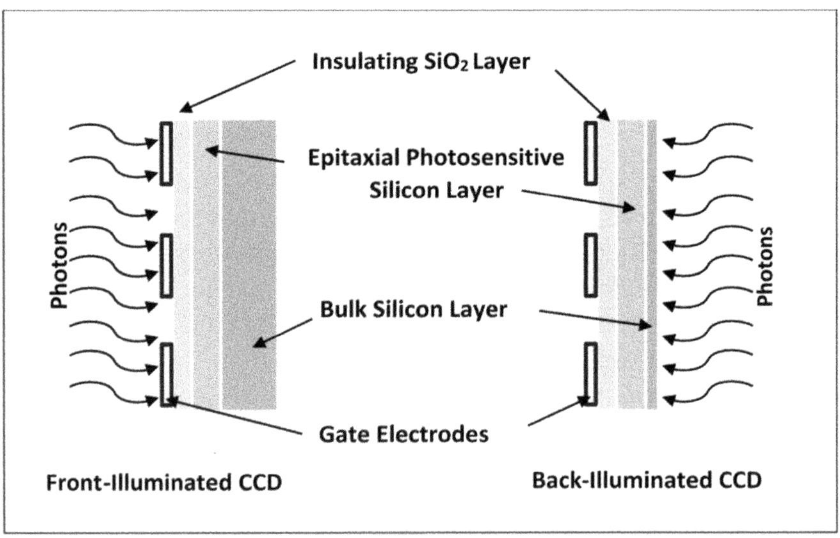

Figure 2.19: Sketch of back and front side illuminated CCDs (adapted from Lesser 2015).

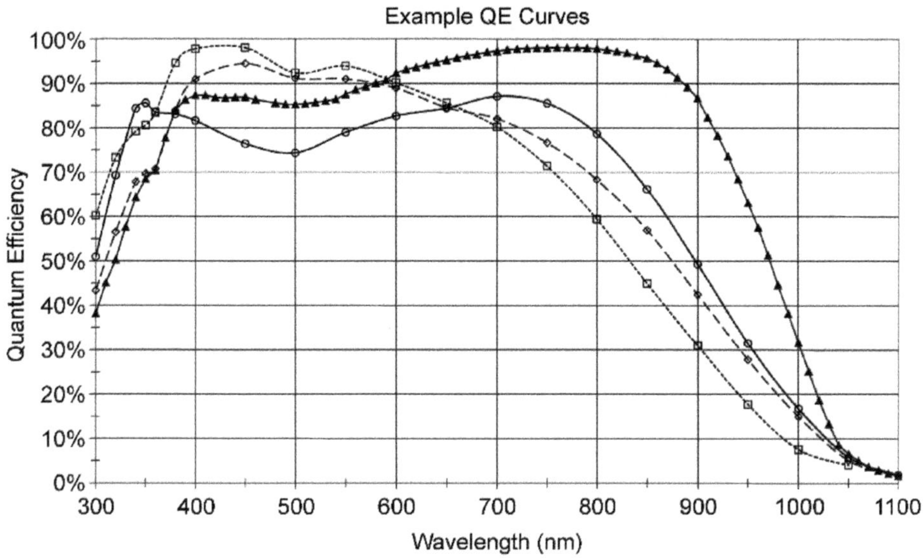

Figure 2.20: QE curves for back-illuminated CCDs with various thicknesses of anti-reflection coatings (from Lesser 2015).

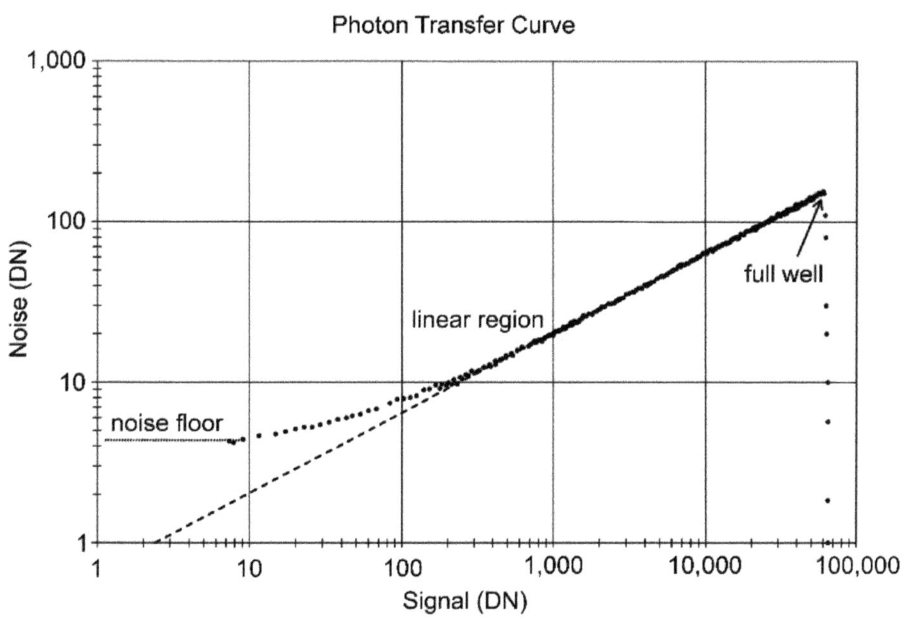

Figure 2.21: CCD noise versus signal in terms of digital numbers, DN (from Lesser 2015). The tip of the linear portion on the right depicts the full-well capacity of the detector.

2.4.7.4 Complementary Metal-Oxide Semiconductor (CMOS): As we have seen above the CCDs are basically metal-oxide semiconductor used as capacitors to accrue charge, transfer, read-out and amplify the signal in each pixel using time-varying sequential electronic gates. In the complementary metal-oxide semiconductor devices (CMOS), each pixel is not treated as an individual entity. The CMOS devices which have now become extremely popular in commercial products more than CCDs, use MOS field effect transistor (MOSFET) amplifier for every pixel. CMOS can thus be classified as an active pixel sensor in contrast with CCD. Further, CMOS have much smaller thickness and pixel sizes than CCDs. Despite its popularity in commercial imaging devices such as cameras and smart mobile phones as well as biomedical devices where imaging is required, as far as astronomy and space science are concerned involving large area and low-noise faint light detection, currently CMOS fails to supplant the CCD. CCDs would remain the chosen detector in future as well (see Lesser 2015).

Table 2.5 gives typical performance parameters of PMT, IPD and CCD such as the wavelength range (λλ in μm), effective pixel size (Px), operating temperature (OT), QE, dark current, readout noise (RO) and readout time (RT), and full-well capacity (FWC).

Table 2.5: Performance parameters for visible radiation detectors.

Detector	λλ (μm)	Px (μm)	OT (units)	Q.E. (%)	Dark (e^-/s/px)	RO (e^-)	RT (s)	FWC (e^-)
PMT	0.2 - 0.8	3 mm*	< 20 C	~40	< 0.001	---	---	---
IPD	0.2 - 0.8	20-40	< 20 C	~40	< 0.001	---	$< 10^{-6}$	$>10^5$
CCD	0.3 - 0.9	9-15	77 K	>70	0.0001	1	$> 10^{-6}$	$>10^5$

2.4.8 Infrared Array Detectors: The efficiency of semiconductor devices used for infrared astronomy depends upon their several characteristics – cut-off wavelength, the quantum efficiency, noise of various origins, full well capacity, read-out time, and array size among others. As discussed earlier, the only semiconductor device used for astronomy in visible wavelengths is the CCD (using silicon chip having a cut-off wavelength of about 1 μm). For the infrared region there are several semiconductor-based detectors that are in use currently. Let us write down a couple of basic relations regarding the cut-off wavelength and operating temperature.

In a semiconductor the band-gap energy E_g is related to the cut-off wavelength as follows,

$$\lambda_c = \frac{hc}{E_g} = \frac{1.24}{E_g(eV)}\ \mu m \tag{2.22}$$

Associated with the cut-off wavelength is also the temperature at which the detector needs to be operated to minimise the thermionic noise (see Hoffman, Love and Rosbeck 2004). This may be evidently obtained as,

$$T_{max} = \frac{200}{\lambda_c(\mu m)}\ K \tag{2.23}$$

Detector arrays are available for longer wavelengths as well by doped semiconductors - for instance, the extrinsic semiconductor arsenic-doped-silicon has a cut-off at 28 μm. In the case of the semiconductor mercury-cadmium telluride ($Hg_{(1-x)}Cd_xTe$), by changing the relative Hg and Cd composition the cut-off wavelength may be manipulated – for x = 0.45 we obtain the cut-off at 2.5 μm while x = 0.3 gives 5 μm. This flexibility in adjusting the cut-off wavelength makes this semiconductor the most versatile among for infrared region from 0.6 to 5 μm (see Rieke (2007) for more details).

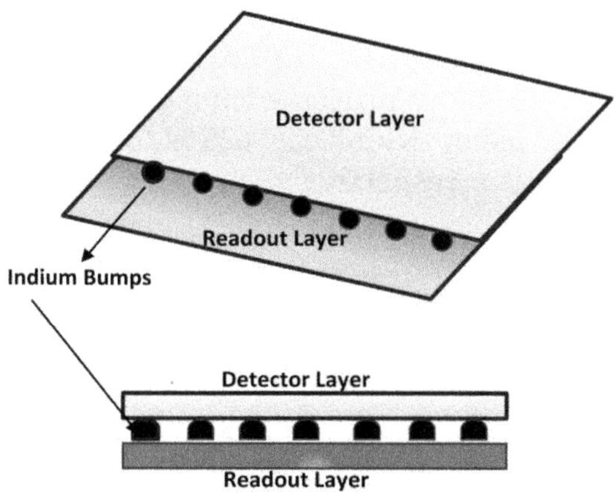

Figure 2.22: Shows schematically the hybrid architecture of an infrared detector array (on top, the lateral view and on the bottom the edge-on view) in which the detector layer (in light shade) is attached to the readout silicon layer (in dark shade) via indium bumps (black dots) that provide the matching of amplifier readout with the detector. Not to scale (adapted from Rieke, 2007).

These infrared arrays are made currently by using a novel technology called molecular beam epitaxy in which the detector is made layer by layer using a molecular beam under vacuum conditions. Another method in vogue is known as liquid phase epitaxy in which the semiconductor material of high concentration in liquid form is used to grow the crystal. Unlike the CCDs the infrared array HgCdTe arrays are manufactured in a hybrid architecture – the detector layer is attached to a readout layer made of silicon by using indium "bumps" that help in matching the amplifier grids to the detector. Figure 2.22 shows this arrangement schematically (adapted from Rieke 2007).

For details on array detectors for infrared region and their characteristic parameters, we may refer to Rieke (2007). Table 2.6 shows a few semiconductor detectors for the near- and mid-infrared regions.

Table 2.6: Infrared Array Detector Characteristics (from Rieke 2007).

Detector	$\lambda\lambda$ (μm)	Px (μm)	OT (K)	RO (e^-)	Dark (e^-/s)	Q.E. (%)	FWC (e^-)
HgCdTe	0.85-2.5	18-20	78	6	<0.1	>70	$>10^5$
	0.6-5.3	18-20	37	6	<0.01	>80	$<10^5$
InSb	0.6-5.5	25	32	6	<0.01	>70	$\sim10^5$
Si:As	5 - 28	18	7	10	0.1	>70	$\sim10^5$

One of the important characterization tests for the array detectors is the linearity of response for different exposure times or different levels of intensity (below saturation level) for a standard light source. Figure 2.23 shows the linearity of digital counts versus exposure time for Mt Abu Infrared Telescope using a HgCdTe HAWAII–1 array detector for K band. Figures 2.24 and 2.25 show processed images obtained from Mt Abu Infrared Telescope using NICMOS3 (256x256 array) and HgCdTe HAWAII–1 array detector respectively.

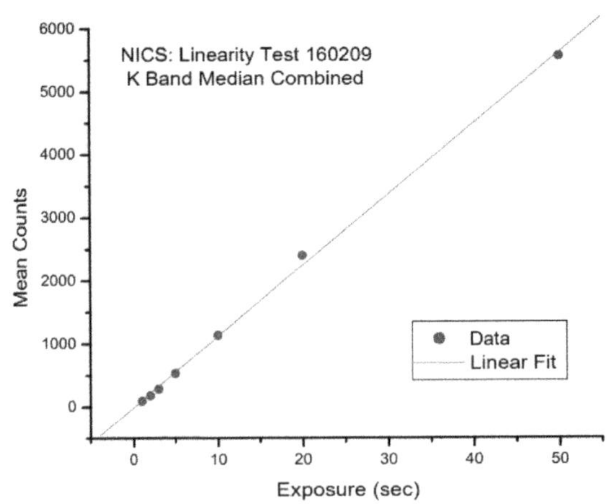

Figure 2.23: PRL Near-Infrared Camera Spectrograph (NICS) characterization curve for linearity.

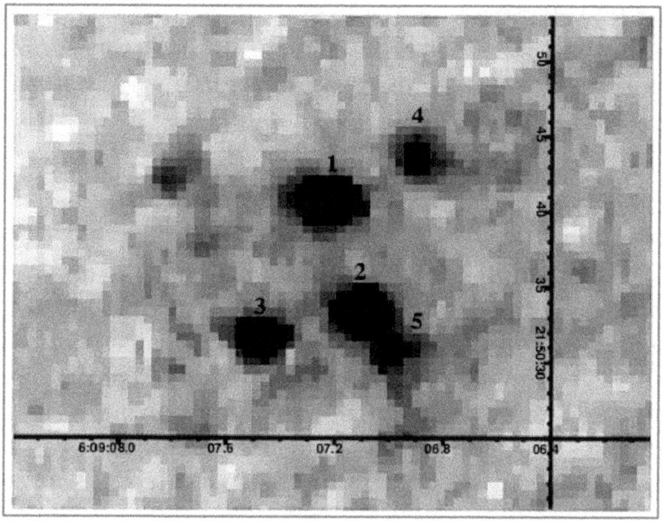

Figure 2.24: K' band image of the young star cluster IRAS 06061+2151 obtained using NICMOS at Mt Abu Observatory (Anandarao et al 2004).

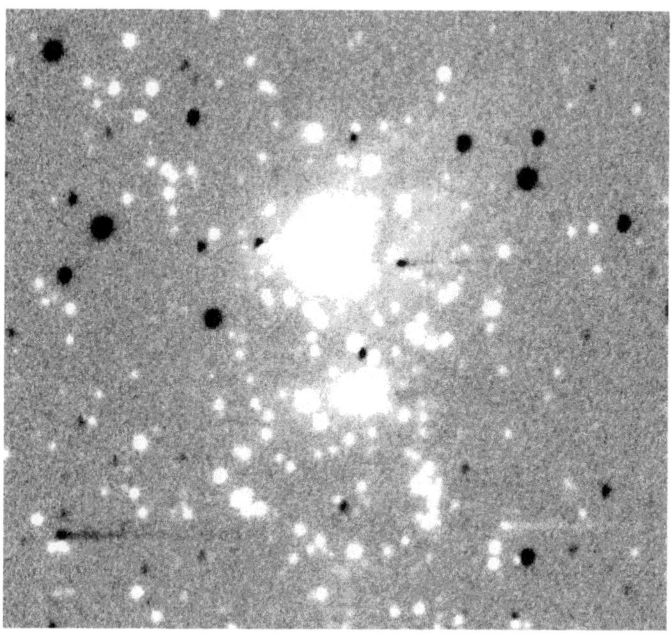

Figure 2.25: A dark and sky-subtracted 'cleaned' image of an extended source in H band obtained using a HAWAII-1 infrared array. The source is a star-forming region namely S235. The observation was made at the PRL Mt Abu Infrared Telescope Facility at Mt Abu. The sky frame is generated by a process called dithering. The dark spots are a result of this process. (Dewangan & Anandarao 2011.)

We close this chapter by introducing an important merit factor for a detector. Bahcall et al (1991) defined what they called 'capability metric' (or Multiplexing capability matrix that applies to both visible and infrared regions) as follows,

$$M_c = N_{pix}\, S_{pix}^2 \tag{2.24}$$

where N_{pix} is the number of pixels in an array detector (n rows x n columns) and S_{pix} is the detector sensitivity per pixel. This figure of merit is applicable to both photometry as well as spectroscopy observations – for the former the figure indicates the number of spatial elements that can be sampled simultaneously (spatial multiplexing M_{xy}) and for the latter it indicates the number of spectral elements that can be sampled simultaneously (spectral multiplexing M_{λ}) in a celestial source.

CHAPTER 3

Radiation Processes and Atomic and Molecular Spectroscopy

Knowledge of the self is the mother of all knowledge. So, it is incumbent on me to know myself, to know it completely, to know its minutiae, its characteristics, its subtleties and its atoms

– Khalil Gibran

Celestial objects, in general, reveal themselves in electromagnetic radiation that they emit (or absorb). The emission has two components: the continuum radiation and line radiation in discreet energies. Knowledge of the mechanism by which these two components arise helps in understanding the underlying physical processes in celestial objects. Various radiation processes for continuum emission, namely, Blackbody, Free-Free, Synchrotron, and 2-photon, prevalent in celestial sources in different circumstances are discussed in this chapter. The observed spectral energy distributions (or simply the flux received at different energies or equivalently at different frequencies/wavelengths) from a celestial body may be interpreted in terms of one or more of these mechanisms. Similarly, the discreet emission from an atom/ion/molecule can be used to derive various physical parameters as well as kinematics/dynamics of the source. Later in this chapter, an introduction to atomic and molecular spectra is given briefly. Most important emission/absorption lines are discussed in more detail. The process of transfer of radiation through circumstellar and interstellar media is discussed briefly. The line formation and broadening processes are discussed. Variety in the spectra of cosmic dust are also discussed.

3.1 Blackbody Radiation

A blackbody is a perfect absorber of radiation having absorptivity of 1. An absorber also emits radiation. A perfect blackbody that satisfies the condition is a cavity made in such a way as it emits electromagnetic radiation in equilibrium with absorption. The brightness of such a blackbody in thermodynamic equilibrium is a function of the equilibrium temperature T, and the frequency of radiation ν. The brightness distribution of radiation with frequency is given by the Planck's equation for blackbody radiation,

$$B_\nu(T) = \frac{2h\nu^3}{c^2}\frac{1}{\left[\exp\left(\frac{h\nu}{kT}\right) - 1\right]} \tag{3.1}$$

It is basically energy emitted per unit time per unit area per unit frequency per unit solid angle and expressed in units of W/m²/Hz/Sterad. In terms of wavelength λ, Planck's equation may be written as

$$B_\lambda(T) = \frac{2hc^2}{\lambda^5}\frac{10^{-6}}{\left[\exp\left(\frac{hc}{\lambda kT}\right) - 1\right]} \tag{3.2}$$

The corresponding units for brightness are W/m²/μm/Sterad. In the above equations k is Boltzmann constant, h is Planck constant. Figure 3.1 gives the radiation distribution for different temperatures.

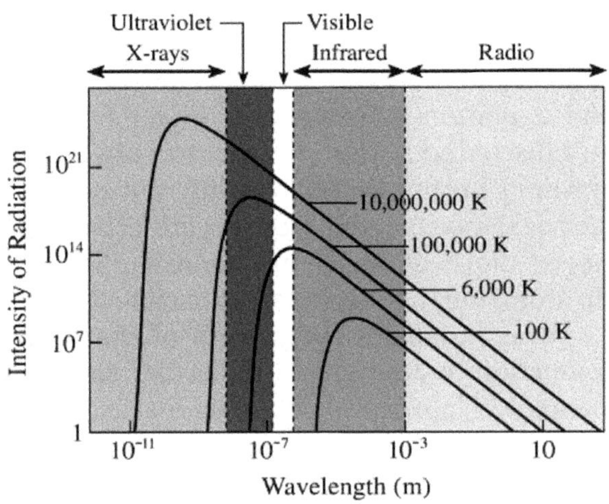

Figure 3.1: Blackbody radiation distribution shown as a function of wavelength for temperatures 100 K to 10^7 K (from K.R. Lang in NASA Cosmos 2010).

As we can see, at any given frequency or wavelength, the Planck's radiation distribution depends only on the temperature of the body in thermodynamic equilibrium. One of the most important properties of a blackbody is its high opacity. Since all main sequence stars do satisfy this condition by having well defined photospheres of high opacity (or large optical depth at UV and optical regions) we may treat these stars approximately as blackbodies. The assumption weakens as we deal with giants whose atmospheres are tenuous. Although treating celestial bodies (stars) as blackbodies is a good approximation, we should be wary of the fact that the optical depth is a function of wavelength and there exist temperature gradients at their surfaces.

The frequency at which the blackbody radiation distribution maximises is given by Wien's displacement law,

$$\nu_m = \frac{3kT}{h} \tag{3.3}$$

Or using in terms of maximising wavelength it is written as

$$\lambda_m = \frac{2898}{T} \tag{3.4}$$

with λ_m in μm. When the radiation distribution is integrated at all frequencies the total emittance is given by

$$B = \sigma T^4 \tag{3.5}$$

This is called the Stefan-Boltzmann law with σ as Stefan-Boltzmann constant. The total luminosity (energy given out per unit time or in units of Watts) of a star can be given by multiplying its surface area with B,

$$L = \sigma T^4 \, 4 \, \pi \, R^2 \tag{3.6}$$

where R is the radius of the star. Two limiting cases may be given here: one is called Rayleigh-Jeans approximation that refers to the region of the distribution where the photon energy is much less than thermal energy the equation for blackbody radiation reduces to

$$B_\nu \, (T) = 2kT\nu^2/c^2 \tag{3.7}$$

and

$$B_\lambda(T) = (2ckT/\lambda^4)10^{-6} \tag{3.8}$$

The Rayleigh-Jeans approximation is applicable for far-infrared and longer wavelengths. Similarly, the second limiting case is for the high energy region where the photon energy far exceeds the thermal energy, the Wien's approximation applies,

$$B_\nu = \frac{2h\nu^3}{c^2} \exp(-h\nu/kT) \tag{3.9}$$

and

$$B_\lambda = (2hc/\lambda^3) \exp(-hc/\lambda kT) \tag{3.10}$$

This approximation applies to the region of high energy radiation.

3.2 Free-Free Radiation

When a free charged particle such as an electron encounters another (such as a free proton, an ion, an atom, or a molecule) then it experiences the Coulomb force field of the latter causing it to accelerate. This process leads to emission of electromagnetic radiation and is called Bremsstrahlung or braking radiation or simply free-free radiation. This process for continuum radiation is important in extended stellar atmospheres (in evolved stars losing mass) where ionized winds extend to large radial distances from the star. The radiation is mostly emitted in the infrared and radio regions. The effective optical depth in plane parallel atmospheres (in which we assume that the thickness of atmosphere is very small compared to the radius of the star) is 2/3 and 1/3 for extended stellar winds (see Lamers and Cassinelli, 1999 for details on free-free radiation). This shows that the radial distance for the effective optical depth is wavelength dependent – being smallest for optical and largest for radio region as illustrated in Figure 3.2.

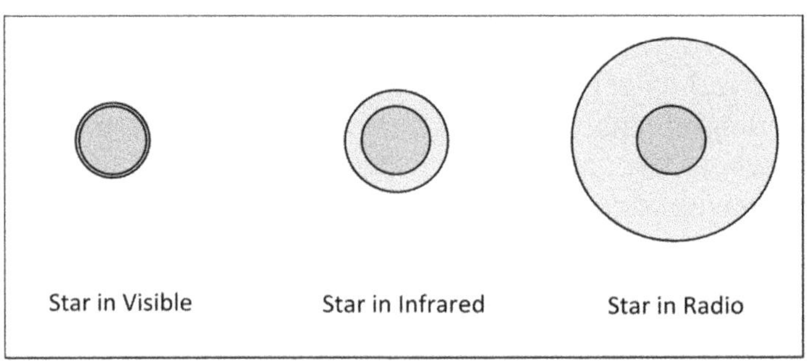

Figure 3.2: An evolved star undergoing mass loss. The lighter colour circles show extended atmosphere at the edge of which the optical depth is 1/3. Figure illustrates different sizes of the star in visible, infrared and radio wavelengths. Free-free radiation contribution is important in infrared and radio. The stellar photosphere is shown in dark region and the contribution due to free-free radiation in lighter colour.

The effectiveness of this process in the interstellar medium or other celestial plasmas depends upon the impact parameter or the distance of closest approach b and the velocity of electron v (see Lamers & Cassinelli 1999). In case of an electron gas at thermodynamic equilibrium at temperature T one can assume Maxwellian distribution for (non-relativistic) electron velocities. In such a case called thermal Bremsstrahlung, the differential power radiated by the gas cloud with volume may be given by (see Woan 2000),

$$\frac{dP}{dV} = 1.7 \times 10^{-40}\, Z^2\, T^{\frac{1}{2}}\, n_i n_e \tag{3.11}$$

In the above equation the units are in W m^{-3}; with Z as ionic charge, n_e and n_i electron and ion number densities. Here in this treatment ions are held at rest. The spectrum is nearly flat at low frequencies and steeply decreases at frequencies greater than (v/b) for non-relativistic electrons. In most cases this process contributes to infrared and lower frequencies.

The reverse process called inverse-Bremsstrahlung or free-free absorption may occur when a free electron moving in the Coulomb field of a proton, or an ion absorbs a photon.

3.3 Synchrotron Radiation

When a charged particle is moving with a velocity v in a magnetic field, then it gyrates around the field and radiates. This radiation is called gyro-radiation or cyclotron radiation. This is also called magneto-Bremsstrahlung. If the velocity of the charged particle (say electron) approach that of light, then it is considered relativistic, and the radiation thus emitted is known as synchrotron radiation. This radiation is highly directional and polarised. In astrophysics there are many situations under which electrons are relativistic and in the presence of magnetic field, synchrotron radiation results. Solar corona, Supernova remnants are just a couple of examples where synchrotron radiation is detected. The frequency at which the emission is maximum is called characteristic or critical frequency,

$$\nu_c = \frac{3}{2}\, \gamma^2 (eB/2\pi m_e) \sin\theta \tag{3.12}$$

where B is the magnetic field, m_e is electron rest mass, e the electron charge, θ the angle between the electron velocity vector and the magnetic field vector. The Lorentz factor γ is given by

$$\gamma = \left[1 - \left(\frac{v}{c}\right)^2\right]^{-1/2} \tag{3.13}$$

where c is the velocity of light.

The emission spectrum for electron is given as follows (see e.g., Woan 2000; Bradt 2008),

$$P(\nu) = 2.34 \times 10^{-25}\, B \sin\theta\, F(x) \qquad in\ W/Hz \tag{3.14}$$

where F is the spectral function given by

$$F(x) = 2.15\, x^{1/3} \qquad \text{for } x << 1 \tag{3.15}$$

$$F(x) = 1.25\, x^{1/2}\, e^{-x} \qquad \text{for } x >> 1 \tag{3.16}$$

with $x = \nu/\nu_c$ (frequency normalised to the critical frequency).

Typical spectral function is plotted in Figure 3.3 showing the above two regimes.

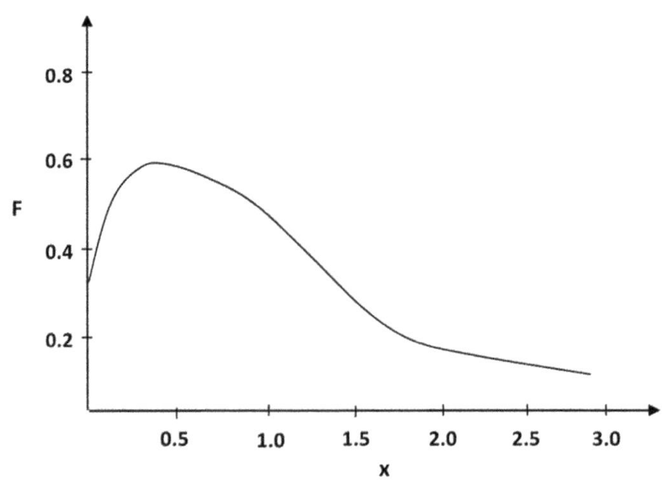

Figure 3.3: Synchrotron spectral function *F(x)* plotted with *x* (see text).

3.4 The 2-Photon Radiation

In atomic Hydrogen the Lyman alpha emission (at 121.6 nm) arises from the spontaneous transition from the *2p* level to the *1s* level (with a lifetime of 1.6 × 10^{-9} s). The transition from *2s* level *1s* level is optically forbidden due to selection rule that angular momentum should be conserved. However, in low density regions, such as the interstellar medium, an electron excited to the *2s* level, having a lifetime of 0.14 s may undergo an optically forbidden transition to the *1s* level via a virtual level with emission of two photons. In Hydrogen atom, such a transition from the optically forbidden metastable level is still possible via a quantum mechanical probability called 2-photon decay (see e.g., Pottasch 1983; Emerson 1996). The electron would make a jump to a virtual state between the *1s* and *2p* levels first and then decays to the ground state. Since the virtual state can in principle occur at any energy between the two states in question the emission leads to continuum radiation. Figure 3.4 shows schematically the energy level diagram for Hydrogen atom. This process is very important contributor to UV flux from tenuous ionized regions such as PNe and HII regions in which the metastable level survives. The continuum energy (in UV) is anywhere between the *1s* and *2s* levels as given below,

$$\nu_c = \nu_1 + \nu_2 = 10\ eV \tag{3.17}$$

The peak emission is found to occur around 350 nm (see Emerson 1999).

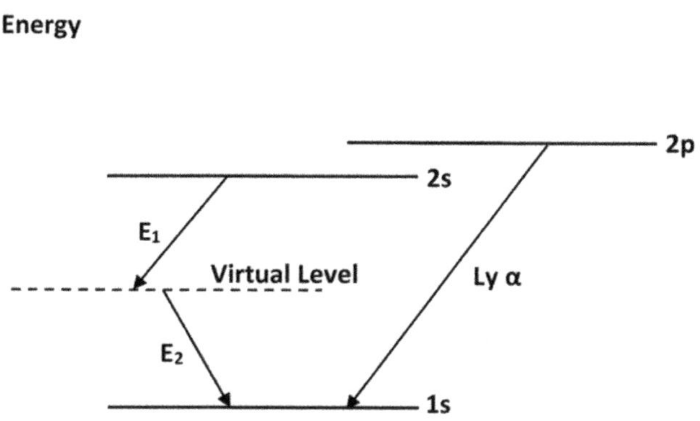

Figure 3.4: Schematic of energy level diagram for Hydrogen atom. The 1s, 2s and 2p levels are shown along with the 2-photon decay from 2s to 1s level via a virtual level which otherwise is forbidden. The energy levels are not to scale. The energy difference between the levels 1s and 2s is 10 eV and that between 1s and 2p (Lyman alpha) is 10.19 eV.

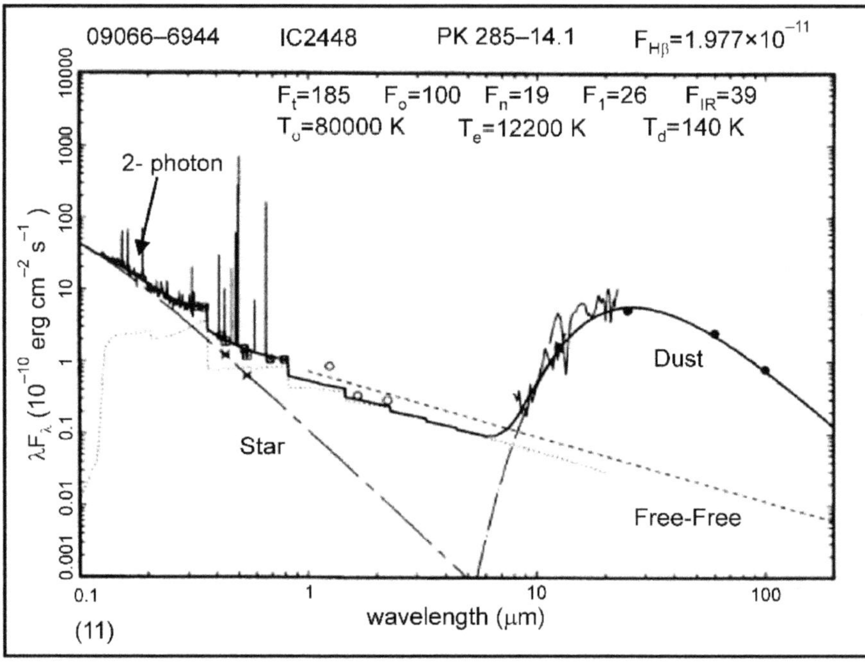

Figure 3.5: Spectral Energy Distribution (as a function of wavelength) in a Planetary Nebula (Zhang & Kwok 1991). The continuum contribution from different mechanisms is marked – 2-photon in UV, Free-Free in Infrared and the Stellar Photosphere and Dust emitting as Blackbodies. The combined spectrum is shown in solid curve superposed by emission spectral lines.

It is observed that about half of the UV emission in HII regions and a substantial part of that in PNe arise from the 2-photon mechanism. Fig. 3.5 shows the spectral energy distribution in a Planetary Nebula showing 2-photon emission in UV as well as free-free radiation component in longer wavelengths.

Figs. 3.5 and 3.6 show examples of observations on PN and SNR respectively, showing contribution from different mechanisms of continuum radiation.

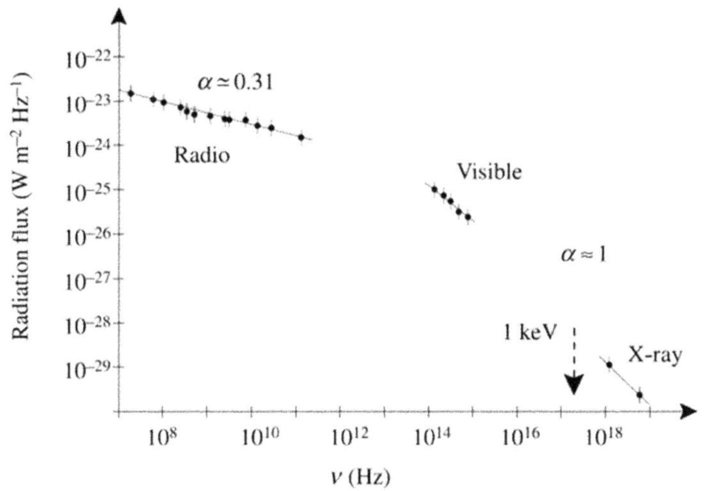

Figure 3.6: Spectral Energy Distribution with frequency in radio, visible and x-ray radiation in the Super Nova Remnant (SNR) Crab Nebula showing continuum radiation due to synchrotron emission. α is the exponent for power law distribution (from Boyd & Sanderson 2003).

In contrast, Fig. 3.6 shows the spectral energy distribution observed in Crab Nebula Super Nova Remnant. The synchrotron components in radio and x-rays are shown along with the exponent for power law distribution expected of the mechanism.

3.5 Basic Atomic Processes

Atomic spectra arise when bound electrons in an atom make transitions from one energy state (or level) to another. The energy levels of electrons in an atom are quantised. These levels are designated by quantum numbers. The principal quantum number n, arising from Schrodinger's equation, designates the quantized bound energy level of the atom. Each of these states are further quantized by the electron orbital angular momentum (l) and spin angular momentum (s). In presence of magnetic field, the levels are further split by magnetic quantum number (m). According to the Pauli's exclusion principle no two electrons can occupy the same quantum state.

The fundamental processes of absorption and emission in an atom arise when it absorbs a photon and emits one respectively. The spontaneous emission occurs when an excited atom makes a transition from the excited state (say, j) to a lower energy state (i). The probability for this transition is given by the Einstein's coefficient for spontaneous emission A_{ji}. Absorption probability for an electron to absorb a photon to jump from one level (say, i) to an upper level (j) is given by the Einstein's coefficient for absorption B_{ij}. These two coefficients are related by the relation,

$$B_{ij} = \frac{c^3}{8\pi h v_{ji}^3} \frac{g_j}{g_i} A_{ji} \tag{3.18}$$

where ν_{ji} is the frequency corresponding to the difference of energy between levels j and i, g_i and g_j are statistical weights representing degeneracies for the levels i and j respectively. It is also possible to designate transition probability by assuming the atom to be a classical harmonic oscillator and defining a term called oscillator strength given by

$$f_{ij} = \frac{4\pi\epsilon_0 m_e}{e^2 \pi} h\, \nu_{ji}\, B_{ij} \tag{3.19}$$

where m_e is the mass of electron, e is its charge, and ε_0 the permittivity of vacuum. The degeneracies and oscillator strengths for two levels are related by

$$g_j f_{ij} = -g_i f_{ji} \tag{3.20}$$

The negative sign for f indicates emission and positive for absorption. It is to be noted that the probability coefficients when inverted give the lifetimes of the transitions τ_{ij}. Figure 3.7 illustrates the basic processes of absorption, and emission.

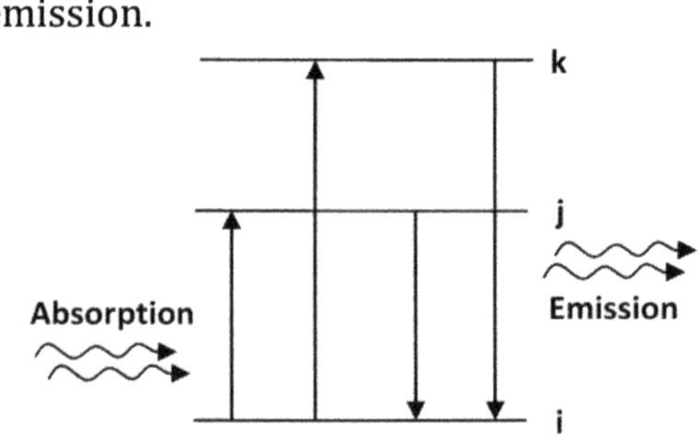

Figure 3.7: Shows energy levels in an atom labelled as i, j and k. When an electron absorbs a photon (shown as wiggly arrows) it transits from level i to level j (or k depending on the energy of the absorbed photon) the transition is called absorption. Emission occurs when electron transits from an upper level to a lower level by releasing a photon.

3.5.1 Spectrum of Hydrogen Atom: Hydrogen is the most abundant atom in the Universe. It is also the simplest and the lightest atom having one single proton in its nucleus and one single electron orbiting the nucleus. To an approximation, we can consider the spectrum of Hydrogen atom as that arising from transitions among quantised energy (sum of kinetic and potential energies) states of electron being inversely proportional to the square of the principal quantum number n. A radiative transition occurs when the electron jumps from a higher to a lower energy state and is characterised by the difference of the two energies given in terms of λ the wavelengths by,

$$\frac{1}{\lambda} = R_\infty Z^2 \left(\frac{1}{n_i^2} - \frac{1}{n_j^2}\right); \tag{3.21}$$

$$with \;\; R_\infty = \frac{\mu e^4}{8h^2 \varepsilon_0^2};$$

$$and \;\; \mu = \frac{m_e M}{m_e + M}$$

Here the subscripts i and j of the principal quantum number indicate respectively the lower and upper quantum states of the transition ($n_j > n_i$). R_∞ is called the Rydberg constant typical for a given atom due to the presence of the reduced mass μ; Z is the atomic number and M the mass of the atomic nucleus. As shown in the Figure 3.8 transitions from higher quantum levels to level $n = 1$ (the lowest level) are called the Lyman series (falling in the UV part of the electromagnetic spectrum). Similarly, the transitions to the level $n = 2$ are called the Balmer series (falls in the visible part); and followed by Paschen ($n = 3$), Brackett ($n = 4$) and so on (occurring in the infrared region). In addition to the principal or radial quantisation, the orbital and spin angular momenta of the electron are also quantised denoted by the quantum numbers l (taking integer values of 0 to n-1) and s (taking values of + ½ or – ½) respectively. The angular momentum states of the atom are denoted by capital letters S, P, D, F etc corresponding to the electronic states $l = 0,1,2,3$ etc. The z-component of the orbital angular momentum arising because of an external magnetic field is also quantised with its quantum number m taking $2l+1$ integer values from $-l$ to $+l$. Further splitting of the energy levels called the "fine structure" occurs due to the coupling of the electron orbital and spin angular momentum vectors. This l-s coupling is important for multi-electron atoms. The total angular momentum quantum number J then is a vector sum of L and S the orbital and angular momentum quantum numbers, $\boldsymbol{J = L + S}$.

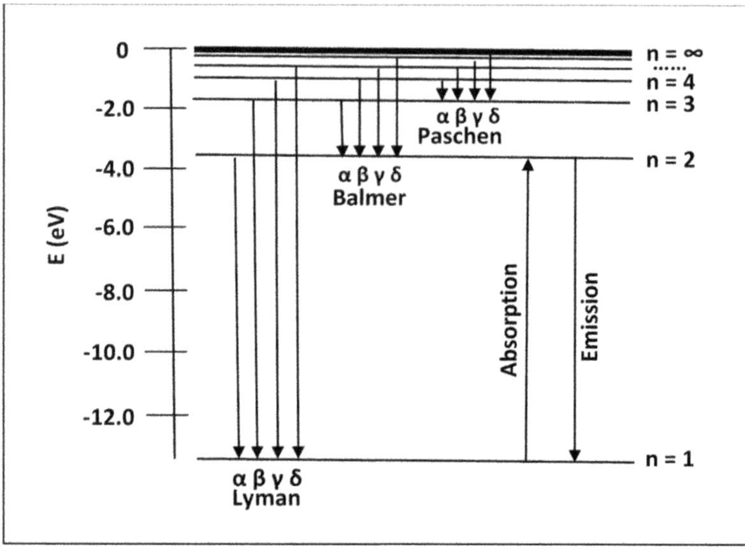

Figure 3.8: Spectrum of atomic Hydrogen (adapted from Carroll & Ostlie 2007).

Figure 3.9 shows the HI Brackett series lines in absorption in the near-infrared band 1.4-1.8 μm from the post-asymptotic giant star IRAS 07134+1005.

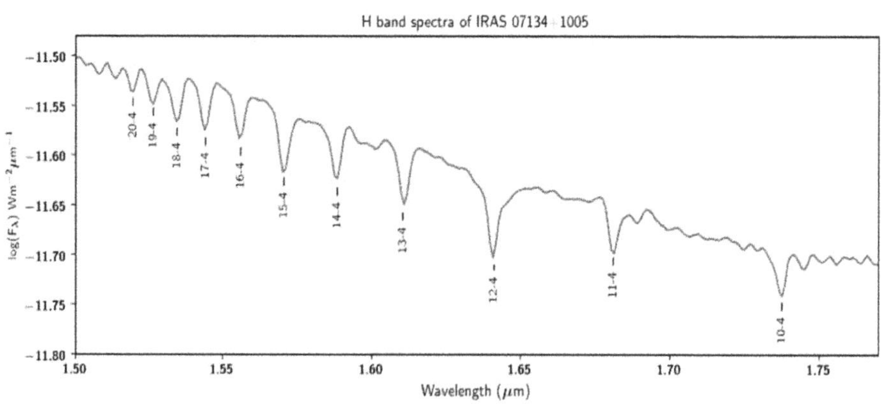

Figure 3.9: Spectrum of atomic Hydrogen Brackett series from 20-4 to 10-4 levels in the infrared region from the source IRAS 07134+1005 taken by NICS at the Mt Abu Observatory (Venkataraman & Anandarao 2008).

Hyperfine structure occurs due to the nuclear magnetic moment interaction accounting for very low energy transitions; the nuclear angular moment quantum number is designated as *i*. The final (including both electron and nuclear) angular momentum is then vector sum $f = j + i$. In Hydrogen atom, the coupling of nuclear angular momentum with that of electron gives rise to the 21 cm emission also called spin-flip transition when the electron flips from -1/2 state to +1/2. This corresponds to *f* going from 0 to 1 state.

It is thus clear that any electron in an atom may be designated with a specific quantum state. According to Pauli's exclusion principle however no two electrons can occupy the same quantum state. Consequently, other quantum numbers being the same, two electrons with spin quantum number of +1/2 and -1/2 may exist. Further transitions from one quantum state to another are restricted by selection rules that ensure conservation of parity or the symmetry of the wave function.

For "optically" allowed transitions (arising from electric dipole configurations) the selection rules are

$$\Delta J = 0, \pm 1; \Delta L = 0, \pm 1; \ \Delta S = 0; \ \ \Delta M = 0, \pm 1;$$

For ΔL and ΔJ transitions from 0-0 are not possible. One electron-jump with Δl=0, +/-1 is also allowed. In addition, parity change is required. The rule ΔS=0 signifies that "inter-combination" transitions are not allowed – e.g., singlet to triplet etc. The rule for magnetic quantum number applies only when the atom is in an external magnetic field.

3.5.2 Collisional Excitation: Apart from photoexcitation described above, atoms may also be excited by collisions with free electrons. Particularly important are the forbidden excitations from collisions. Collisional excitation rate for an optically thin forbidden transition is a function of electron temperature (T) and density (N_e) and is given by (see Emerson 1996)

$$C_{ij} = \frac{8.6 \times 10^{-12}}{T^{\frac{1}{2}}} N_e \frac{\Omega(i,j)}{g_i} e^{-\left(\frac{E_{ij}}{kT}\right)} \tag{3.22}$$

Here g_i is the statistical weight called Gaunt factor for level i and $\Omega(i,j)$ is the collision strength and E_{ij} is the energy difference between levels i and j.

3.5.3 Critical Density: We may derive a critical density for matter above which the radiative transitions arise in thermodynamic equilibrium (thermal origin) and below which de-excitation by collisions becomes important (collisional origin). This may simply be derived by equating the rate of spontaneous emission to the rate of collisional de-excitation. The critical density for transitions of an atom/ion between states i and j is given by (Tennyson 2005),

$$n_c = \frac{\sum_{i<j} A_{ji}}{\sum_{i\neq j} C_{ji}} \tag{3.23}$$

The summation above is for all possible transition routes for recombination lines. As defined earlier, A_{ji} is the Einstein coefficient and C_{ji} is the rate of collisional de-excitation.

The (thermal) case for which the density is above the critical density the transitions are thermal, and this fact may be used to derive the equilibrium temperature. Likewise, the (collisional) case for which the density is below the critical density, i.e., the collisions with electrons control the dynamics, may be used to derive the electron density. Thus, the study of spectra in both cases may be used to obtain temperature, density, and nature of the radiation field of the matter.

3.5.4 Forbidden Lines: In tenuous (low-density) interstellar medium, collisional de-excitation is not very important for recombination lines (for instance H line series); while for heavier atoms, it is the only probable mechanism via the slower electric quadrupole and magnetic dipole transitions that are forbidden by the selection rules for electric dipole transitions. For transitions between levels of same parity are forbidden by selection rules for electrical dipole transitions *(for the latter).*

Such transitions are extremely slow having longer lifetimes or much smaller probability coefficients - by several orders of magnitude - compared to those for allowed transitions. Typical lifetimes of various transitions are (see Tennyson, 2005):

$$\tau_{\text{electric-dipole}} \sim 10^{-8}\ \text{sec};\ \tau_{\text{electric-quadrupole}} \sim 10^{-3}\ \text{sec};\ \tau_{\text{magnetic-dipole}} \sim 1\ \text{sec}.$$

Forbidden emissions arise from ground levels of the atoms such as O, N, C etc. These forbidden transitions are possible to be observed in astrophysical nebulae which are highly tenuous and the de-excitation by collisions is possible unlike in denser environments as is the case, e.g., in laboratory plasmas. These lines are very important diagnostic tools for ISM.

The selection rules are the same for forbidden transitions as for allowed transitions except that the parity change is not required (see Lang 1974, for details).

The forbidden lines important in astrophysical context arise in the atoms and ions of ground state configurations $2p^q$ or $3p^q$ with $q = 2, 3, 4$ (Lang 1974). These transitions are excited by electron impact. Excitation of these forbidden transitions throw light on the electron temperature and density by observing ratios of intensities of certain carefully chosen transitions.

For temperature we need lines from same species with different excitation energies: e.g., [OIII] forbidden lines ratio I(λ4959+λ5007)/ I(λ4363); while for density we need species with nearly same excitation energy but different probability coefficients: [SII] I(λ6717)/I(λ6731) and [OII] I(λ3726)/I(λ3729) ratios (the wavelengths are in Angstrom units).

3.5.5 Stimulated Emission: In this case, photon emission is stimulated by another photon of a particular energy depending upon the transition, as shown in Figure 3.10 (see for details Tennyson 2005). Then two photons would emerge in the same direction coherently. This process requires a population inversion deviating from the Boltzmann distribution for level population in Local Thermodynamic Equilibrium. This process leads to the Light Amplification with Stimulated Emission of Radiation (Laser and by similar process Maser in microwave region). Einstein had derived the probability coefficient for this process and is called Einstein's coefficient for stimulated emission.

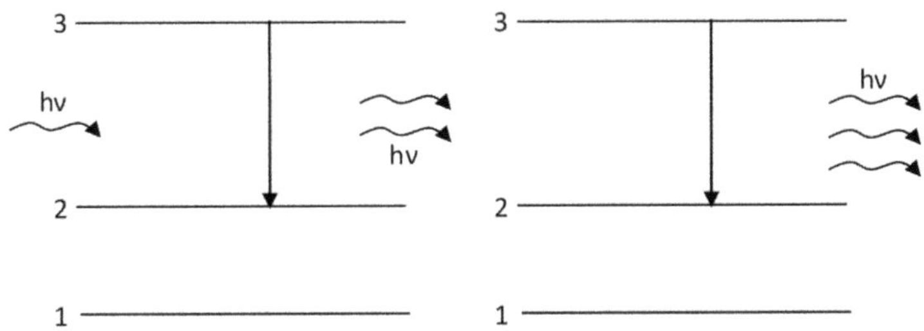

Figure 3.10: A schematic showing energy levels for laser action. Photon of appropriate energy stimulates the laser/maser emission as shown in the figure.

In several astrophysical sources stimulated emission in vibration-rotation levels of certain molecules is noticed in microwave wavelength region (Microwave Amplification by Stimulated Emission of Radiation – MASER, akin to LASER). In massive protostars methanol (CH_3OH) and OH masers are pumped (stimulated) by far-infrared radiation and are considered as very useful probes to unravel the physical mechanisms at play (see Stahler & Palla 2004). Similarly, H_2O maser emission (at 1.35 cm) stimulated by collisional interaction is also found in proto-stellar sources of both low and high masses (Stahler & Palla 2004). Maser lines of OH are observed from sources representing stellar evolution such as OH-IR stars (post-AGB stars) (see Pottasch 1983).

3.6 Molecular Spectra

Molecules are made of atoms bonded to each other by mainly two types of bonds called ionic and covalent bonds. The covalent bond forms when atoms share valence electrons: two valence electron per bond, e.g., H_2, CH_4. In ionic (and metallic) bonds the electrostatic bonding keeps the atoms together – positively and negatively charged atoms. For example, the molecule NaCN ($Na^+ + CN^-$) has ionic bonding between atoms having large electro-negativity. Similarly, in the molecule $FeSO_4$, a metallic bonding forms by the electrostatic attraction between conduction electrons (anions) in one atom (in this case the SO_4 radical) interacting with positive metallic ions (Fe, the cations). We consider here a few specific molecules of import in astrophysical context and describe their spectra briefly. Molecules may be divided into three main categories as follows.

Diatomic Molecules: These are molecules having two atoms bonded together. In homo-nuclear molecules both the atoms are the same such as molecular Hydrogen, Nitrogen and Oxygen. These atoms do not possess dipole moments being homo-nuclear. Among the other diatomic molecules of importance, we discuss here is Carbon Monoxide (CO). The spectra of CS, CN and OH are also important for diagnostics of ISM.

Polyatomic Molecules: These molecules are formed by bonding of more than two atoms. Some of the most important ones among such molecules are Methane (CH_4), Ammonia (NH_3) among others. To date more than a few hundred complex molecules have been identified in ISM from infrared, sub-millimetre, and radio observations.

Large Molecules and Dust Grains: As the size of molecules increases the vibrational and rotational modes would have restricted degrees of freedom and in such large molecules as in the Poly-cyclic Aromatic Hydrocarbons (PAH), transitions due to stretching and bending occur.

3.6.1 Types of Molecular Spectra – Electronic, Vibrational and Rotational Spectra

In molecules the electrons do not possess a central force as in an atom. The spectrum therefore becomes very complicated. In addition to the electron motion that is very fast the nuclei too would have freedom of rotation and vibration at a much slower rate. This allows us to make the approximation, known as the Born-Oppenheimer approximation, that the three may be separately treated. This is a reasonably good approximation owing to the large difference between electron mass and the nuclear mass. For an advanced treatment, called Frank-Condon principle, we may refer to Bransden & Joachain (2003). Consequently, molecular spectra have three separate components: electron transitions; vibrational transitions; and rotational transitions. The energy ranges of these transitions are depicted schematically in Figure 3.11.

Based on the energies involved in the three types of transitions we can write the total energy of the molecular system having electronic (E_{el}), vibrational (E_{vi}) and rotational (E_{ro}) energies as

$$E = E_{el} + E_{vi} + E_{ro} \tag{3.24}$$

with the inequality,

$$E_{ro} \ll E_{vi} \ll E_{el}$$

Here we deal briefly with the spectra of certain molecules that hold very high significance to astrophysical objects. Molecules containing more than two atoms (called polyatomic) usually form and survive in environments at equilibrium temperatures of less than 4000 K while diatomic molecules could survive up to 8000 K, due to their dissociation energy being higher than for the former.

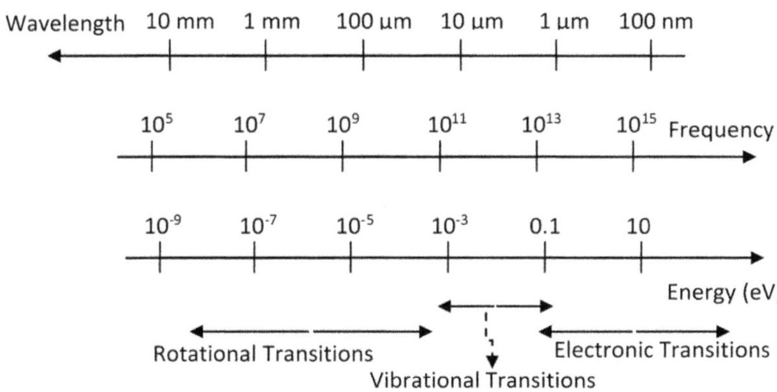

Figure 3.11: Wavelength/Frequency and energy ranges of molecular spectra.

The Schrodinger wave functions representing quantum state energies of electrons and the nuclei may be separated completely according to the Born-Oppenheimer approximation. Further the vibrational and rotational wave functions also may be separated to somewhat lesser degree of certainty. Hence, we may write the total wave function for a molecule as the product of electronic and nuclear wave functions as follows (the subscripts *e* and *n* denote electrons and nuclei),

$$\Psi(R_A, R_B, r_i) = \psi_e(r_i)\, \psi_n(R_A, R_B) \tag{3.25}$$

Here R_A and R_B are the vectors connecting the molecular centre of mass with nuclei A and B respectively; r_i are vectors between the electrons and with each electron with the two molecules. The nuclear wave function can further be written as the product of the vibrational and rotational wave functions as

$$\psi_n = \psi_v(R)\,\psi_r(\theta,\phi) \tag{3.26}$$

Here R is the inter-nuclear distance and θ and φ are rotational degrees of freedom representing rotation around two mutually orthogonal axes.

The solutions to the Schrodinger equation for rotational energy levels may be obtained as

$$E_r = \frac{\left(\frac{h}{2\pi}\right)^2}{2\mu R^2} J(J+1) \tag{3.27}$$

J is the rotational angular momentum quantum number and takes integer values $J = 0, 1, 2, \ldots$ and R represents the vector giving the inter-molecular separation and its orientation with respect to a reference axis. μ is the reduced mass given by (for a diatomic molecule),

$$\mu = \frac{M_A M_B}{M_A + M_B} \tag{3.28}$$

M_A and M_B are the masses of the atoms constituting the molecule. With $R = R_0$ the inter-molecular separation and under the assumption that a molecule is rotating like a rigid body, the equation for energy levels reduces to

$$E_r = \frac{\left(\frac{h}{2\pi}\right)^2}{2I_0} J(J+1) = B_0\, J(J+1) \tag{3.29}$$

B_0 is the rotational constant for a molecule, with the subscript representing the (ground) vibrational level. Notice that for $J = 0$, $E_r = 0$, meaning that there is no zero-point energy for rotational levels. Further, at faster rotations (for J being large) molecules tend to move apart and hence the rigid body rotation would not be appropriate in such cases.

Assuming that the vibrations of a molecule may be represented by a harmonic oscillator, we can write down the vibrational energy levels as

$$E_v = \frac{h\omega}{2\pi}\left(v + \frac{1}{2}\right) \tag{3.30}$$

Here v represents the vibrational quantum number taking integer values 0, 1, 2, ... and the levels are equally spaced. ω is the angular frequency of oscillations (vibrations). For small values of v, the electronic potential $V(R)$, in which the nuclei of the molecule move, may be assumed by a parabolic function (see Figure 3.12). The factor $h\omega$ is expressed in units of wavenumber i.e., cm^{-1}. The vibrational frequency is related to reduced mass as

$$\omega = \left(\frac{k_f}{\mu}\right)^{1/2}$$

Here k_f is known as force constant representing the curvature of the electronic potential. This dependence of frequency on reduced mass is significant for molecules having different isotopic composition.

In real world the molecular vibrations cannot be strictly represented by a harmonic oscillator. For higher energy levels the potential would deviate from parabolic shape and also the energy levels would not be equally spaced as shown in Figure 3.12. The vibration-rotation energy levels for an anharmonic oscillator may be written as (see e.g., Tennyson 2005),

$$\begin{aligned} E_{VR} = \omega_e\left(v + \frac{1}{2}\right) + B_e J(J+1) - \omega_e x_e\left(v + \frac{1}{2}\right)^2 \\ - D_e J^2(J+1)^2 \\ -\alpha_e\left(v + \frac{1}{2}\right)J(J+1) + \ .... \end{aligned} \tag{3.31}$$

Here B_e is the rotational constant, $\omega_e x_e$ is the leading anharmonic correction, D_e is the centrifugal distortion and α_e represents the variation of B_e with energy state. Along with ω_e, all these constants are called spectroscopic constants expressed in units of cm^{-1}. The rotational constant for vibrational state v is given by,

$$B_v = B_e - \alpha_e\left(v + \frac{1}{2}\right) + \ ....$$

It should be noted that within each vibrational state, there are rotational transitions.

3.6.1.1 Selection Rules and Molecular Spectra: Molecules possess a permanent dipole moment owing to the asymmetric charge distribution.

For such molecules electric dipole transitions are possible. The strength of such spectral lines depends upon the square of the dipole moment. However, for homonuclear molecules the permanent dipole moment is zero and hence the dipole transitions are forbidden. Further, for the linear symmetric molecules such as CO_2 the dipole transitions are forbidden; only under some conditions when such molecules are distorted, dipole transitions become possible.

Pure Rotational spectra of diatomic molecules in Σ symmetry electronic ground states have selection rules that allow dipole transitions for which,

$$\Delta J = \pm 1$$

Transition between two zero J states is not possible.

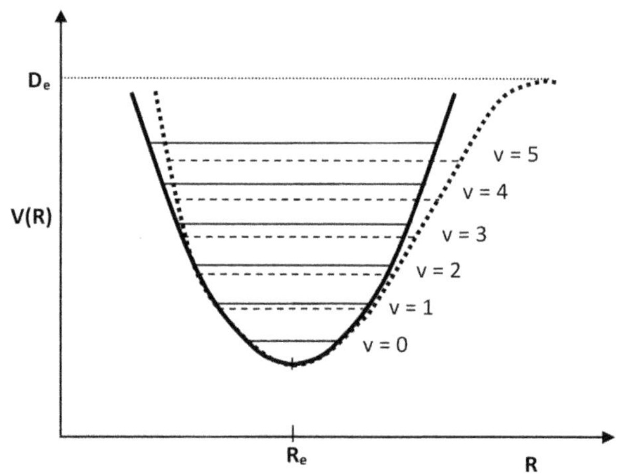

Figure 3.12: A schematic showing electronic potential V(R) for a molecule. The dark thick curve shows the case for harmonic oscillator approximation and the dashed thick curve for realistic anharmonic potential. The vibrational levels 0 to 5 are shown for both the potentials.

For homonuclear molecules the rules for the electric quadrupole transitions are

$$\Delta J = 0, \pm 2$$

Again, the transitions between two zero J states are not allowed.

Vibrational spectra have selection rules,

$$\Delta v = \pm 1$$

As mentioned earlier, with each vibrational state corresponding rotational states are possible.

3.6.2 Hydrogen Molecule: Molecular hydrogen is the most abundant molecule in astrophysical sources (see Shull & Beckwith 1982 for a detailed review). Hydrogen molecule was first detected in the ISM through these UV electronic transitions by rocket-borne probes in the early 1970's (Carruthers 1970). As already mentioned, molecular hydrogen is homo-nuclear and hence does not possess a permanent dipole moment. For this reason, the electric dipole transitions are forbidden. However, weak electric quadrupole transitions are possible and through such spectra it is possible to detect H_2 in astrophysical sources in the near- infrared (vibration lines) and mid-infrared (rotational lines) wavelength regions (see Sternberg & Dalgarno 1989). The excitation of these spectra is facilitated by the famous Lyman and Werner electronic transitions in the ultraviolet region as shown in Figure 3.13. The dissociation energy for this molecule is 4.5 eV and corresponds to the vibration state of $v = 14$. The Lyman and Werner transitions descend to the ground electronic state and have 85% probability to excite various vibration levels and their associated rotation levels, while dissociation may result for 15% of the instants (Shull & Beckwith 1982).

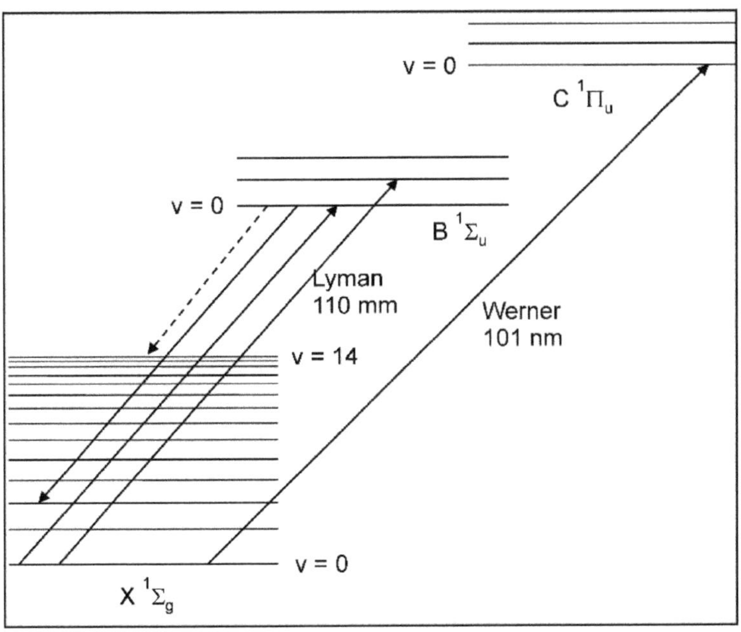

Figure 3.13: Werner and Lyman transitions in H_2 molecule. The figure shows vibrational levels v = 0 to 14, for the ground electronic state.

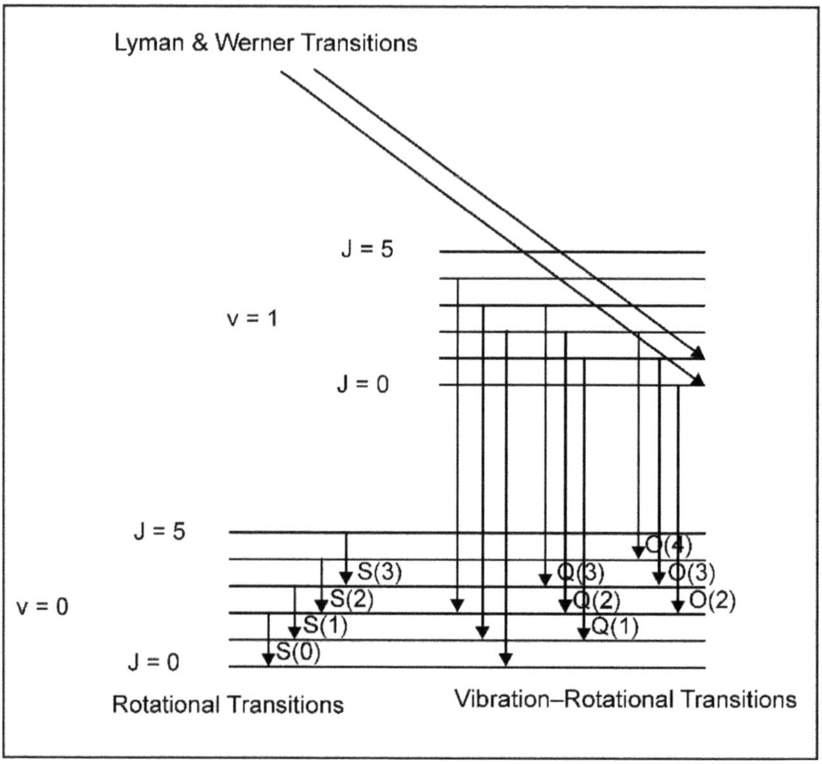

Figure 3.14: Infrared Spectral lines arising from Werner and Lyman transitions to the ground vibration state in H_2 molecule.

Figure 3.14 depicts the excitation of the infrared spectra in the ground and first vibration states and the corresponding rotation transitions. The figure shows vibration-rotation and pure rotation spectra. With the advent of infrared detectors during the second half of the last century, the vibration-rotation spectra of the molecule (in the infrared region) were detected in colder environment of interstellar medium (ISM) and since have become powerful diagnostic tools for probing the ISM and protostars.

The ratio of vibration-rotation lines may be used to delineate the mechanism of excitation namely thermal (collisional) excitation or non-thermal excitation by fluorescence (Shull & Beckwith 1982; Draine et al 1983; Black & van Dishoeck 1987; Sternberg & Dalgarno 1989; Draine & Bertoldi 1996; Luhman et al 1998; Walmsley et al 2000). For reference the line ratios had been compiled by Glass (1999). Figure 3.15 shows molecular hydrogen vibrational-rotational spectra from a star forming region.

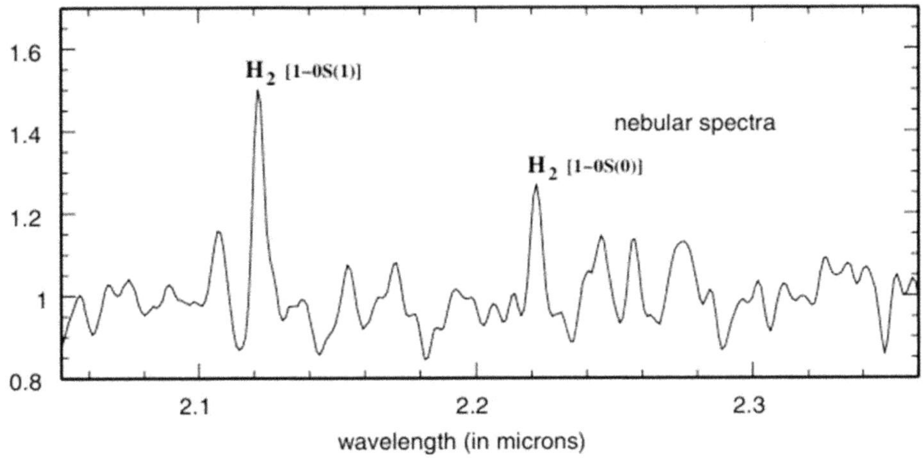

Figure 3.15: Infrared spectra of H_2 molecule in a star forming region (from Anandarao et al 2004).

3.6.3 Carbon Monoxide: CO is the second most abundant molecule after H_2. Its dissociation energy is 11.1 eV and hence once formed in the ISM or circumstellar environments it stays longer than most of the molecules (see e.g., Emerson 1996; Tennyson 2005). This fact combined with its rich prominent rotational lines in the ground vibration state renders this molecule one of most important tools to map the cold ISM regions. Figure 3.16 shows the vibration-rotation spectral transitions. Particularly important are the transitions $J = 1\text{-}0$ at 2.60 mm and $J = 2\text{-}1$ at 1.30 mm in the radio region. Infrared lines arise from vibrational-rotational transitions – most important being the fundamental $v(1\text{-}0)$ line at 4.6 μm (see Figure 3.16) for which the first ($\Delta v = 2$) and second ($\Delta v = 3$) overtone bands lie in the near-infrared observable using the ground-based telescopes. Figure 3.17 shows the first and second overtone vibrational-rotational lines in absorption in the K band and H band respectively from two post-AGB stars observed using NICMOS at the Mt Abu Observatory.

Table 3.1 provides a list of the prominent and diagnostically important transitions in the visible and near-infrared regions for various atoms, ions, along with the molecular vibrational-rotational transitions from H_2 and CO. The table includes both allowed and forbidden transitions. Besides H, O, N, and S listed in the table, the elements He, C, Ca, Ne, Na, Mg, Fe, and many others at various stages of ionization occur in the spectra of various astronomical sources. Apart from the vibrational-rotational transitions of H_2, the molecule has prominent pure rotational lines as well - at ground vibrational level $v = 0$, with $\Delta J = 2$ from S(0) to 0-0 S(5) as shown in Fig 3.14 - in the mid-infrared wavelength region 6.9 to 28.2 μm.

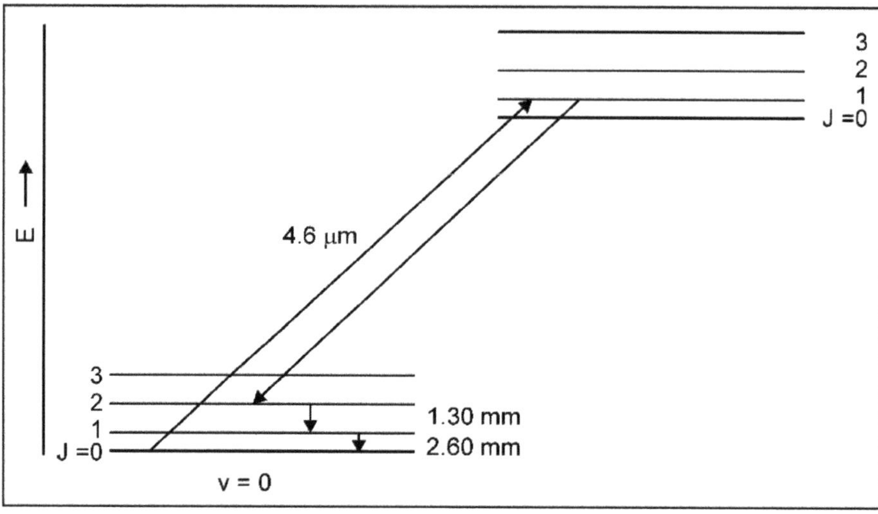

Figure 3.16: Infrared Spectral lines from CO molecule.

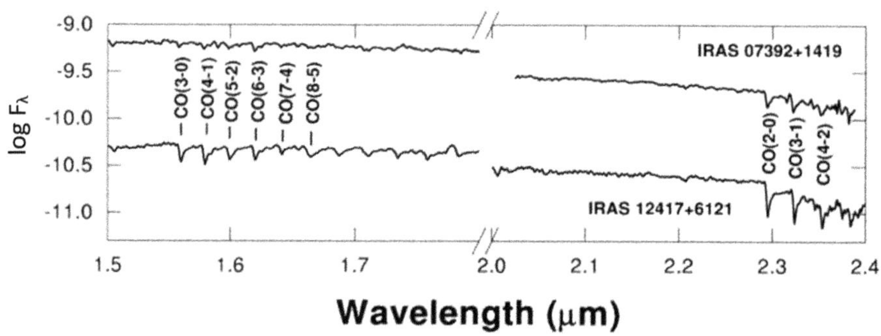

Figure 3.17: Infrared Spectra of two post-AGB stars showing overtone vibrational-rotational lines in absorption from CO molecule (from Venkataraman & Anandarao 2008).

3.6.4 Other Important Molecules: The other important molecules in the astrophysical context are CS, CN, CH_4, CO_2 that may be detected in both infrared and radio regions. The transitions may be excited by collisions or by radiation processes.

If we assume that the excitation of the molecules occurs by collisions with molecular hydrogen (considering cold molecular environment in ISM), then it is possible to constrain the density of the ISM at which a particular molecule gets excited most efficiently.

Table 3.1 Prominent Atomic/Ionic/Molecular Lines in Astrophysical Sources in the visible and Infrared wavelength regions.

Species	Transition	Wavelength (μm)
HI	Ba δ	0.4102
	Ba γ	0.4340
	Ba β	0.4861
	Ba α	0.6563
	Pa δ	1.005
	Pa γ	1.094
	Pa β	1.282
	Pa α	1.875
	Br γ	2.165
HeI	$2p^3P^0$ - $2s^3$ S	1.083
	$4d^3D$ - $3p^3P^0$	1.699
[OI]	$^2D_{5/2}$ - $^4S_{3/2}$	0.3229
	$^2P_{1/2}$ - $^2D_{5/2}$	0.7319
	$^2P_{1/2}$ - $^2D_{3/2}$	0.7330
[OIII]	$^1D_2$ - $^3P_2$	0.5007
	$^1D_2$ - $^3P_1$	0.4959
	$^1S_0$ - $^1D_2$	0.4363
[NII]	$^1D_2$ - $^3P_2$	0.6583
	$^1D_2$ - $^3P_1$	0.6548
	$^1S_0$ - $^1D_2$	0.5755
[SII]	$^2D_{3/2}$ - $^4S_{3/2}$	0.6731
	$^2D_{5/2}$ - $^4S_{3/2}$	0.6716
H_2	(1-0) S(3)	1.957
	(1-0) S(2)	2.034
	(1-0) S(1)	2.121
	(1-0) S(0)	2.223
	(2-1) S(1)	2.247
CO	v (1-0) Fundamental	4.7
	$\Delta v = 2$ 1st overtones	2.30
	$\Delta v = 3$ 2nd overtones	1.65

Following a simple model described by Smith (2004) the mean free path for a molecule (defined as the distance that a given molecule can travel before colliding with a hydrogen molecule-auto-collisions may be very rare and may be neglected) can be given by

$$\lambda_p = \frac{1}{\sigma_p n(H_2)} \tag{3.32}$$

σ_p is the collision cross-section for the molecule in question and $n(H_2)$ the density of hydrogen molecule in the region of ISM under consideration. The mean-free path as shown by the above equation is larger if the density is very low and smaller if the density is high. The cross-section depends on the impact parameter of the size of a molecule. Further the rate of collisions C is inversely related to the collision timescale as

$$C = \frac{1}{\tau_C} = \sigma_p n(H_2) v_{th} \tag{3.33}$$

v_{th} is the thermal velocity related to temperature as $T^{1/2}$. If the molecule decays radiatively, the rate is given by the Einstein's coefficient of spontaneous emission A. The two timescales namely C^{-1} and A^{-1} should be comparable for a critical density n_c given by

$$n_c = \frac{10^{11} A}{T^{1/2}} \tag{3.34}$$

A is the coefficient for spontaneous emission and T the temperature of the ISM.

If for a molecule $C << A$ (cf. eqn. 3.23), then the collisional excitation process is very inefficient and only radiative transitions become important. For the case of a cold ISM at T = 10 K, the critical density for CO molecule for J = 1-0 rotational transition (at 115 GHz) for which $A = 7\text{x}10^{-8}$ s^{-1} is ~ 10^3 cm^{-3}. In comparison for the J = 2-1 transition (at 98 GHz) of CS molecule for which $A = 2\text{x}10^{-5}$ s^{-1} the critical density is ~ 10^6 cm^{-3}. Therefore, the CO transition may be used to trace the low-density regions and the CS transition for high density regions of ISM.

3.7 Very Large Molecules: Polycyclic Aromatic Hydrocarbons

Infrared spectra of interstellar medium showed several broad unidentified features in emission (e.g., Sellgren et al 1983). These features occurring at 3.3, 6.2, 7.7, 8.6 and11.3 μm, were attributed to chains of hydrocarbon molecules named polycyclic aromatic hydrocarbon (PAH) molecules by Leger & Puget (1984) and Puget et al (1985). The laboratory spectra of several such molecules corroborated with the observed features (Allamandola et al 1987). These are planar sheets of large molecules (or small grains) composed of several hexagonal Benzene rings of sp^2-hybridized carbon atoms (each C atom having three σ bonds - a double bond and a single bond in one plane - and a π bond perpendicular to the plane). The carbon atoms are connected to each other with H, N, O atoms, and radicals saturating the outer bonds of peripheral carbon atoms (e.g., Kwok 2004) forming aromatic networks and aliphatic bridges. With linear sizes typically $\leq$ a nanometre, these molecules belong to the class of nanoparticles. With more and more observations on infrared spectra pouring in over the years, it is now evident that these molecules are ubiquitously present in a variety of astronomical sources, such as, protostars, star formation regions, evolved stars such as post-AGB stars, HII regions, and external galaxies. For comprehensive account of PAH molecules, we may refer to Puget & Leger (1989), Kwok (2004), Whittet (2003), Tielens (2008), and Pendleton & Allamandola (2002). Typical structure of a PAH molecular complex is shown in Figure 3.18. The figure depicts various features marked by aromatic complexes and aliphatic bridges (taken from Pendleton & Allamandola, 2002; see also Kwok, 2004). Figure 3.19 shows schematically the emission mechanism in PAH molecules triggered by the absorption of a FUV photon – infrared emission occurs when transitions take place from the excited S_2 level to T_1 level via inter-system crossing mechanism (for details see Tielens 2005). Emission in various stretching modes – in-plane and out-of-plane stretching modes in the C-H and C-C bonds and the combinations modes is shown in Figure 3.20 (from Whittet 2003) and Figure 3.21 (from Venkataraman et al 2017). Table 3.2 gives important infrared emission features detected in ISM.

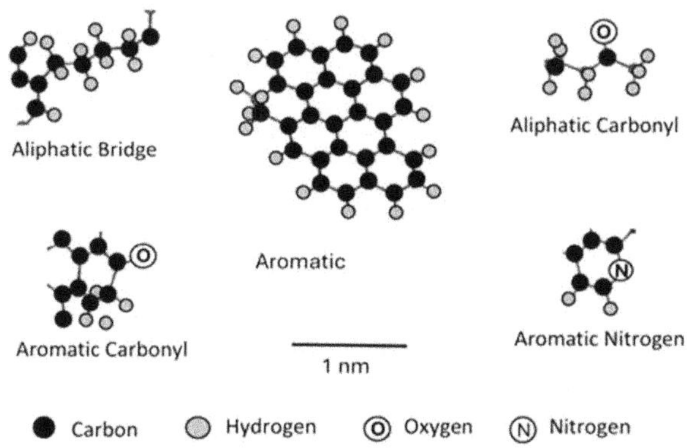

Figure 3.18: Structures in a Polycyclic Aromatic Hydrocarbon molecule complex. Various bonding networks are shown in the five panels: Aromatic network (centre); Aliphatic bridge (top left); Aliphatic carbonyl (top right); Aromatic carbonyl (bottom left); and Aromatic nitrogen (bottom right). (Adapted from Pendleton and Allamandola 2002).

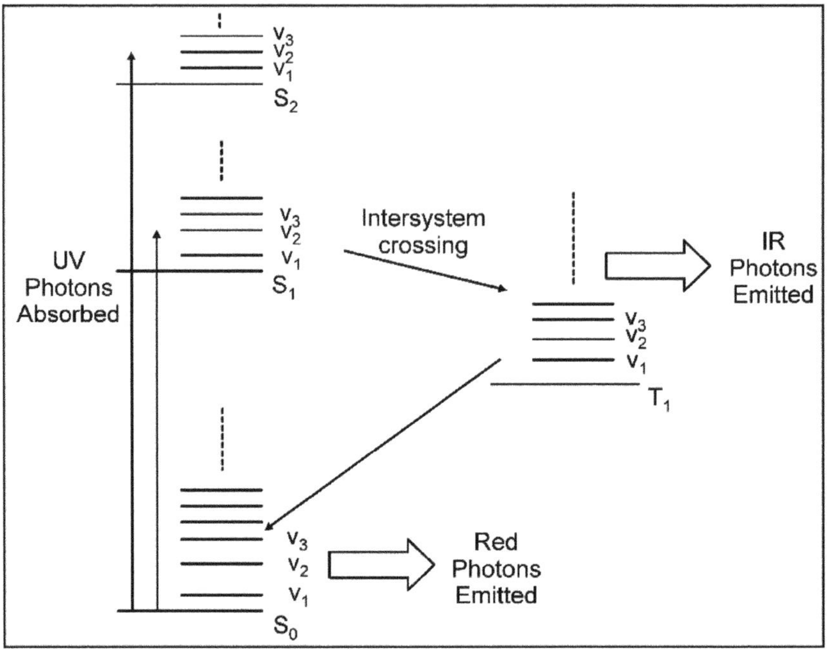

Figure 3.19: Excitation mechanism in neutral Polycyclic Aromatic Hydrocarbon molecules (adapted from Allamandola et al 1987).

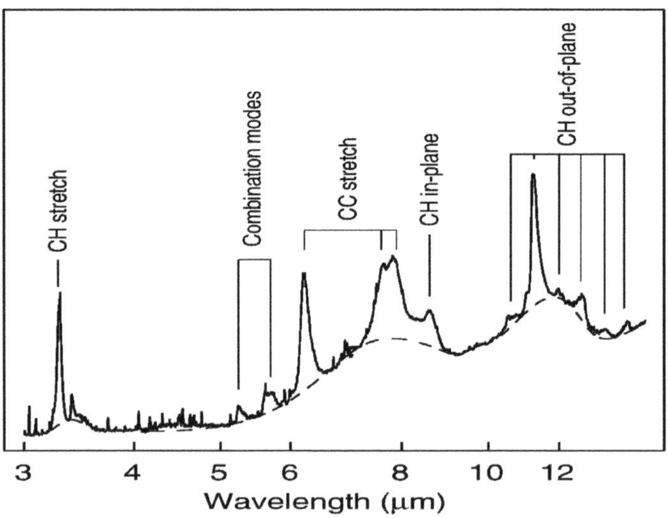

Figure 3.20: Spectrum of a PAH molecule showing various modes (from Whittet 2003).

Table 3.2: Interstellar Dust and PAH spectral features.

Species	Wavelength (μm)	Comments
Silicate	9.7, 18.5	SiO
Silicon Carbide	11.3	
Graphite	0.2175 (UV)	Alternatively, PAH
PAH features	3.3	Aromatic C-H stretch
	3.4	Aliphatic C-H stretch
	5.2 5.65	Combination of C-H bend and C-C stretch
	6.2	Aromatic C-C stretch
	6.9	Aliphatic C-H bend
	7.6 7.8	C-C stretch C-H in-plane bend
	8.6	C-H in-plane bend
	11.0	C-H out-of-plane bend solo - cation
	11.2	C-H out-of-plane bend solo - neutral
	12.7	C-H out-of-plane bend trio - cation
	13.6	C-H out-of-plane bend
	14.2	C-H out-of-plane bend
	16.4	C-C-C in-plane and out-of-plane bend

3.8 Cosmic Dust Grains

Dust is an all-pervading feature of cosmos – not just only in the terrestrial environment but it is universally present in many a situation. This makes the topic one of the most important in modern day astrophysics (see e.g., Kwok 2013).

Stars following the exhaustion of core hydrogen undergo mass loss in a variety of ways discussed briefly as it were in the Introduction (Chapter 1). Main contributors to the mass loss and hence to the enrichment of ISM are the AGB stars (with progenitors of low-intermediate mass) and Supernovae (having massive stars as their progenitors). In addition to these sources star forming molecular clouds also offer cites for formation of dust grains in their relatively higher density regions. It is basically a cyclic process, so to say, as shown in Figure 3.22. Basically, mass thus lost to the ISM cools and various molecules form in the circumstellar environs. Dust grains form in the ISM and circumstellar matter when molecules stick together, in a process called coagulation, to form large grains and large planar molecular chains of sizes typically of 1 nm, the PAHs (discussed in the previous section). The sizes of the dust particles vary from a small fraction of a micrometre to a few micrometres. The distribution of sizes usually assumed to follow a power law (as ascertained from laboratory experiments).

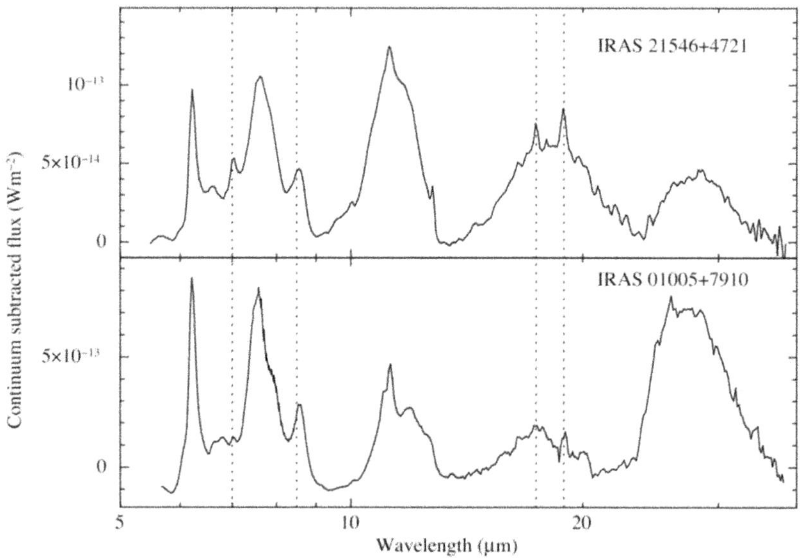

Figure 3.21: Continuum-subtracted spectrum of a PAH molecule along with other lines observed by Spitzer satellite in two post-AGB stars (from Venkataraman et al 2017).

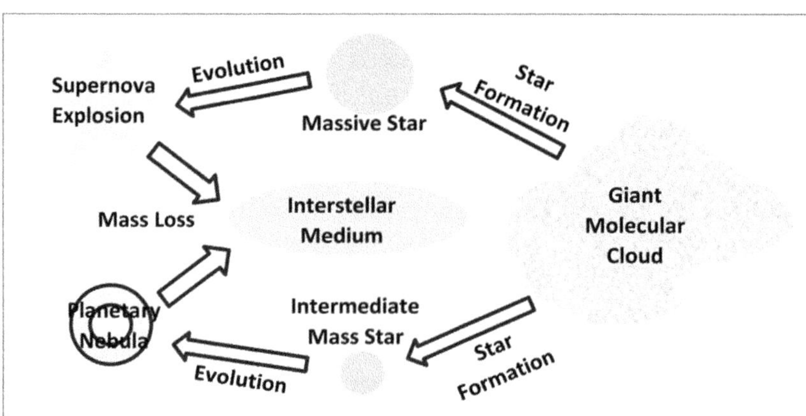

Figure 3.22: Cycle of Dust in ISM (adapted from Tielens & Allamandola 1987).

Figure 3.23 shows contribution of various components of ISM to the radiation field. In the shorter wavelengths hot gas is the principal contributor while in the infrared region emission from dust contributes most.

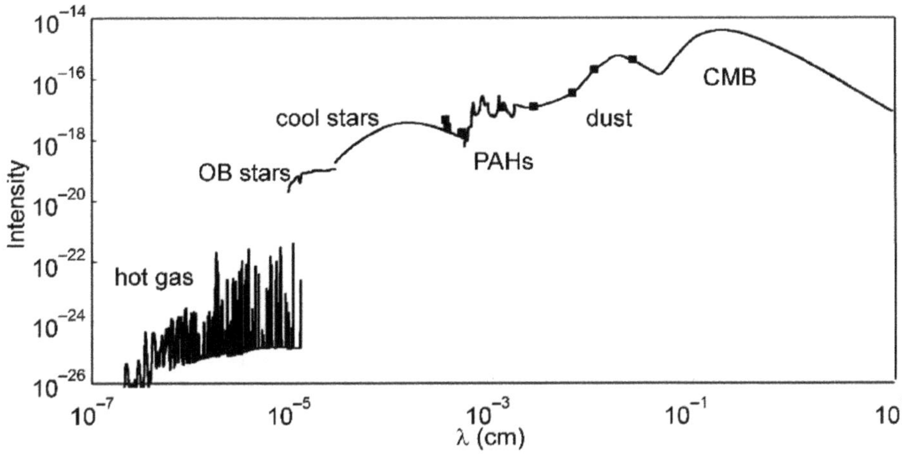

Figure 3.23: Contribution from various components to the interstellar radiation field (from Tielens 2005; see also cross references).

3.8.1 Spectral Features of Dust Grains and PAH Molecules: Dust grains are identifiable from the broad emission and absorption features arising from stretching and bending modes of constituent molecular bonds: for example, the stretching modes of the dust constituents such as SiO (silicate dust) and SiC (silicon carbide dust) and the modes in the C-C and C-H bonds in PAH molecules (see Tables 3.2 and 3.3). In addition to the prominent emission features listed in the table, PAH molecules also exhibit broad nearly flat features aptly called "plateaus" in the regions 3.2-3.6, 6-9, 11-14, and 15-19 μm. These plateaus are attributed to combination and blending of many modes involving C-C stretch, C-H bending (both in-plane and out-of-plane) and C-C-C bending (Tielens 2005). Due to the overlap of these plateaus with the other prominent modes listed in the table the overall appearance of the mid-IR spectra is like a rising continuum.

3.9 Extinction

One of the earliest discovered and the most important aspects of the presence of dust grains is their absorption and scattering effects as the radiation passes through intervening medium towards an earth-based observer. Extinction is a physical parameter that accounts for both absorption and scattering that photons undergo while traversing the interstellar or circumstellar medium. It is therefore the sum of absorption and scattering effects of the constituent matter in the ISM. The particles that scatter and absorb the photons are not only gas particles (atoms and molecules) but also dust grains of different composition. We would consider here only the effects due to dust grains. Towards this effort we can apply here the famous Beer-Lamberts law. Thus, the fraction of intensity dI absorbed and scattered by a column of dust grains in the line of sight in a particular direction (observer-object line of sight) can be written as,

$$\frac{dI}{I} = -n_d \, C_{ext} \, dL \tag{3.35}$$

Here I is the intensity of photons emerging from an object before encountering a cloud of dust grains and n_d is the number density of grains per unit cross-section in the line of sight; and dL is the elemental column length in the line of sight. $C_{ext}$is the extinction cross-section that is a function of dust properties such as its composition, its complex refractive index, size, and shape as well as the wavelength of the photon that is scattered or absorbed. Now integrating the above equation, we get with τ_λ as the optical depth,

$$I = I_0 \exp(-\tau_\lambda)\,; \; \tau_\lambda = n \, L \, C_{ext} \tag{3.36}$$

$$C_{ext} = C_{abs} + C_{sca} = \frac{Q_{ext}}{\pi a^2} \; ; with \; Q_{ext} = Q_{abs} + Q_{sca}$$

Extinction cross section is the sum of scattering and absorption cross sections as shown above and is the ratio of the extinction coefficient Qext and the geometric cross section, the surface area of the dust grain of size a (πa^2). L is the entire column length between the object and the observer.

Figure 3.24 shows extinction cross-section for radiation of wavelengths corresponding to the range between infrared and x-rays (cf. Figure 3.23). We may notice the bump at 10 and 18 μm corresponding to silicate dust absorption and the famous bump at 217.5 nm that is usually attributed to graphite absorption, but other alternatives such as PAH are also suggested (Blasberger etal 2017). The steep drop called the Lyman ledge is due to the absorption of radiation at 91.6 nm corresponding to the Lyman continuum by the Hydrogen atoms.

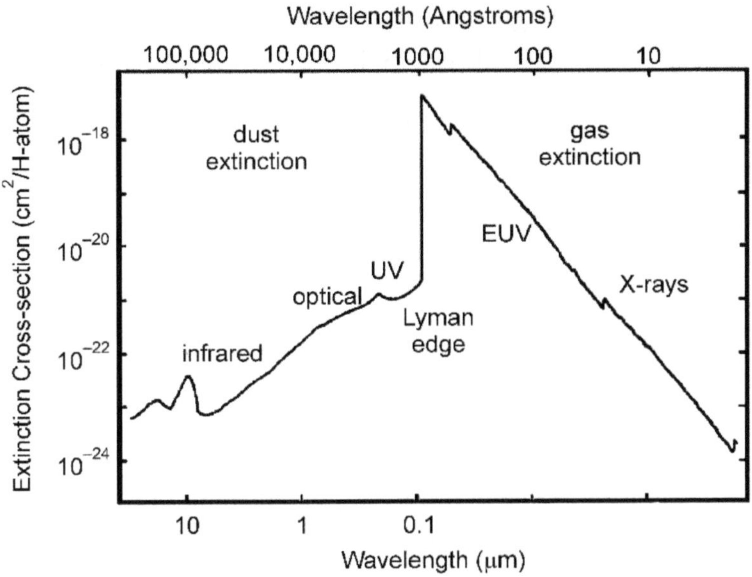

Figure 3.24: Extinction Cross-section as a function of wavelength in interstellar medium (C. Ryter given in Smith 2004).

As shown in the figure, the extinction towards the left of the Lyman edge towards longer wavelengths is predominantly caused by the dust in the interstellar medium while that in the shorter wavelengths is by warm and hot gas particles. The deep gorge marked "Lyman Edge" in the figure is caused by the bound-free absorption of photons at 91.2 nm by H atoms leading to ionization.

We may write the optical depth in terms of magnitude scale by converting natural logarithm to logarithm to the base of 10. We get,

$$A_\lambda = -\frac{2.5}{2.3}\ln[exp(-\tau_\lambda)] = 1.087\,\tau_\lambda \tag{3.37}$$

A_λ represents the extinction (in magnitude scale) that photons of a particular wavelength suffer due to passage through ISM (and/or circumstellar matter). This extinction parameter as defined above represents the difference in magnitudes between an un-reddened star and a distant (reddened) star of the same spectral type, as the radiation suffers extinction. Such a star shows colours that differ from an un-reddened star. A parameter called colour excess can then be defined as

$$E(\lambda) = \frac{Q_e(\lambda) - Q_e(\lambda_1)}{Q_e(\lambda_1) - Q_e(\lambda_2)} = \frac{A(\lambda) - A(\lambda_1)}{A(\lambda_1) - A(\lambda_2)} \tag{3.38}$$

Here λ denotes any given wavelength while λ_1 and λ_2 denote two wavelengths that are considered as standards used to eliminate the distance term in the individual extinction equations. The colour excess that indicates the presence of dust is given by the excess in *B*, *V* colours,

$$\frac{E_{\lambda-V}}{E_{B-V}} = \frac{A_\lambda - A_V}{E_{B-V}} = R_V\left(\frac{A_\lambda}{A_V} - 1\right) \tag{3.39}$$

where,

$$R_V = \frac{A_V}{E(B-V)} \tag{3.40}$$

Here A_V is considered as the extinction at the selected wavelength corresponding to visual band (545.5 nm) and E is the B - V colour index. R_V is usually $\sim$ 3 but may vary in different lines of sight and in other galaxies. This ratio is expected to indicate grain formation and growth due to coagulation (see Whittet 2003).

In general, for large particles, i.e., size a is of the order of one wavelength or larger as is the case for infrared wavelengths, on an average the extinction coefficient may be written down as a parameterized quantity,

$$Q_\lambda \propto \lambda^{-p}\,;\; \tau_\lambda \propto \lambda^{-p}$$

The exponent p above is found to be 1.62 - 1.85 by different authors (see Whittet 2003 and Glass 1999 for details), for the Galactic ISM. It shows therefore that the longer the wavelength the smaller would be the extinction; the extinction at the visible (V) band is about a factor of 10 more than that at the near-infrared K band. This is the reason why the regions enshrouded in dust and gas are better explored via infrared bands than the optical bands.

For the optical and near-infrared regions, the colour index may then be written as,

$$\frac{E_{\lambda-V}}{E_{B-V}} = 1.19\,\lambda^{-p} - R_V \tag{3.41}$$

The ratio R_V may also be expressed in terms of visible-IR colour index as follows,

$$R_V = 1.1\,\frac{E_{V-K}}{E_{B-V}} \tag{3.42}$$

From this, we can find the extinction law (a plot of colour index versus λ^{-1}) for ISM. The law may be different for different environments (e.g., our Galaxy versus other galaxies). For the Galactic ISM (in or close to the plane of the Galaxy), in general for the V band, the average rate of extinction A_V/L with distance is given by 1.8 mag/kpc (see e.g., Whittet 2003). In comparison, the extinction rate towards the Galactic centre is estimated to be ~ 4 mag/kpc.

Figure 3.25 shows the average extinction curve of the ISM (average curve for various lines of sight in the Milky Way galaxy taken from Fitzpatrick 1999; 2004) for the UV through visible and near-IR regions. The figure clearly illustrates the four discernible attributes namely, the IR region power law, the optical "knee", the near-ubiquitous absorption bump at 217.5 nm and the steep rise in the far-UV region. The UV bump is attributed to small graphite grains of sizes ≤ 0.2 μm, while the rise in the UV region is attributed to very small particles (VSG, even smaller than graphite grains) (see Whittet 2003). For a detailed discussion on the average extinction curve, we may refer to Draine (2003), Whittet (2003), and Fitzpatrick (2004).

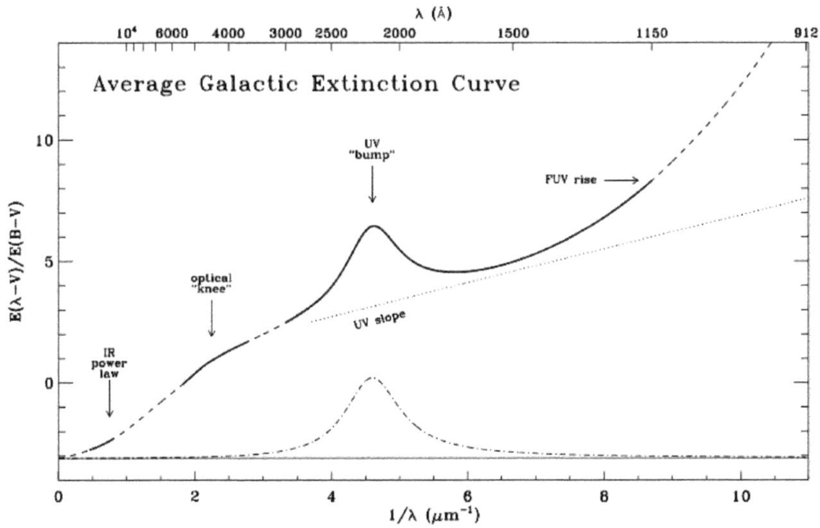

Figure 3.25: Average Milky Way Extinction Curve showing various features from UV to Infrared region (for R = 3.1) (from EL Fitzpatrick 1999, 2004).

3.9.1 Sources of Dust: The principal source of dust is the ISM in all its phases comprising of Supernovae remnants, Novae, AGB stars including Carbon stars, Protostars apart from the Solar system objects (in the interplanetary medium).

3.9.2 Composition of Dust Grains: Clearly, since dust must form from available elements present in the ISM, we expect depletions of elemental abundances relative to standard reference abundance and attribute the depleted abundance to be embedded in dust grains. Thus, we not only require relative abundances (relative to H) in ISM but also dust composition so that we would be able to quantify the elemental abundance embedded in dust particles. Thus, the parameter depletion is defined as under-abundance of an element in gas phase relative to the standard.

The abundance N_X of an element X in gas phase relative to its solar value may be defined as (in terms of logarithmic values normalised with the abundance of hydrogen H),

$$\left[\frac{X}{H}\right] = log\left\{\frac{N_X}{N_H}\right\} - log\left\{\frac{N_X}{N_H}\right\}_{\odot} \tag{3.43}$$

The symbol $\odot$ refers to the Sun. In a similar fashion we can also define a depletion index as follows. It may be noted here that the present-day metallicity of the ISM is consistent with that of the solar system to within uncertainties (see Whittet 2003, for a detailed discussion).

The depletion index for an element $D(X)$ is defined as follows with N_X and N_H denoting abundances for element X and Hydrogen (H),

$$D(X) = \log\left[\frac{N_X}{N_H}\right] - \log\left[\frac{N_X}{N_H}\right]_{ISM} \tag{3.44}$$

The suffix for the second term denotes the standard reference for ISM such as the Solar abundance. One more important parameter is the fractional depletion defined as,

$$\delta(X) = 1 - 10^{D(X)} \tag{3.45}$$

When $\delta(X) = 0$ it means that all atoms are present in gaseous state (no depletion or 0% depletion), and when equals to 1, it means that all atoms are in solid state (depletion is 100%).

Another important parameter that needs to be considered is the condensation temperature T_C, defined as the temperature at which 50% of the atoms of an element condenses into solid phase under thermodynamic equilibrium conditions. Figure 3.26 shows a plot of depletion index for elements as a function of their condensation temperature. A correlation of the parameters plotted would indicate that the dust grain condensation occurs in equilibrium conditions (Whittet 2003).

Within errors of observations, the plot in Figure 3.26 shows two distinct groups, namely volatile and refractory elements: those having $T_C < 1000$ K are called volatile elements and those having $T_C > 1000$ K are called the refractory elements. The refractory elements show a strong correlation indicating predominant depletion as compared to near zero depletion for volatile elements.

It is abundantly clear that spectroscopic observations are very important for knowing the composition of dust grain and then to attempt to correlate or reconcile with elemental abundances.

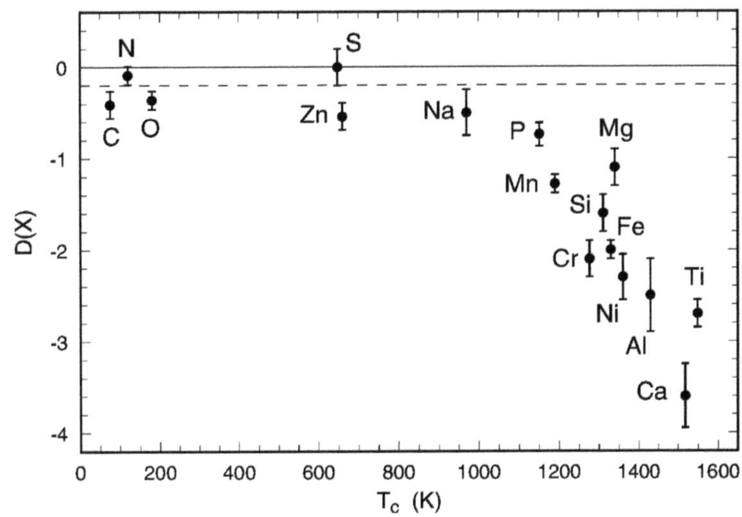

Figure 3.26: Depletion index for elements as function of their condensation temperature (from Whittet 2003). The D(X) zero line represents the Solar values and the dashed line the OB stars.

3.9.3 Gas-to-Dust Ratio: The correlation of gas and dust in ISM was established by observations on atomic and molecular hydrogen (see Whittet 2003, for a review). Bohlin et al (1978) showed that in general gas and dust are well mixed in ISM. Hence it is important to know the concentration of dust with respect to gas. Thus, one of the most important factors that characterise the ISM is the ratio of gas and dust by mass. With ρ_d representing the dust mass density (summed over all refractory elements) and ρ_H that of Hydrogen (being most abundant, representing the gas), the dust-to-gas ratio is given by (see for details, Whittet 2003),

$$Z_d = 0.71 \frac{\rho_d}{\rho_H} \sim 0.004 - 0.009 \tag{3.46}$$

The ratio indicates the fraction by mass of elements trapped in solid state as dust grains.

Elements (apart from Hydrogen) that are presumably form dust grains are divided into three categories by Whittet (2003): (i) elements such as C, N, O, S, Zn with low depletion indices 0-60%, generally irrespective of environmental physical conditions; (ii) elements like Mg, Si, P with high depletion 80-100% in diffuse clouds and with lower depletions in inter-cloud medium – hence density dependent; and (iii) elements like Fe, Ti, Ca, Mn, Cr and Ni with density dependent high depletion indices 80-100% that survive in extreme conditions and represent rather indestructible form of dust grains. The major elements by fraction of mass that are depleted into dust grains in ISM are O, C, Mg, Si and Fe.

Thus, the real-world dust grain models (that predict the complex refractive index as a function of wavelength) need to take the above results into consideration.

In general, the dust grains are identified to be silicates (of magnesium and iron), graphite, silicon carbide, and amorphous carbon.

3.9.4 Dust Grain Models: Basic dust grain models use theory of light scattering and absorption by small spherical/spheroidal dust particles (see Bohren & Huffman 1983). From the theoretical treatment the optical properties namely absorption and scattering coefficients at a given wavelength are obtained for a given grain composition, of given shape and size. These data are then applied to a model that calculates the light output as a function of wavelength for a light source enshrouded in a cloud of dust particles. The light source (assuming a blackbody of a certain equilibrium temperature) may be a star of a given spectral type and luminosity class, for instance. DUSTY is such a modelling code (Ivezic & Elitzur 1997; Ivezic et al 1999) extensively used by astronomers.

Any grain model deals with the scattering and absorption processes for electromagnetic radiation due to a dust grain. The grain is assumed to be sphere or (prolate or oblate) spheroid while, in reality, it is irregular, fluffy, and porous. Any model or a set of models should be able to understand and explain the observed results on interstellar extinction curve (as discussed earlier), polarization of grains, the spectral features (both absorption and emission) in various spectral regions and excess continua, abundances and depletions of condensable elements putatively trapped in dust grains and formation and destruction of dust particles in various astrophysical environments (see Whittet 2003). Also important are the ratio of O and C in O- and C-rich grain environments, average grain sizes and grain size distribution (and their possible effect on the extinction curve).

For most of the grain models that reconcile well with the observations the grain size distribution is a power law of the following type,

$$n(a) \propto a^{-q} \quad ; where\ q = 3.5 \tag{3.47}$$

This size distribution is known as MRN distribution (Mathis et al 1977).

Table 3.3 gives important parameters and observed features for dust grains in interstellar medium. In the table VSG denotes very small particles which are thought to be responsible for the rise of ISM extinction in the UV region. The spectral features as listed in the table serve as an important input to the composition of dust grains and helps the models reconcile with observations. The most versatile grain models are discussed in Draine (2003). The latest model known as discrete dipole scattering model (DDSCAT) due to Draine is also available for users (Draine & Flatau 1994, 2003). This model may be used for any composition of grains and the shapes may be manipulated by using appropriate number of dipoles creating voids to simulate the more realistic porous grains.

Table 3.3: Dust Grain Properties and observable features (adopted from Whittet 2003).

Property	Silicate Dust	Carbonaceous Dust
Composition	Am. $MgSiO_3$ Fe inclusions	Am. C Graphite PAH (VSGs)
Origin	O-rich Star Dust	C-rich Star Dust; ISM
Size (μm)	< 1.0	< 0.02 - 0.001
Elements	O, Mg, Al, Fe, Si	C
Extinction	UV-Vis-IR	UV-Vis-IR; Far-UV
Polarization	UV-Vis-IR	UV-Vis-IR and None for PAH /Graphite
Absorption Features	9.7; 18.5 μm (Silicates); 11.3 μm (SiC)	217.5 nm (graphites); None for PAH
Emission Features	9.7; 18.5 μm (silicates); 11.3 μm (SiC); Mid/Far-IR Continuum	3.3, 6.2, 7.7, 8.6, 11.2 μm (PAH); Mid-IR Plateaus (PAH); Mid/Far-IR Continuum (thermal)

3.10 Radiation Transfer and Line Formation and Broadening

In the foregoing pages we described briefly the processes of absorption and emission in an atom. Absorption and emission take place in astrophysical sources such as an interstellar cloud of gas powered by a central source of energy such as a star. A star itself may be surrounded by matter either by accretion during its formative stage or by shedding of mass during its evolution. As the radiation traverses through circumstellar or interstellar medium towards the observer it undergoes absorption and scattering as described earlier. The process of how the radiation gets transferred through the medium is a complex topic in astrophysics.

Further, a spectral line contains a wealth of information embedded in it. Its shape, width and shift reveal physical conditions in which the line is formed. Its ionization state and excitation level reveal the energetic of the region being investigated. We deal with this subject briefly here. We describe these aspects also in the subsequent subsections.

3.10.1 Radiation Transfer: Let us consider a cold cloud of matter surrounding a hot source of radiation. The cloud absorbs the radiation depending on the physical properties of its material. The fraction of specific intensity (intensity at a particular frequency or wavelength) that the cloud absorbs in the line of sight is given by (see Emerson 1996 for full details),

$$\frac{dI_\nu}{I_\nu} = -\kappa_\nu \, \rho \, ds \tag{3.48}$$

where κ_ν is the specific absorption coefficient per unit volume and is equal to the fractional decrease in intensity per unit length along line of sight and ρ is the mass density and *ds* is the incremental length in the line of sight. Upon integration on both sides, we obtain,

$$I_\nu = I_0 \exp[-\int \kappa_\nu \, \rho \, ds\,] \tag{3.49}$$

with I_0 as the unimpeded intensity from the source. As already introduced in the earlier subsections, we can now define a new quantity known as the optical depth as,

$$\tau_\nu = \int \kappa \, \rho \, ds \tag{3.50}$$

such that

$$I_\nu = I_0 e^{-\tau_\nu} \tag{3.51}$$

The intervening cloud of matter between the source and the observer may be characterized by two cases: one, called optically thin cloud, is defined by $\tau_\nu \leq 1.0$ and the second, called optically thick cloud, arises when $\tau_\nu \geq 1.0$ (causing extinction of light by more than about 65%). If we now suppose the cloud is sufficiently hot and optically thin, then there will not be absorption but only emission takes place; in which case we can write the contribution to the total emission (integrated along the line-of-sight *s*) from a layer of the cloud as

$$dI_\nu = \int j_\nu \, \rho \, ds \tag{3.52}$$

where j_ν is the specific emission coefficient per unit mass, per unit frequency, per unit solid angle.

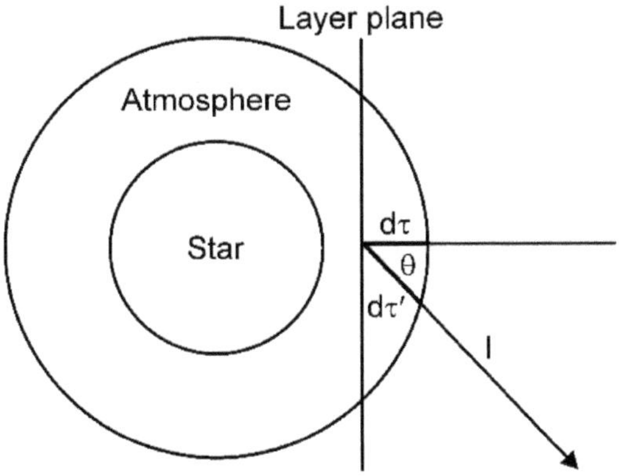

Figure 3.27: Schematic of a star and its atmosphere with intensity I shown making an angle θ with the perpendicular to the plane of an atmosphere layer. dτ represents optical depth: dτ = dτ' cos θ.

However, if the cloud is hot but also optically thick then it will not only emit but also absorb radiation. Then we can write down the fraction of intensity due to increase by emission and decrease by absorption as,

$$dI_\nu = \int j_\nu \, \rho \, e^{-\tau_\nu} \, ds = \int \frac{j_\nu \rho}{\kappa_\nu \rho} \, e^{-\tau_\nu} \, (\kappa_\nu \rho \, ds) \tag{3.53}$$

$$= \int S_\nu \, (\tau_\nu) \, e^{-\tau_\nu} \, d\tau_\nu$$

where S_ν $(= j_\nu/\kappa_\nu)$ is known is the source function. Figure 3.27 illustrates the geometry of a star surrounded by a spherical shell of matter (cloud). The figure shows the direction of specific intensity making an angle θ with direction of the net flux (integrated over specific intensity in all directions) towards the observer. The sphericity of the geometry of a star and its circumstellar matter makes the transfer angle dependent.

Now combining the absorption and emission parts we obtain a more general equation for transfer of radiation as

$$dI_\nu = -\kappa\,\rho\,I\,ds + j_\nu\,\rho\,ds$$

or,

$$\frac{dI_\nu}{d\tau_\nu} = I_\nu - S_\nu(\tau_\nu) \tag{3.54}$$

The sign is considered as positive towards the observer (out of the source) for intensity and as positive towards the source for τ and *s*.

When we consider the curvature of the atmosphere the equation becomes more complex as follows,

$$\mu\frac{dI_\nu}{d\tau_\nu} + \frac{1-\mu^2}{\tau_\nu}\frac{\partial I_\nu}{\partial\mu} = I_\nu - S_\nu \tag{3.55}$$

where $\mu = \cos\theta$.

3.10.2 Plane Parallel Atmosphere: For a special case of what is known as "plane parallel" stellar atmospheres, wherein the thickness of the atmosphere is much smaller compared to its radius of curvature, the curvature effect may be neglected (i.e., $\mu = 1$, whence the second term on the left-hand side of the eqn. 3.55 drops out). This is a reasonably good approximation for most stars in most evolutionary stages. This allows us to consider the atmospheric layers as plane parallel sheets perpendicular to the radiation flow direction and hence the intensity may be treated as independent of the angle of propagation of radiation. This will greatly simplify the computations. Under this approximation we may write the above equation of transfer of radiation as,

$$\mu\frac{dI_\nu}{d\tau_\nu} = I_\nu - S_\nu \tag{3.56}$$

3.10.3 Source Function: The source function, defined as the ratio of emission coefficient to the absorption coefficient, is usually a Planck Blackbody distribution of radiation under local thermodynamic equilibrium condition at a temperature *T*. It also depends on the density and radiation field of the background. Thus,

$$S(\nu, T) = j_\nu/\kappa_\nu = B(\nu, T) = \left(\frac{2h}{c^2}\right)\frac{\nu^3}{(e^{h\nu/kT} - 1)} \tag{3.57}$$

or in terms of wavelength the function may be written as

$$B(\lambda, T) = \left(\frac{2hc^2}{\lambda^5}\right)\frac{1}{(e^{hc/\lambda kT} - 1)}$$

3.11 Line Broadening

A spectral line is ideally purely monochromatic similar to a delta function, but it is rendered broad due to several processes that occur in the emitting region. Here we discuss these broadening mechanisms.

3.11.1 Natural Broadening: An atom absorbs a photon whereby one of its electrons in an energy level is excited to jump to an upper state. The energy difference between the two levels *i* and *j* will determine the photon energy required for this process to occur. An atom emits a photon when an excited electron at an upper level makes a downward transition to level *i*. This is called spontaneous emission with Einstein's coefficient A_{ji}. The Heisenberg's uncertainty principle demands, however that,

$$\Delta E\ \Delta t \sim h/2\pi \quad \text{or} \quad \Delta\nu \sim 1/2\pi\Delta t \quad \text{or} \quad \Delta\nu \sim A_{ji}\ /2\pi$$

where ΔE and Δt $(= 1/A_{ji})$ denote the uncertainties in the energy and lifetime of the electron state in question. The uncertainty in the frequency of the transition is $\Delta\nu$. This means that if the lifetime of the electronic state is long then its energy uncertainty or spread is small and vice versa. As is shown in the above equations the frequency spread due to this quantum mechanical uncertainty is proportional to the Einstein's coefficient for spontaneous emission for the transition from level *j* to *i*. The intensity profile in this case may be shown to be,

$$I(\nu) = \frac{A_{ji}}{4\pi^2}\frac{1}{(\nu - \nu_0)^2 + (A_{ji}/4\pi)^2} \tag{3.58}$$

where ν with subscript zero denotes the line central frequency of maximum intensity. The natural broadening predominantly contributes to the wings of a line profile but normally its overall contribution is very small. Full width for a given transition at central wavelength λ and spontaneous emission coefficient A_{ji} is given by $\Delta\lambda = (\lambda^2 A_{ji})/(2\pi c)$ where c is the velocity of light. For example, the Hydrogen Balmer α line at 656.3 nm with $A_{ji} = 4.7\times10^{9}$ s^{-1}, has a natural (full-)width of 0.001 nm while the forbidden line of [OIII] at 500.7 nm with $A_{ji} = 2\times10^{-2}$ at 500.7 nm has negligibly small width. In general, while natural widths are negligible compared to other mechanisms for broadening, for allowed transitions these are much larger than those for forbidden transitions due to the large difference in A_{ji}.

3.11.2 Doppler Broadening: Inside an emitting source at local thermodynamic equilibrium at a temperature T the atoms are in random motion and the distribution of their velocities is given by Maxwell-Boltzmann distribution. The number of atoms with velocity between v_r and, $v_r + dv_r$ is given by,

$$n(v_r)dv_r = N\left(\frac{m}{2\pi kT}\right)^{1/2} e^{-\frac{mv_r^2}{2kT}} dv_r \tag{3.59}$$

where N is the number of particles of mass m at a temperature T. Now substituting velocity with frequency of the line radiation (noting that, $\Delta v/v = \Delta\nu/\nu$) and integrating over all the frequencies then we obtain,

$$I(\nu) = \frac{1}{\sqrt{\pi}\,\Delta\nu_D} e^{-\left[\frac{\nu-\nu_0}{\Delta\nu_D}\right]^2} \tag{3.60}$$

where $\Delta\nu_D$ is the Doppler width given by

$$\Delta\nu_D = \left(\frac{\nu_0}{c}\right)\left(\frac{2kT}{m}\right)^{1/2} \tag{3.61}$$

or in terms of wavelength this may be written as

$$I(\lambda) = \frac{1}{\sqrt{\pi}\,\Delta\lambda_D} e^{-\left[\frac{\lambda-\lambda_0}{\Delta\lambda_D}\right]^2} \tag{3.62}$$

where the Doppler width is given by

$$\Delta\lambda_D = \frac{\lambda_0}{c}\left(\frac{2kT}{m}\right)^{1/2} \tag{3.63}$$

In the above equations the rest frequency and the rest wavelength are denoted by ν_0 and λ_0 respectively. It can be noticed that the intensity distributions above are Gaussians. If we consider microscopic or macroscopic turbulent motions in the gas, then since such a distribution will also be Gaussian (being random) the turbulence term may be simply an additive term in quadrature to the Doppler width. The width then may be written as sum of two terms as

$$\Delta\lambda_{total} = (\Delta\lambda_D^2 + \Delta\lambda_T^2)^{1/2}$$

The subscript T denotes turbulence at micro- and macro-scales. Usually, the micro-turbulent term representing random motions in the atoms or molecules is typically a few km/s typically for stellar atmospheres. Macro-turbulence is due to the random motion of blobs or packets of matter and can cause the line to broaden significantly. The overall turbulence contribution to the line width can be determined by observing profile widths of two lines that are arising at the same position or optical depth in the object.

3.11.3 Pressure or Collisional Broadening: The emitting atom may undergo collisions with other atoms depending upon temperature and density of the gas. This will cause the energy levels to spread, and it is called the pressure or collisional broadening. This is similar to the natural broadening, but its contribution can be more significant. The normalized intensity distribution due to pressure broadening is given by,

$$I(\nu) = \frac{\Gamma}{4\pi^2} \frac{1}{\left\{(\nu - \nu_0)^2 + \frac{\Gamma}{4\pi^2}\right\}} \tag{3.64}$$

where Γ is proportional to the number of collisions per unit time and is given by

$$\Gamma = 2\, N_c\, \sigma\, V$$

N_c is the total number of colliding particles, σ is the collision cross-section and V is the average velocity of the particles. The profile is called Lorentzian and is similar in shape to the natural broadening. The broadening due to collisions will cause the wings of the profile to expand thus rendering the shape of the profile a non-Gaussian at the wings. The combined Gaussian and Lorentzian profiles are called Voigt profile (described in section 2.11.4).

It may be mentioned here that the Stark broadening (linear case) occurring due to the splitting of lines by application of external electric field has an effect like the pressure broadening.

3.11.4 Rotational Broadening: Stars have spin or rotation around their own axis. The line emanating from the star will be blue-shifted when the portion that moves towards the observer is blue-shifted while the portion that moves away is red shifted with respect to the central region of the star that has rotational velocity perpendicular to the line of site of the observer hence there will not be a component. The sum effect of this is that the line gets broadened. The one unknown parameter in this scenario is obviously the angle of inclination of the rotation axis with respect to the sky, *i*, being 90^0 when the rotation axis is in the plane of the sky and 0^0 when it is perpendicular or in the direction of observer's line of sight. Figure 3.28 shows the schematic diagram of a rotating star: on the left is shown the general case of a star with its rotation axis inclined to an observer by an angle *i*; on the right, the star is shown as seen from its rotation axis, now considered perpendicular to the observer (or the line of sight). The larger circle represents the equatorial plane (at the radial distance *R* equal to the radius of the star) and the smaller circle represents a plane (perpendicular to the rotation axis) at a radial distance of *r*. In the figure, the direction of *r* (meeting the line *p*) makes an angle θ with the line of sight (towards the observer). The observer sees the same Doppler shift along the line $p = r \sin \theta$ (for details see Emerson 1996). For a rigid body, the rotational velocity is proportional to the radial distance.

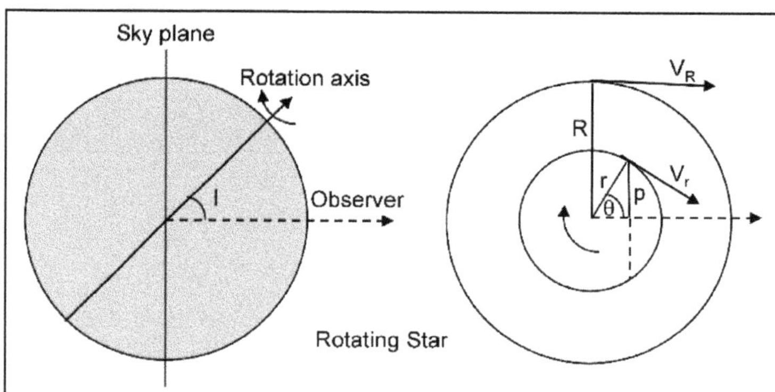

Figure 3.28: Left: A rotating star will broaden the line profile. The angle of inclination of the rotation axis with respect to the observed is given by i. The angle of inclination is unknown quantity. If the axis is towards the observer, then i = 0 hence rotation cannot be detected. Right: Rotating star seen from the rotation axis perpendicular to the observer direction. V_R, and V_r are the rotational velocities at the equator R and at radial distance r respectively. Observer direction is shown in dashed lines with arrows. See text for details.

If the star is assumed to be a rigid body, then the rotational velocity can be written as

$$V_r = \frac{r}{R} V_R \tag{3.65}$$

where V_R and V_r are the rotational velocities at the equator and at radial distance r from the centre of the star respectively.

At the radial distance r the rotational velocity component in the direction of the observer is (see Figure 3.28) simply $[(r/R) \sin\theta \ V_r]$. All the points at $p = r \sin\theta$, the perpendicular distance from r onto the line-of-sight, will have the same velocity or Doppler shift towards the observer. Now considering the general case of rotation axis at an angle i with the line of sight we can re-write the Doppler shift as

$$\delta\lambda = (\lambda_0/c)\,(V_r \sin i)\,(p\,/R) \tag{3.66}$$

The shift at a position opposite to the point r will have the sign changed in the above equation. Thus, the line profile can be calculated from contributions from all such points as

$$\frac{I_p}{I_0} = \left[1 - \left(\frac{\delta\lambda}{\delta\lambda_m}\right)^2\right]^{1/2} \tag{3.67}$$

with $\delta\lambda_m = (\lambda_0/c)\,(V_R \sin i)$, the maximum width. We note here that there is no direct method to accurately determine the angle of inclination (see Emerson 1996).

The normalised profile then may be written as

$$I(\lambda) = \left(\frac{2}{\pi\delta\lambda_m}\right)\left[1 - \left(\frac{\delta\lambda}{\delta\lambda_m}\right)^2\right]^{1/2} \tag{3.68}$$

The FWHM of the profile is given by,

$$\Delta\lambda = \sqrt{3}\,\delta\lambda_m$$

The shape of the profile is more like rectangular and can easily be distinguished from the Doppler broadened profile (Gaussian). This does not depend upon line strength unlike the thermal and micro-turbulent broadening. Macro-turbulent component however may be difficult to distinguish from rotational broadening (Emerson 1996).

3.11.5 Voigt Profile: The combined thermal (Gaussian) and pressure (Lorentzian) broadened profile is given by convoluting the two and is called the Voigt profile. Denoting the Gaussian function by $G(x)$ and the Lorentzian function by $L(x)$, we write here the final Voigt function (for details we may refer to Emerson 1996) as the convolution of the two ($V = G * L$):

$$V(a,u) = \frac{a}{\pi^{3/2}\,\Delta\nu_D} \int_{-\infty}^{\infty} \frac{e^{-x^2}}{[(u-x)^2 + a^2]}\,dx \tag{3.69}$$

with $x = u = \Delta\nu/\Delta\nu_D$ *and* $a = \Gamma/(4\pi\Delta\nu_D)$.

Voigt function is numerically evaluated and tabulations for various input values of Doppler widths ($\Delta\nu_D$) and Lorentzian widths (Γ) are available in literature. While the central region or the core of a Voigt profile is dominated by the Gaussian component, its wings are predominantly comprised of Lorentzian part.

3.11.6 Zeeman Splitting: The magnetic quantum number in an atom has $2J+1$ levels possible where J is the total electron angular momentum. These levels are degenerate when an atom is under the influence of an external magnetic field and are split into sub-levels causing broadening of the corresponding spectral lines (see e.g., Schulz 2005),

$$\Delta\lambda_m = \pm\left(\frac{e}{4\pi m_e c^2}\right)\lambda^2 gB = 4.67\times10^{-7}\,\lambda^2 gB\ m\text{Å}\ (kG)^{-1} \tag{3.70}$$

where, B is in kilo Gauss (kG), c is the velocity of light, and m_e is electron mass. Using high-resolution spectroscopy, the Zeeman split can be resolved and hence the strength of the magnetic field may be found out.

3.12 Line Formation

In order to obtain physical parameters from a measured line profile it is necessary to know how the line is formed in the atmospheres of a star. For this in turn we need model atmospheres as to how the temperature, pressure and density vary with radial distance. Models have been developed based on the equations of stellar interiors and assuming opacity functions in the atmospheres. We may refer to Kurucz's pioneering work on this (Kurucz, 1979).

From the previous section on transfer of radiation, we can write the intensity of the emergent radiation as,

$$I_\nu = \int_0^\infty S_\nu\,(\tau_\nu)\, e^{-\tau_\nu/\mu}\, \frac{d\tau_\nu}{\mu} \tag{3.71}$$

The flux received from a star of radius R at a distance d is then given by

$$F_\nu = 2\pi \frac{R^2}{d^2} \int_0^1 I_\nu(\mu)\, \mu\, d\mu \tag{3.72}$$

These equations are applicable for the case of a plane parallel atmosphere – the thickness of the atmosphere is assumed much smaller than the stellar radius. This assumption fails in the case of extended sources or extended stellar atmosphere due to stellar wind.

The absorption coefficient is given by

$$\kappa_\nu = \kappa_{L\nu} + \kappa_C \tag{3.73}$$

where subscript L_ν denotes line absorption term at frequency ν (corresponding to the transition i-j) and C denotes continuum absorption term at that frequency. The line absorption term may be expanded as,

$$\kappa_\nu = \frac{e^2}{4\,\varepsilon_0 m_e\, c}\, f_{ij}\, \phi_{ij}(\nu)\, N_a(i) + \kappa_C \tag{3.74}$$

where f_{ij} is the oscillator strength which is the classical parameter denoting the strength of the line i-j, $N_a(i)$ is the number of absorbing atoms in level i, φ_{ij} is the line profile function (Gaussian/Lorentzian), ε_0 is electrical permittivity of space, and e and m_e are the charge and the mass of electron. Here the absorption coefficient corresponds to a single transition, but in general, we need to take sum of all the transitions that have significant contribution at the given frequency (see Emerson 1996, for details).

By definition, the source function S_ν is the ratio of emission and absorption coefficients. As mentioned earlier in this chapter, under local thermodynamic equilibrium (LTE) condition the source function may be considered as the Planck's blackbody radiation function $B(T)$. Hence,

$$S_{\nu} = \frac{j_{\nu}}{\kappa_{\nu}} = (\kappa_{L\nu}\, S_L + \kappa_C S_C)/(\kappa_{L\nu} + \kappa_C) \tag{3.75}$$

where S_L and S_C are the source functions for line and continuum respectively. Under LTE conditions both may be approximated by Planck's function.

The line formation is a complex process, and we need to make certain assumptions to make the problem tractable. Line formation is considered in statistical equilibrium under Local Thermodynamic Equilibrium condition. Under this condition the population of ionization states is given by Saha equation, and the population of excitation levels is given by Boltzmann equation. What is not known very clearly is the atmosphere of a star. Basically, the successful line formation theory (that agrees with observed lines) depends upon how realistically well we can model stellar atmospheres with uncertainties in the radial profiles of temperature, opacity.

3.12.1 Equivalent Width: We need to characterise a resolved spectral line profile to derive relevant physical parameters from it. A parameter called Equivalent Width is often used for this purpose both for absorption and emission lines. It basically defines the strength of the line with respect to the continuum level. The EW is a function of the number of absorbing or emitting atoms, respectively for an absorption profile or an emission profile. Usually in the case of an emission line we characterise it by the full-width at half maximum (FWHM). In both the cases the profile is treated as a Gaussian. With F_λ (or F_ν) and F_C the fluxes on the line and at the nearby continuum, the equivalent width (EW) for an absorption profile is defined as,

$$W = \int_{\lambda_1}^{\lambda_2} \frac{F_C - F_\lambda}{F_C}\, d\lambda = \frac{\lambda^2}{c} \int_{\nu_1}^{\nu_2} (F_C - F_\nu)/F_C \, d\nu \tag{3.76}$$

The integral above actually defines the depth of the absorption line at a given frequency or wavelength and is integrated over all frequencies or wavelengths relevant for the line. The EW may be seen as the equivalent rectangular area representing actual area under the line profile – i.e., EW = depth times the full-width at half-minimum (or height times the full-width at half-maximum for emission line).

3.12.2 Curve of Growth: Curve of growth represents the variation of EW with the variation of physical parameters that define the strength of a spectral line, principally the number of contributing atoms/ions/molecules. Using the equivalent width (EW) as defined in the foregoing section we can find out the growth of the EW as the number of absorbing atoms increase with increase of optical depth under different broadening conditions such as thermal or Doppler and pressure or collisional process. We can notice that as the optical depth is increased beyond ~ 1 (optically thick case) the line gets saturated, and the width is no longer linearly related to the number of atoms (or molecules) as for optically thin condition. The curve so formed has three very distinct regions. Let us examine these.

We consider a cloud having a uniform source function and absorption takes place only at wavelengths relevant to the line. In this case since μ can be considered 1 (for plane parallel atmosphere), we have the emerging intensity given by

$$I_\nu = \left(\int_0^{\tau_{L\nu}} S_\nu \, e^{-\tau_{L\nu}} \, d\tau_{L\nu}\right) + I_C \, e^{-\tau_{L\nu}} \tag{3.77}$$

We are dealing with a weak absorption line meaning that $I_C < S_\nu$.

Under certain reasonable assumptions made in the line formation theory, Milne & Eddington derived the curve of growth for stellar absorption lines (see Emerson 1999, for details). Firstly, a constant ratio of line-to-continuum absorption coefficient is assumed independent of depth in the atmosphere: $\kappa_L/\kappa_C = \eta_\nu$. Further the source function is assumed to be linear with continuum optical depth $S = (a + b\,\tau_C)$. The dimension-less EW may then be written as,

$$\frac{W}{\lambda} = \frac{\pi e^2}{4\pi\varepsilon_0 m_e c^2} R_c \, \lambda \, f \frac{N_a}{\kappa_C} \tag{3.78}$$

where R_C is the maximum possible depth of the absorption line below the continuum. The number of absorbing atoms N_a is given by,

$$N_a = \frac{N_i}{N_{ion}} \frac{N_{ion}}{N_{el}} \frac{N_{el}}{N_H} \tag{3.79}$$

where N_{el}/N_H is the abundance of element being considered with respect to that of Hydrogen atom; N_{ion}/N_{el} is the fraction of ionized element to a given degree of ionization (as given by the Saha equation) and N_i/N_{ion} is the fraction of excited ions over the weighting function g (from the Boltzmann equation).

In the above equation the variation of the dimensionless EW with the factor ($\lambda\, g\, f\, N_a$) or equivalently the line profile width, is usually called the curve-of-growth (C-o-G). For weak lines where the number of absorbing atoms is small, the Gaussian nature dominates the shape of the line profile.

In case of weak lines, we can show that the EW is linearly related to the number of absorbing atoms (or $\Delta\lambda_D$ the Doppler or Gaussian profile width) as,

$$\left(\frac{W}{\lambda}\right)_{linear} = 0.501\, R_c\, \Delta\lambda_D \tag{3.80}$$

This case is shown in Figure 3.29 as solid curve and the initial rising linear portion of the curve of growth shown in Figure 3.30.

As we increase the density or the depth the number increases tending to saturate the line (see the dashed curve in Figure 3.29 and the middle part of the C-o-G in Figure 3.30).

For saturated lines for which the Gaussian profile is still majorly contributing, we can write the EW as

$$\left(\frac{W}{\lambda}\right)_{Sat} = 2R_c\frac{\Delta\lambda_D}{\lambda}\left[\ln\left(\frac{\kappa_0'}{\kappa_C}\frac{1}{\sqrt{\pi}}\frac{\lambda^2}{c}\frac{1}{\Delta\lambda_D}\right)\right]^{1/2} \tag{3.81}$$

where κ'_0 is the frequency integrated absorption coefficient.

If we further increase the density or the depth of atmosphere, then in such a case, the wings of the already saturated line profile become the dominant contributor to the EW requiring the Lorentzian to define the profile (see dotted curve in Figure 3.29 and the third part of the C-o-G in Figure 3.30). In such a case the EW is given by (from Milne-Eddington model; see Emerson 1996),

$$\left(\frac{W}{\lambda}\right)_{Strong} = R_c\frac{\lambda}{c}\left[\frac{\kappa_0'}{\kappa_C}\frac{\Gamma}{4}\right]^{\frac{1}{2}} \tag{3.82}$$

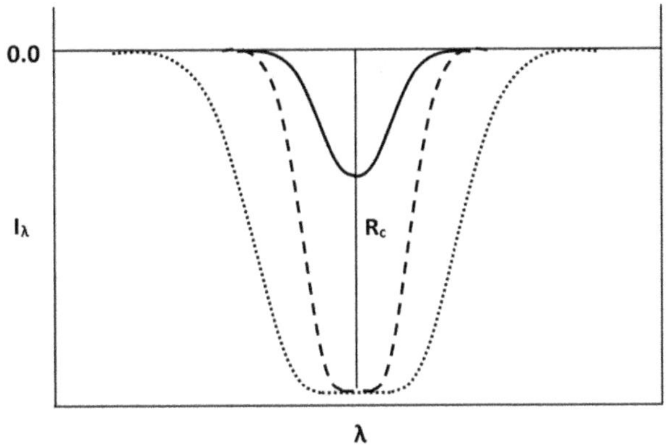

Figure 3.29: Absorption line profiles of a star shown schematically (not to scale). The full profile shown as full curve represents a weak line; the profile in dashed curve shows saturated line; and the dotted profile shows the extremely strong absorption characterized by growing wings. The vertical line at the core of the profiles is the maximum possible depth of the line Rc.

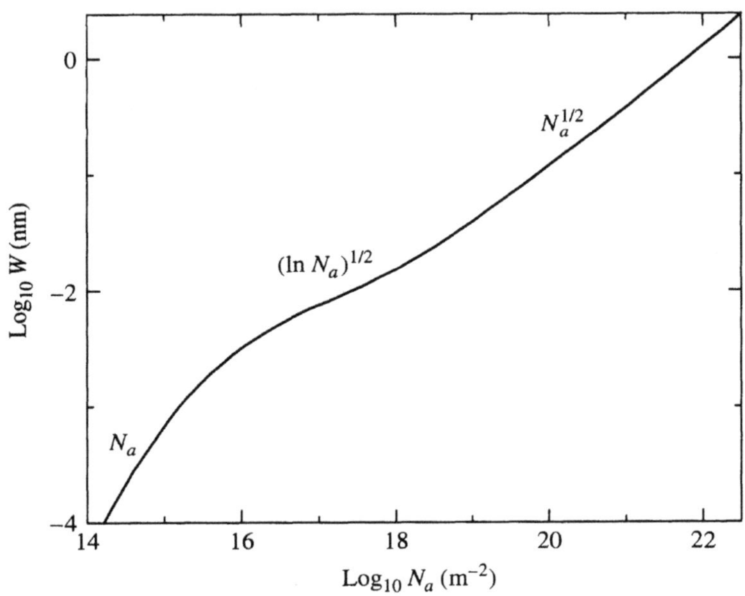

Figure 3.30: Curve of Growth for line formation under different conditions of opacity. Log (*W*) is plotted against log(N). Figure taken from Carroll and Ostlie (2007).

Essentially one may understand that when the layers of stellar atmosphere are optically thin i.e., optical depth τ_ν is much less than 1 the C-o-G is linear with the N_a (Figure 3.30). When the optical depth becomes more than 1 up to about 6-7 (i.e., optically thick case), the slow part of the curve rules where the EW becomes proportional to $(ln\ N_a)^{1/2}$. Finally in situations where the optical depth is very high then the EW varies as $N_a^{1/2}$.

Summarizing then the dimensionless EW is given in terms of N_a for various conditions as

$$W \propto N_a\ (\tau_\nu \ll 1);\ W \propto (ln\, N_a)^{\frac{1}{2}}\ (1\ < \tau_\nu < 7);\ W \propto N^{\frac{1}{2}}\ (\tau_\nu \gg 1)$$

3.13 Stellar Spectra

Stars exhibit a variety of spectra at various stages of their birth and evolution. The stellar spectra vary due to the fact that the effective temperature being different for different stars. Spectra vary from absorption spectra of main sequence stars to emission spectra in hot stars that are losing mass. For a comprehensive discussion on analysis of stellar spectra and their interpretation, we may refer to Gray (2021).

Here we present a few examples of these spectra.

3.13.1 Absorption Spectra: Absorption spectra arise in stellar atmospheres in photosphere when an atom (or molecule) absorbs a photon coming from hotter inner regions of a star. Upon absorbing a photon an electron goes to an excited state as shown in the Figure 3.31 (top left). De-excitation may happen after a while to a lower state. For instance, atoms or molecules may absorb UV radiation and re-emit in the infrared region.

3.13.2 Resonance Scattering: In this process a photon is absorbed as mentioned above; but de-excitation occurs to the same state as if the photon undergoes a change in direction (see top right of Figure 3.31). This phenomenon is important in the case of stellar wind or stars with significant mass loss.

3.13.3 Recombination: An ion captures a free electron; the electron then cascades downwards to several states emitting photons on its way (as illustrated in the bottom right of Figure 3.31). For instance, Hydrogen ion (proton) captures a free electron resulting in series of recombination lines such as Balmer series etc.

3.13.4 Collisional Excitation: An atom or ion may be excited also by collisions (thermal process) or by absorption of a photon (a non-thermal process called photoexcitation). As shown in the Figure 3.31 (bottom left) this process gives rise to emission of a photon. This process is important in hot outer atmospheres of stars such as the chromosphere and corona of the Sun and other sun-like stars.

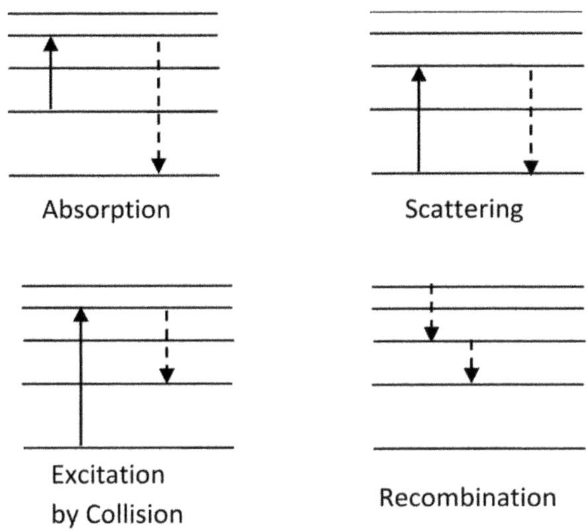

Figure 3.31: Schematic showing the processes of absorption, resonance scattering, excitation by collisions, and recombination.

CHAPTER 4

Photometric Techniques

Measure what can be measured and make measurable what cannot be measured

– Galileo Galilei

The purpose of photometric observations of a celestial object is to understand the continuum emission mechanisms. Photometric technique involves measurement of flux of a celestial source in a few selected passband filters. A photometric passband is chosen such that its width is smaller than its central wavelength by a few times – typically the half-power bandwidths are a few percent of the central wavelength. Photometric observations made at several wavelength bands are then used to construct what is known as the spectral energy distribution (SED). SED is simply a plot of observed fluxes at various bands versus the central wavelength of the bands. Suitable modelling of the SED is then done to find out the physical process or processes that accounts for the observed continuum radiation. Quite often, more than one mechanism (as discussed in Chapter 2) needs to be invoked to explain the observed SED. We need to make precise measurements of the fluxes at various passbands for accurate modelling. Standard sets of filters are recommended for photometry in visible and infrared wavelength regions to minimise subjectivity in data acquisition. The measured fluxes are usually expressed in magnitude scale for both visible and infrared bands. For SED modelling, we need to convert all data (in magnitudes) into the more physical flux units. The observations need to take account of all extraneous factors contributing to the noise that reduces the quality of data. To obtain a meaningful and positive observation, the signal-to-noise ratio (*SNR*) is required to be more than 3.

The optical arrangement of a typical photometer is described in the following section. In the earlier times before the advent of array or panoramic detectors (such as CCDs), aperture photometry (with a sky sample in a few seconds of arc) with a photomultiplier as detector was employed.

4.1 Optical Layout of a Photometer

Typically, the optics for a photometer as shown in the Figure 4.1, consists of an aperture (A) to restrict the field of view to the desired size, and a collimator (C) to make the converging telescope beam (T) parallel. The collimated beam then passes through a filter wheel (F) and an imaging lens (I) to image the object on to the detector (D). A Fabry lens or a field lens, placed after the field aperture, is generally used to image the exit pupil of the telescope on the detector so that the image would not wander out of the detector field (due to atmospheric effects). Usually in order to keep the star at the required position on the detector (centre) a guiding star facility (G) is employed with the help of a 45 degrees mirror (M). Alternatively, a beam-splitter may be used, instead of M, with reflected light carrying only a small fraction of the total light from the field star. In modern astronomical observations a panoramic detector such as a CCD is employed to image the object – especially for extended sources. But even for a single point source imaging detectors are used instead of a single element detector. Photometric calibration is done by observing non-variable standard stars that are stable and whose colours are well established (see Ch.1 and Appendix 3 and the ensuing sections in this Chapter). After obtaining the apparent magnitudes of the programme stars the flux calibration is made using the A0V star Vega (α Cygnii).

The sequential steps involved in observations and analysis till we get a meaningful photometric flux for a programme object are: (i) Observations need to be made on a dark, cloudless and clear night, called a "photometric night"; (ii) Each programme object needs to be observed for sufficient amount of exposure time in order to get a fairly high signal-to-noise ratio; (iii) Observations on a few standard stars with accurately known apparent magnitudes and colours need to be made on the same night preferably with similar zenith angles and similar colours as the programme object (if these are already known); (iv) sufficient number of dark frames, flat filed frames, and sky frames are required to correct respectively for dark current, inter-pixel sensitivity difference for imaging detectors such as CCDs, and sky-background elimination. This step is needed for the standard star observations as well; (v) The signal from programme stars as well as standards need to be processed (in terms of a voltage, a current or counts/second) and *SNR* needs to be established; (vi) Then using these processed signals and the magnitudes of the standard stars, the processed signals from the programme stars need to be calibrated and magnitudes obtained; (vii) Finally the magnitudes of programme stars thus calibrated need to be converted to flux units by using the data from a standard (non-variable) A0 V star such as Vega for which the magnitudes in all the visible and near-infrared bands are considered zero. If required, the steps (ii) and

(iii) (and (iv)) may be repeated to improve *S/N*, since the exposure time should not be too long lest the atmospheric conditions may change during the exposure.

In what follows we define various methods of photometry observations and analyses depending upon the type of observations required. We discuss various calibration procedures that need to be undertaken for accounting for various extraneous factors that contribute to the signal. We define a point source and an extended source. We discuss certain photometric tools to study astronomical sources.

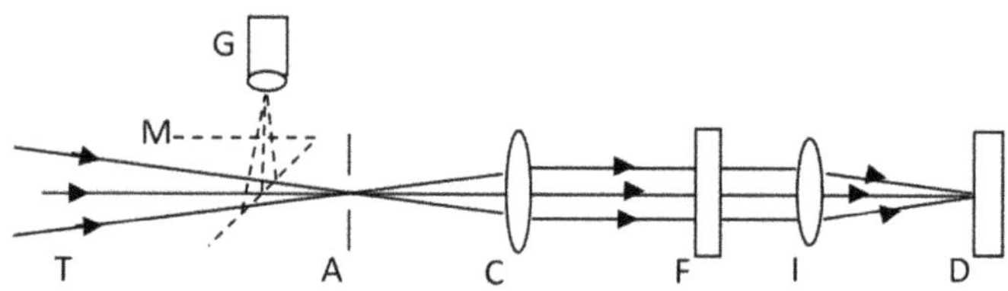

Figure 4.1: Schematic of photometer optics: T is telescope beam; M is flipping mirror for guiding star unit G; A is the field aperture; C and I are collimating and imaging lenses; F is the filter and D the detector. In addition, a field lens (not shown here) may be kept at the diaphragm at A.

4.2 Integration or Exposure Time and Signal-to-Noise Ratio

Integration or exposure time is the time required to expose the detector to the incoming signal from an astronomical object to obtain a reasonably good signal-to-noise ratio (*SNR*). Exposure time depends on the sensitivity of the detector used. In an ideal photon-counting detector, the photons follow the Poisson statistics and hence the uncertainty in the signal is equal to the square root of the signal itself. In the real world however, we need to consider various sources of noise – such as that from the detector and the sky-background. The general expression for *SNR* may be written as (e.g., Harlow 1992; Chakraborty et al 2005),

$$\frac{S}{N} = \frac{K_0(\lambda)\, F(\lambda)\, \delta\lambda\, t}{\{K_0(\lambda)\,[F(\lambda) + A_{bg}B(\lambda)]\,\delta\lambda\, t + n_{pix}[N_D t + N_{ro}^2]\}^{1/2}} \tag{4.1}$$

where $K_0(\lambda) = T_0(\lambda)\, A_T/E_{ph}(\lambda)$ with $T_0\,(\lambda)$ total transmittance, A_T the area of the telescope primary and E_{ph} the energy of photon of wavelength λ. The exposure time is denoted by t; bandwidth by $\delta\lambda$; flux (in W/m²/μm) from the object and the background by F and B respectively; the number of pixels on the detector array sampling the sky and object by n_{pix}; and the dark current and read-out noise by N_D and N_{ro} respectively; and A_{bg} is the area of background (in square of arcsec) sampled. The object must be sampled so as to include the first bright fringe surrounding the Airy maximum for a diffraction-limited observation. Normally, the sampling is done to include the region surrounding the star that equals the area of background sampling. We may note that if the noise due to dark and read-out is negligibly small compared to the background, then the *SNR* is proportional to the square root of exposure time *t*. The SNR expression above is typically for a CCD detector and needs to be used for other detectors appropriately, e.g., for a single-element detector such a PMT the readout noise is zero and $n_{pix} = 1$.

Three general cases of deciding appropriate exposure times are discussed here:

1. For a required *SNR*, if exposure times for program star and standard are t and t_0 respectively, then their magnitudes satisfy the following equation.

$$m - m_0 = 2.5 \log_{10}(F_0/F)$$

or

$$m - m_0 = 2.5 \log_{10}(t/t_0) \tag{4.2}$$

For a required *SNR*, the above equation gives the limiting magnitude m for an exposure time of t.

2. Alternatively, for a given program star of magnitude m and required *SNR*, the integration or exposure time is given by,

$$t = t_0 10^{0.4(m-m_0)} \tag{4.3}$$

3. Finally, for a case where the exposure time is the same for program and standard stars, then the *SNR* for program star is

$$m - m_0 = 5 \log_{10}\left(\frac{C_0}{C}\right)$$

or

$$C = C_0\, 10^{0.2(m_0-m)} \tag{4.4}$$

where $C = SNR$. Note that the factor 5 is because $C \sim t^{1/2}$.

4.2.1 Background-limited and Detector or Photon-limited Performance: The observation is said to be under background-limited performance (BLIP) when the background noise is a dominant term in the above equation for *SNR* compared to the detector noise terms. The observation is said to be photon- or detector-limited when the background noise is negligible compared to detector related dark and readout noise. These specific cases simplify the SNR equation (eqn. 4.1) given above as follows:

Case 1: When signal is so strong (for a bright object) that both the sky-background and detector noise are negligible, we can see from the following simplified equation that the SNR depends on the square root of the signal and the square root of the exposure time (cf. Poisson statistics).

$$\frac{S}{N} = \{K_0(\lambda)\, F(\lambda)\, \delta\lambda\, t\}^{1/2}$$

Case 2: Background-limited Performance: The sky-background is much larger than the detector noise. In this case also the *SNR* depends upon the square root of the signal and the square root of the exposure time if the signal term is smaller than the back-ground term,

$$\frac{S}{N} = \frac{K_0(\lambda)\, F(\lambda)\, (\delta\lambda\, t)^{1/2}}{\{K_0(\lambda)\, [F(\lambda) + A_{bg} B(\lambda)]\}^{1/2}}$$

Case 3: Photon/Detector-limited Performance: The detector noise dominates over the sky-background. Then, the *SNR* takes the form,

$$\frac{S}{N} = \frac{K_0(\lambda)\, F(\lambda)\, \delta\lambda\, t}{\{K_0(\lambda)\, [F(\lambda)]\, \delta\lambda\, t + n_{pix}[N_D t + N_{ro}^2]\}^{1/2}}$$

4.2.2 Zero-Point Magnitude: The calibration of a photometer is done by observing several steady (non-variable) standard stars of known magnitude in the filters of interest. The instrumental (apparent) magnitude thus obtained needs to be corrected by adding a constant that gives the actual apparent magnitudes of the stars. This constant evidently is different for different filters. The equation for the constant called the 'zero-point magnitude" for any filter, centred at an effective wavelength λ, can be written as

$$m_\lambda = -2.5\log(I) + ZP \qquad (4.5)$$

Here the instrumental magnitude obtained by using the standard photometric analysis package such as IRAF or GNU Astronomy Utilities (GNUAU) is denoted by I. The Zero Point (*ZP*) magnitude (= $-2.5 \log (I/I_r)$, with I_r the reference brightness) is simply the magnitude when $I = I_r = 1$. It accounts for all the instrument-specific parameters through the reference brightness I_r.

4.3 Sky Background Across the Electromagnetic Spectrum

Before we go into the details of methods of photometry, we need to examine the contribution of background sky in the visible and infrared wavelength region. The sky background is mainly due to scattered light from atmospheric molecules (undergoing Rayleigh scattering). In addition to this, emission from some atmospheric constituent atoms and molecules also contributes to the brightness. Atmosphere acts also like a blackbody, and the emission typically peaks at 10 μm wavelength and is important factor to reckon within the near-infrared and mid-infrared regions.

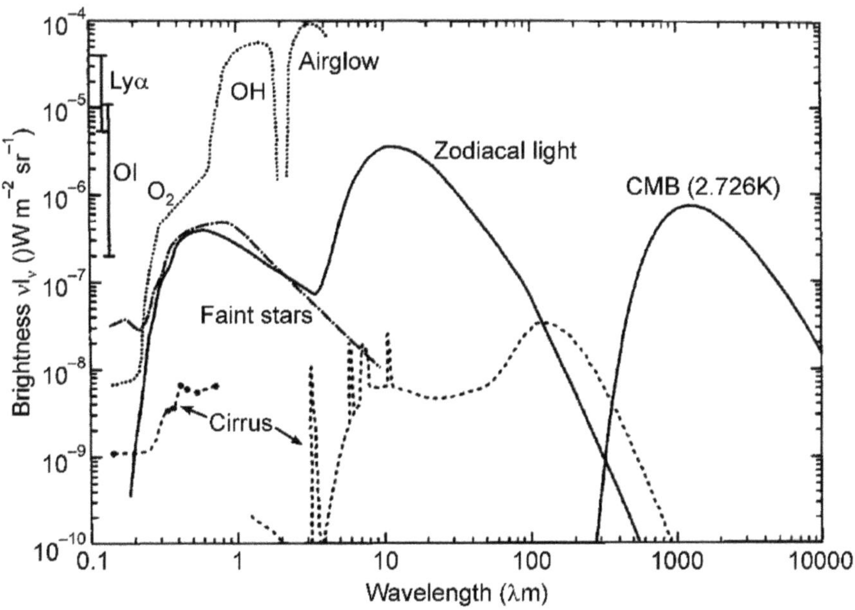

Figure 4.2: Sky brightness due to various components (from Glass 1999).

In Figure 4.2 is shown sky brightness (νI_ν) as a function of wavelength from UV to beyond Far-IR up to millimetre region as well as various sources of background. The figure shows that in the visible region it is the faint stars and oxygen molecular emission and to some extent the OH Meinel bands contribute. In the near-infrared region the main contributor to the sky brightness is the OH Meinel band emission and beyond 2 μm the sky emission as a blackbody starts dominating. In both the regions the zodiacal light contributes though not very significantly in the near-infrared. The contribution of cirrus clouds is negligible in comparison with the other factors as mentioned above. We discuss more in detail about the atmospheric emission in the near-infrared region in the later sections.

It is important to consider various factors that contribute to the sky background at different wavelength regimes.

Sky background in the visible and infrared regions (see Figure 4.2) is comprised of the following factors (see e.g., Tokunaga 2000):

1. Airglow from OH molecular emissions chiefly in the J and H bands; the emission occurs at altitudes of Mesosphere of Earth.
2. Zodiacal light in the ecliptic pole with equivalent blackbody temperature of 5800 K with maximum emission occurring in the visible region.
3. Zodiacal light from interplanetary dust grains at the ecliptic pole with equivalent blackbody temperature of 275 K making it an important factor for mid- and far-infrared regions (IRAS measurements).
4. Telescope and its environment thermal emission at ambient temperature, affecting emission in the mid-infrared peaking around 10 μm.
5. Thermal emission of the atmosphere in wavelength region beyond about 2-3 μm.
6. Molecular emissions in visible and infrared regions: Molecular and atomic Oxygen for visible region; H_2O, CO_2, N_2O, CH_4 and O_3 for near- and mid-infrared.

In order to perceive the combined contribution from sky, telescope and instrumental background emission at various near- and mid-infrared bands (for the 3.0 m IRTF at Mauna Kea, Hawaii) we refer to Tokunaga's (2000) Table 7.2.

4.3.1 Sky Background in the Optical Region: Figure 4.3 shows the sky background emission in the visible region. A few strong emission lines from atmospheric constituents (from O, O_2 and Na in the visible region and OH in the near-infrared) may be seen in the figure. This is especially an important consideration for spectroscopic observations and analysis.

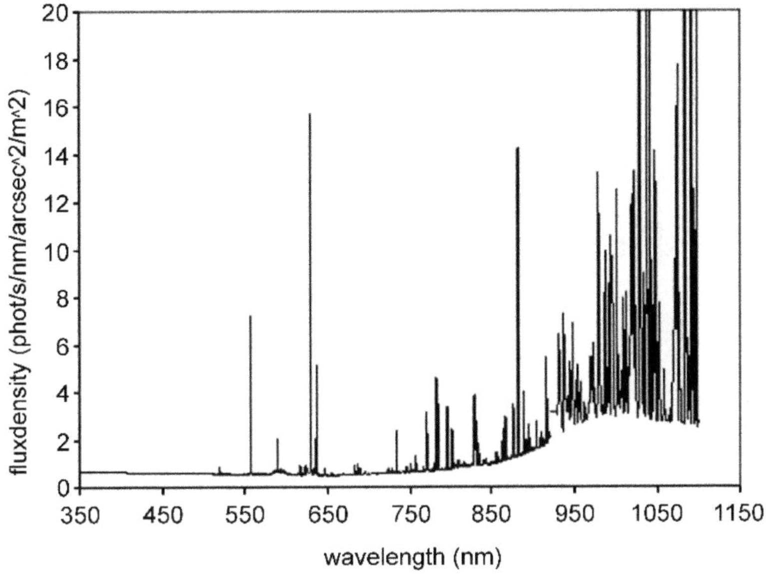

Figure 4.3: Atmospheric Emission in the optical region (from 450 nm) and near-infrared region (up to 1100 nm). (Gemini Observatory Web site).

4.3.2 Atmospheric Emission in the Near- and Mid-Infrared Regions: The OH Meinel bands are shown in Figure 4.4 in the near-infrared region from 1 to 2.5 μm. It may be noted that the thick forest of OH emission lines serve the purpose of wavelength calibration for spectroscopy (see Chapter 5). As shown in Figure 4.5 the continuum emission as well as absorption from water vapour dominates in the region beyond about 3 μm since the atmosphere emits like a blackbody.

To account for atmospheric contamination, observations in the infrared region must always include sampling of background and infrared standard stars in the immediate vicinity of a programme object. This task must be carried out for both for photometry as well as spectroscopy. Even with these correction measures, ground-based observations beyond 3 μm are extremely difficult not only because of severe atmospheric absorption and emission but also by the proximity environmental emission. As much as it is necessary to account for it, it is truly a daunting task since the atmosphere varies temporally and spatially (with zenith angle) as well as from site to site and with altitude.

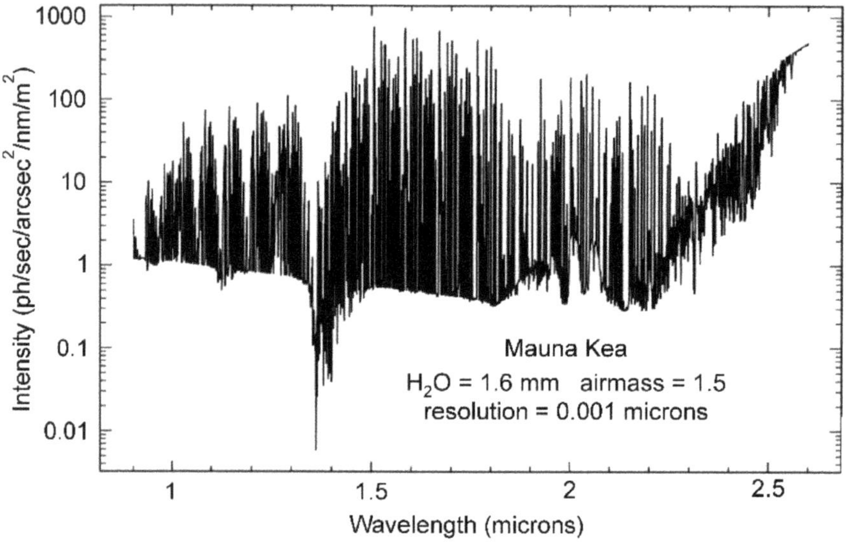

Figure 4.4: Atmospheric Emission in the near-infrared region 1-2.5 μm. Copious OH band emission is seen in the region (Gemini Observatory Web site).

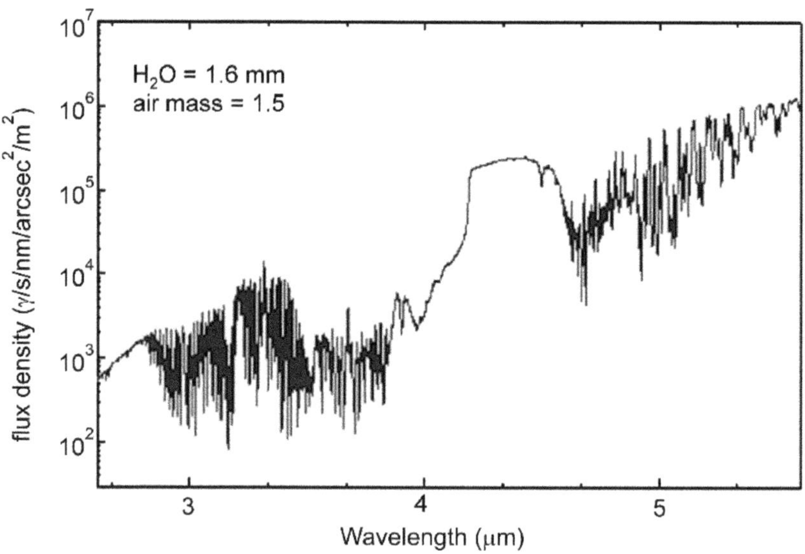

Figure 4.5: Atmospheric Emission in the near-infrared region of 2.5 – 5.5 μm. (Gemini Observatory Web site).

4.4 Photometric Standard Filters in the Visible Region

Based on their pass-band widths, photometric filters are standardized as wide-band (width >/= 30 nm), intermediate band (width between 10 – 30 nm) and narrow band (width ~ a few nm) filters. Considerable efforts had been made in the past to standardize filters used for photometry to reduce subjectivity. Most photometric studies of stars use standard broad-band filters in both visible and near-infrared regions. Especially in the infrared region standard filters underwent considerable changes with continued efforts to reduce subjectivity in filter characteristics. This is because of the atmospheric absorption that defines the regions of wavelength observable from the ground-based telescopes, and it depends upon the site of observation. These efforts in filter designing to avoid as far as possible the absorption (and emission) features due to atmospheric constituents are continuing even today with the advent of modern methods to make filters with high transmission and excellent profile definition.

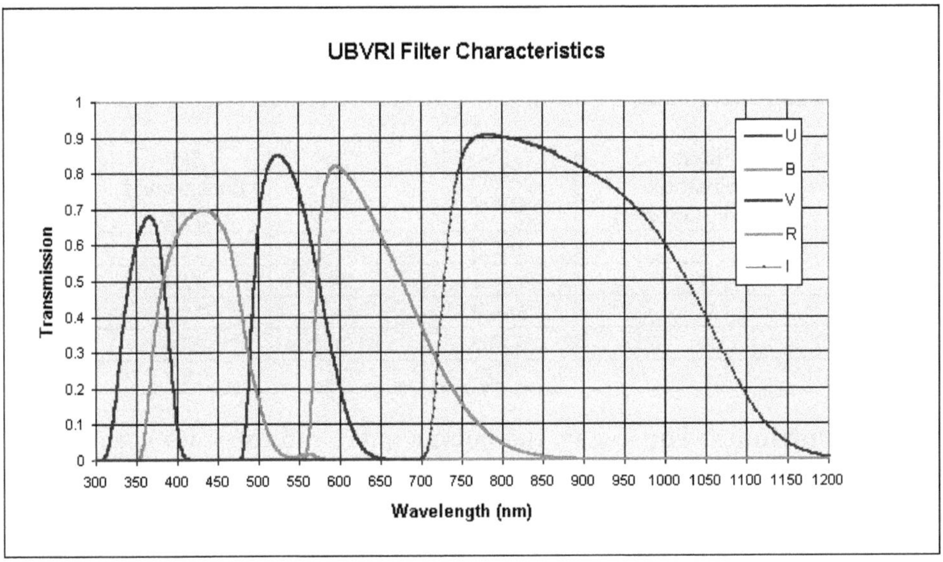

Figure 4.6: Standard filter characteristics for optical photometric observations. (from Santa Barbara Instruments Group – Johnson-Cousins System, 2007).

Standard sets of filters for the visible and near-infrared wavelength ranges were defined and designed (Johnson 1966; Bessell & Brett 1988; Bessell 2005) for using with photometers to facilitate the realistic depiction of spectral energy distribution of a source to understand the underlying physics of the continuum source. Figure 4.6 shows such standard filters in the range 300 – 1200 nm in the visible bordering on near-UV on shorter wavelength side and near-IR on the longer wavelength side. As we can see each of these filters has spectral width of a few ten nm. These filters had been used traditionally for the past several decades in observatories the world over. Photometric filters were 149 standardized by astronomers from meticulous observational campaigns conducted at sites that have minimum extraneous affects, such as Mauna Kea, Hawaii and Chile. Table 4.1 gives characteristic parameters, such as the effective wavelength, bandwidth, and sky background at standard sites, for standard filters used for photometry in visible region.

Basically, for a photometric filter, the effective wavelength may be defined as

$$\lambda_{eff} = \frac{\int \lambda\, F(\lambda) R_f(\lambda) d\lambda}{\int F(\lambda) R_f(\lambda) d\lambda} \tag{4.6}$$

where $R_f(\lambda)$ is the spectral response of the filter and $F(\lambda)$ flux received from the source.

Table 4.1: Johnson-Cousins Photometric Filter Characteristics.

Filter Designation (Johnson-Cousins)	λ_{eff} (nm)	$\Delta\lambda$ (nm)	Sky Background (mag.)
U	366.3	65	21.3
B	436.1	89	22.1
V	544.8	84	21.3
R	640.7	158	20.4
I	798	154	19.7

4.4.1 Colour Index: For a star, the colour index indicates its spectral and luminosity classes. It is defined as the difference between the magnitudes in two photometric bands. The monochromatic flux received from a celestial object, at a particular wavelength λ within the wavelength region defined by the filter transmission function, may be written as (see Carroll & Ostlie 2007),

$$F_\lambda\, d\lambda = \frac{L_\lambda}{4\pi d^2}\, d\lambda = B_\lambda(T) \frac{R^2}{d^2} \tag{4.7}$$

where R is the radius of the photosphere of the observed star and d its distance from the observer. L_λ and B_λ are the specific luminosity and the Planck function (at an effective temperature T) respectively at the wavelength λ. Integrating the flux over the filter response function we get at two effective wavelengths,

$$U = -2.5\log_{10}\int_{\lambda_1}^{\lambda_2} F_\lambda\ S_U\ d\lambda\ +\ C_U \tag{4.8}$$

$$B = -2.5\log_{10}\int_{\lambda_1}^{\lambda_2} F_\lambda\ S_B\ d\lambda\ +\ C_B \tag{4.9}$$

The limits on integration λ_1 and λ_2 refer to the blue and red ends of the filter bandwidth at zero amplitude (height). C_U and C_B are constants. Then the colour index is defined as

$$U - B = -2.5\log_{10}\frac{\int F_\lambda S_U d\lambda}{\int F_\lambda S_B d\lambda}\ +\ C_{U-B} \tag{4.10}$$

where R is the radius of the object and d is its distance; S is essentially a function that defines overall photometric response including filter response as a function of wavelength. Here, the constant $C_{U\text{-}B} = C_U - C_B$. The constant may be estimated from photometric standard stars.

Similarly, we may have the colour $(B - V)$ as

$$B - V = -2.5\log_{10}\frac{\int F_\lambda S_B d\lambda}{\int F_\lambda S_V d\lambda}\ +\ C_{B-V}$$

Let us now examine the use of these colours as defined above in photometric study of astronomical sources. Here, the constant $C_{B\text{-}V} = C_B - C_V$.

4.4.2 Colour-Colour and Colour-Magnitude Diagrams: When we plot one colour index of a celestial source as a function of another, we obtain the Colour-Colour Diagram (C-CD). Likewise, when we plot the (absolute) Magnitude as a function of a colour index the resultant plot is called Colour-Magnitude Diagram (C-MD). As discussed earlier, in the visible region the colours usually and most appropriately chosen are $(U - B)$ versus $(B - V)$ for C-CDs and V versus $(B - V)$ for C-MDs. Objects of different physical parameters would occupy different positions in these spaces. Using this simple criterion, these diagrams have traditionally been in use since a long time for classification of observed stars. Thus, photometry is a tool for delineating the various types of sources in the C-CD and C-MD space. Physical parameters such as colour temperature (determined from the colours and closely approximated for effective temperature) and the luminosity class of stars can be determined from the C-CDs while we can determine the spectral class from the C-MDs as well as possible age of evolution. As a matter of fact, the C-MD of V vs $(B - V)$ is considered to represent the traditional Hertzsprung-Russell Diagram (HRD) a plot of luminosity vs effective temperature.

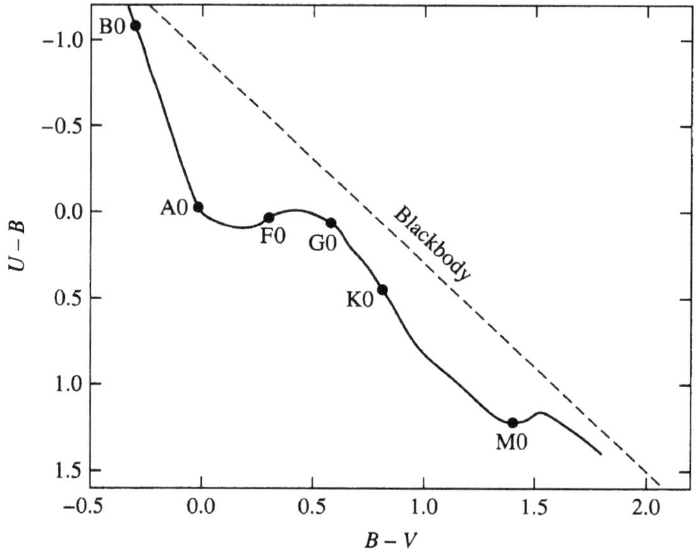

Figure 4.7: The colour-colour diagram showing the colour index *(U - B)* vs *(B - V)* for various spectral types of main sequence stars of different temperatures along with blackbody locus (from Carroll & Ostlie 2007).

Figure 4.7 shows a typical colour-colour diagram in which the colour (U - B) is plotted as a function of the colour (B - V) for various main sequence stars of different effective temperatures. Also plotted in the figure for comparison is the locus of the colours of a blackbody at different equilibrium temperatures corresponding to dwarf stars from M to B types. We can see the departure for stars hotter than M0 and cooler than B0. The departure is due to the photospheric absorption lines of various elements – especially Hydrogen recombination series (Balmer) – we may notice the maximum departure in the A0 type stars in which the H line absorption is most prominent (see Figure 1.6 in Chapter 1). Even with this mismatch, we can delineate the spectral types of the stars using the colour-colour diagram *(U - B)* vs *(B - V)* diagram. The colour *(B - V)* may be used to estimate the effective temperature of a main sequence star as follows:

$$T_{eff} \sim \frac{8540}{(B-V)+0.865} \quad K$$

This formula is an empirical derived from the $(B - V)$ colour indices for various stars of temperatures less than 10000 K or fainter than an A0 V star.

Especially while dealing with a cluster of stars, the colour-magnitude diagram is very useful as compared to colour-colour diagram. Figure 4.8 shows the colour-absolute magnitude for a Globular Cluster spectacularly depicting positions of coevolving stars of various evolutionary stages (obtained from Hipparchus catalogue, see Schulz 2005).

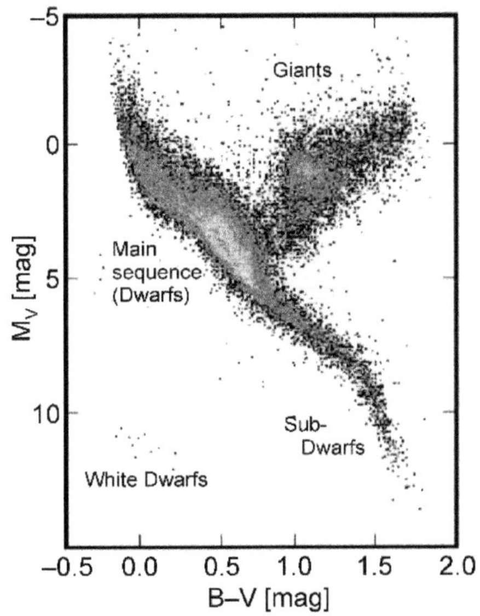

Figure 4.8: $(B - V)$ Colour – Absolute Magnitude Diagram of 40000+ single stars from Hipparcos catalog (from Schulz, 2005).

4.4.3 Extinction of light: As discussed in Chapter 3, light undergoes absorption and scattering as it traverses the distance from object to the observer by intervening extra-terrestrial matter, namely gas and dust grains. Figure 4.9 shows schematically the process of extinction due to circumstellar and interstellar matter intervening the line-of-sight between the observer and the object. Photometric and spectroscopic observations need to be corrected for these effects to find the true fluxes from the object in each wavelength band.

4.4.4 Colour Excess: Due to the terrestrial atmosphere absorption as well as the circumstellar and interstellar absorption (as shown in the previous section), the colour of a particular main sequence star is over-estimated. Terrestrial atmospheric effects being the same for all stars for the same zenith angle, the excess due to interstellar and/or circumstellar extinction can be estimated by observing a standard star with known colour. Then the colour excess of a programme (unknown) is defined as

$$E = (B - V) - (B - V)_0 \quad (4.11)$$

E representing the colour excess in $(B - V)$. The subscript "0" refers to an unreddened star or a star with zero extinction.

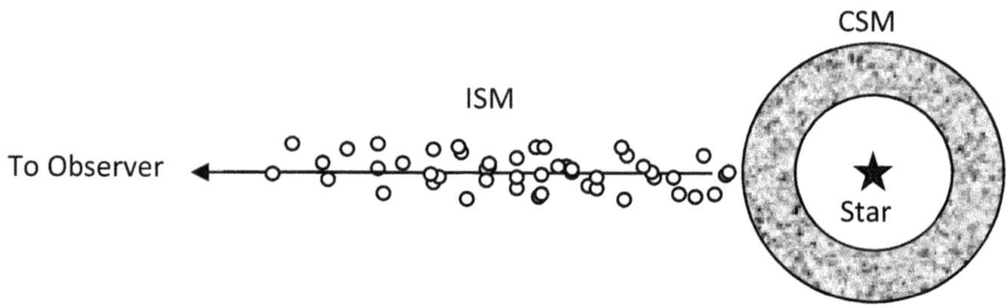

Figure 4.9: Schematic of Extinction Effect from Circumstellar and Interstellar Media.

We discuss below a few alternate sets of standard filters used for special or specific photometric observations. These filters have bandwidths typically smaller than the Johnson-Cousins system given in the previous subsection.

4.4.5 Stromgren Standard Filters: Intermediate-band filters in the visible region had been proposed by Stromgren (1966) called the *u, v, b, y* system (see Table 4.2 for their characteristics). These were initially specially designed for photometry of F, A and B type stars but later extended to all spectral types. These filters are used for specific purpose of measuring certain important spectral features such as the Balmer decrement (Balmer decrement occurs at the wavelength 364.7 nm where continuous bound-free absorption by recombination to the $n = 2$ level sets in H atoms; the position of the wavelength however shifts to longer wavelengths when the gas pressure is higher. It Is measured by taking the ratio of equivalent widths of Balmer alpha and beta lines). Similarly, there is Paschen ($n = 3$) discontinuity at 820.6 nm but too shallow to observe. Observations of these features are used to find the pressure broadening effect, and the strength of Balmer β absorption line.

Table 4.2: Stromgren Photometric Filter System.

Filter Designation	λ_{eff} (nm)	$\Delta\lambda$ (nm)
u	352.0	31.4
v	410.0	17.0
b	468.8	18.5
y	548.0	22.6
β_w	489.0	15.0
β_n	486.0	3.0

As may be seen from the table, the last two rows in the system are designed to delineate (find the strength of) the HI Balmer β absorption line centred at 486.1 nm by taking the ratio of the two bands, β_n and β_w, centred on-line and off-line respectively. The colour indices defined for this system are,

$$m_1 = (v - b) - (b - y) \tag{4.12a}$$

$$c_1 = (u - v) - (v - b) \tag{4.12b}$$

The first of these colours namely m_1 measures the depression due to metal lines around 410.0 nm and the colour index c_1 measures the Balmer discontinuity or decrement. It may be noted that the colour index *(b - y)* approximates to *(B - V)* colour in broad-band photometry but the latter is less sensitive to metallicity. Also, the band y approximates to V band for most of the non-M type stars. Grundahl et al (2002) showed that precision Stromgren photometry (to better than 0.01 mag, required for determining effective temperature, surface gravity and metallicity) can be done using CCDs with good standard stars. We may refer to Bessell (2005) for a very detailed description of photometric systems in both visible and infrared regions; and their application to various physical problems associated with stellar photospheres.

4.4.6 Wing's Narrow-band Photometric Filters: Wing (1967) designed several narrow-band filters to delineate absorption features from late type stars such as K and M types and evolved giants such as Carbon stars (see Bessell 2005 for detailed description and references). Table 4.3 below gives the filter characteristics.

Table 4.3: Wing's Narrow-band Filters for Late-type Stars.

Band	λ_{eff} (nm)	$\Delta\lambda$ (nm)	Spectral Features
	688.0	6	CaH
1	712.0	6	TiO
2	754.0	5	Continuum
3	781.0	4	Continuum
4	812.0	5	CN
	991.0	5	FeH
5	1039.5	5	Continuum
6	1054.0	6	VO
7	1081.0	6	Continuum
8	1097.5	7	CN

As is clear from the two tables 4.2 and 4.3, and the description provided above, the intermediate- and narrow- band filters are for very specific purpose. We must keep in mind while using these filters that their calibration and standardisation are rather involved and require high precision photometry. For this reason, most observatories use the immensely popular, well-calibrated and time-tested UBVRI broadband photometric system in the visible region. Standardisation of these broadband filters for various observatory sites may be found in Bessell (2005).

4.5 Near-Infrared Photometric Filters

We have seen in Chapter 1 that the visible light from cosmic sources arrives at the ground without undergoing severe absorption by terrestrial atmospheric gases. But the infrared radiation suffers deep absorption from the lower atmospheric constituents. Figure 4.10 shows atmospheric transmission in the wavelength region 1-30 μm from a standard benchmark site, Mauna Kea in Hawaii. It is very evident that the atmospheric gases absorb the radiation at several bands. The chief absorbing gases are H_2O, CO_2, CO, CH_4 and O_3. The deep gorge around 10 μm is the work of O_3. As may be seen on the figure there are good portions of the wavelength region which are relatively free from the absorption and free to reach unhindered a good elevated observing site. Such free bands are designated as J, H, K, L, M, N and Q bands as shown in the figure. In the bands centred around 3 and 5 μm the absorption is chiefly from water vapour. One may see the water vapour effect in the beginning of the K band also. The bands J, H, and K are also contaminated by copious emission from OH radical of the atmosphere. In recent times, Simons & Tokunaga (2002) and Tokunaga et al (2002) designed near-infrared filters considering all the above factors that limit the transmission in this region of wavelengths.

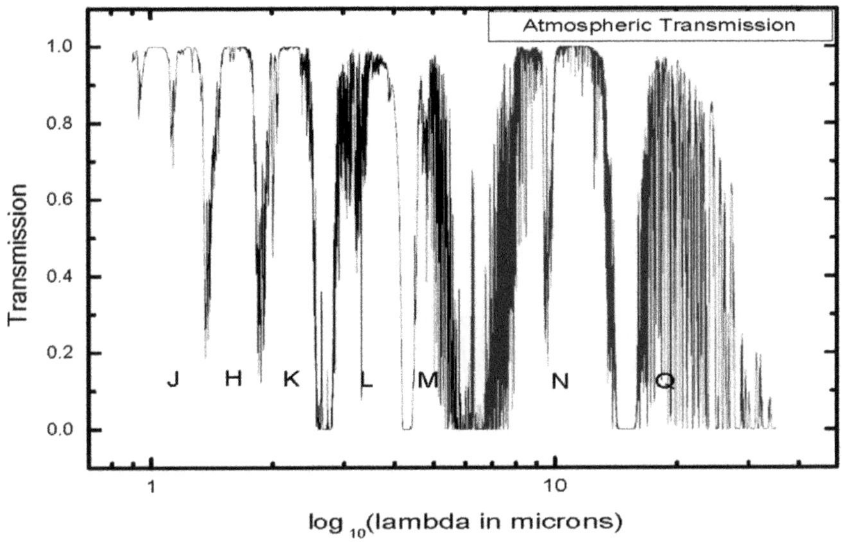

Figure 4.10: Near- and Mid-infrared absorption by the earth's atmosphere at a standard site (Hawaii). The atmospheric windows are marked J, H, K, L, M, and N bands. (from Gemini Observatory Web site).

Table 4.4 lists standard filters in the near-infrared region 0.8 - 5.0 μm along with sky-background in magnitude for a good site (e.g., Mauna Kea). The K band has a variant, namely K_S or K' with a narrower width than the original to avoid the H_2O absorption bands. Similarly, the L and M bands also have variants L' and M' for the same purpose. Figure 4.11 shows these filter definitions in the region 0.8 - 2.5 μm superposed by the atmospheric absorption features for two standard sites Mauna Kea Observatory and United Kingdom Infrared Deep Sky Survey. The Z and Y bands in the lower panel of the figure are designed for far red visible region to avoid the H_2O absorption around 0.93 μm.

In the table the wavelengths and bandwidths are in μm and *S*, the surface brightness is given in mag/arcsec^2 (values are from Shure et al (1994) for IRTF 3.0 m Telescope). Standardisation of these broadband filters for various observatory sites may be found in Bessell (2005).

Table 4.4: Near-Infrared Filters, Wavelengths, Bandwidths and Sky Background.

Band	λ (μm)	$\Delta\lambda$ (μm)	S (mag)
J	1.26	0.31	15.9
H	1.62	0.28	13.4
Ks	2.15	0.35	14.1
K	2.21	0.39	13.7
L	3.50	0.61	4.9
L'	3.78	0.59	4.5
M	4.85	0.62	-0.7
M'	4.78	0.22	0.3

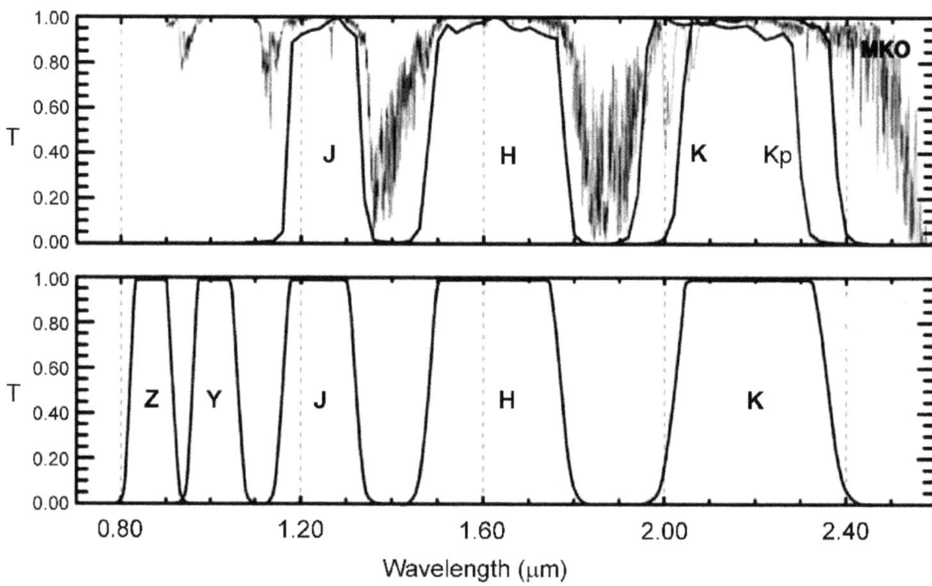

Figure 4.11: MKO (top panel) and UKIDSS (bottom panel) Standard filter characteristics (smooth curves in both panels) for near-infrared region (from Bessell 2005). The spiky pattern in the top panel are the actual atmospheric transmission showing absorption bands that force the design of the filter profiles suitably. Compared to earlier defined/designed JHK filters (top panel), curves shown in the bottom panel are modern and well-defined.

4.5.1 Infrared Colour Indices: Similar to the visible colours we may define infrared colour indices taking the standard JHK bands. For obtaining colour indices, currently we use several infrared bands such as the JHKK'ML bands as well as bands defined in space-based observations (e.g., Spitzer). From the near-infrared photometry we use *(J - H)* and *(H - K)* colour indices to delineate stars of different spectral and luminosity classes. The colour index *(V - K)* is also a good indicator of giant stars. These infrared colour indices are useful in identifying stars having circumstellar extinction. With the advent of satellite data (e.g., SPITZER), other colour indices involving L and M, or longer bands are also being used for better accuracy in classification.

The extinction vector that represents the locus of stars having circumstellar extinction is defined as

$$(J-H)-(J-H)_0 = 1.74\,[(H-K)-(H-K)_0] \tag{4.13}$$

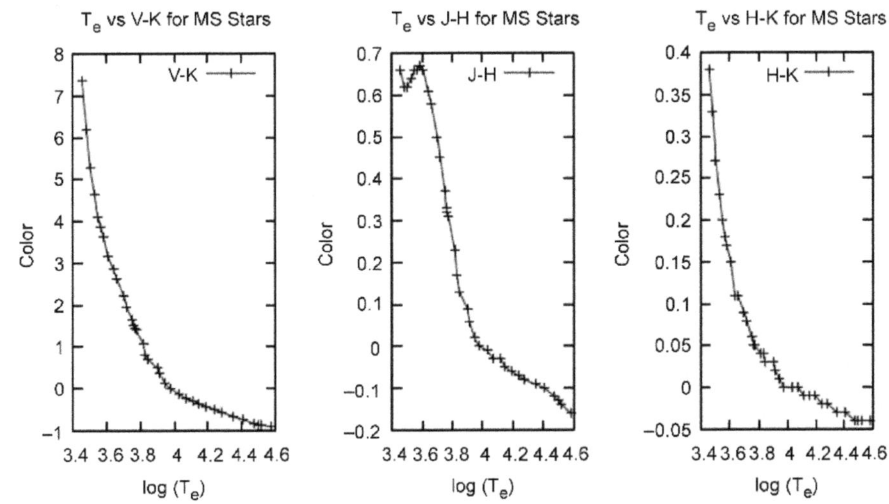

Figure 4.12: Intrinsic Colours of Main sequence Stars (Dwarfs of Luminosity Class V from M type to O type) plotted against their effective temperatures T_e (Data taken from Tokunaga, 2000).

Figure 4.12 shows intrinsic colours *(V – K)*, *(J – H)*, and *(H - K)* as a function of effective temperature for main sequence stars (also called dwarfs of luminosity class V) from M type to O type. For any given star if the colour is in excess of what is given above for dwarf stars it is considered to have a colour excess. The colour excess is due to extinction from circumstellar matter as well as the interstellar matter.

Appendix A4 gives some visible and near-infrared colour indices for the brightest stars of luminosity classes I, III, and V.

4.5.2 Colour-Colour and Colour-Magnitude Diagrams in Infrared: The C-CD and C-MD have been in use also in the near-infrared and even at longer wavelengths wherein objects (especially those with circumstellar envelopes) having different physical conditions could be delineated clearly. Stars in the main sequence, stars in the giant phases and stars having circumstellar matter such as evolved stars or protostars occupy different positions in the C-CD space. One can also approximately determine the extinction towards the sources under investigation from the extinction vectors (loci of extinction) over-plotted on the C-CDs and C-MDs.

The standard colour-colour diagram (C-CD) using JHK photometry is the *(J - H)* vs *(H - K)* for main sequence and giant stars with zero extinction (cf. the *(U - B)* vs *(B - V)* diagram in the visible region shown earlier) is shown in Figure 4.13. The two dashed-lines are the extinction vectors, showing that as the extinction of a star of a given spectral type increases, its position shifts towards the top-right direction. Using this diagram, we can obtain colour temperature for stars up to an effective temperature of 6000 K. The C-CD helps in delineating stars of different classes - for example, AGB, PAGB and YSOs both have circumstellar matter but, in the C-CD space these different objects occupy different regions and thus can be identified.

Figure 4.14 shows C-CD (top panel) and C-MD (bottom panel) diagrams in JHK bands for a young star cluster in the star forming region IRAS 21413+5442 (Anandarao et al 2008). The objects occupying the region to the right-top of the diagram are stars with circumstellar matter (having positive values of both colours). The C-M diagram showing likely spectral types of young stellar objects derived from their absolute magnitudes (from the distance modulus; see for details, Anandarao et al 2008).

C-CD and C-MD are also made using SPITZER satellite photometric bands in the near- and mid-infrared region in combination with the ground-based observations on JHK bands to effectively separate various sources in star clusters (e.g., Dewangan & Anandarao 2011). Such plots are not only complementary to those in the traditional visible and near-infrared bands but also help in bringing out additional information and in many cases unambiguously.

An excellent example of colour-colour diagrams using far-infrared bands centred at 12, 25 and 60 μm from IRAS photometry was given by Pottasch (1993) that demarcates various celestial sources very clearly (see Figure 4.15). The diagram depicts objects having cool (increasing *y*-axis) and warm (increasing *x*-axis) dust environments (see also van der Veen & Habing 1988).

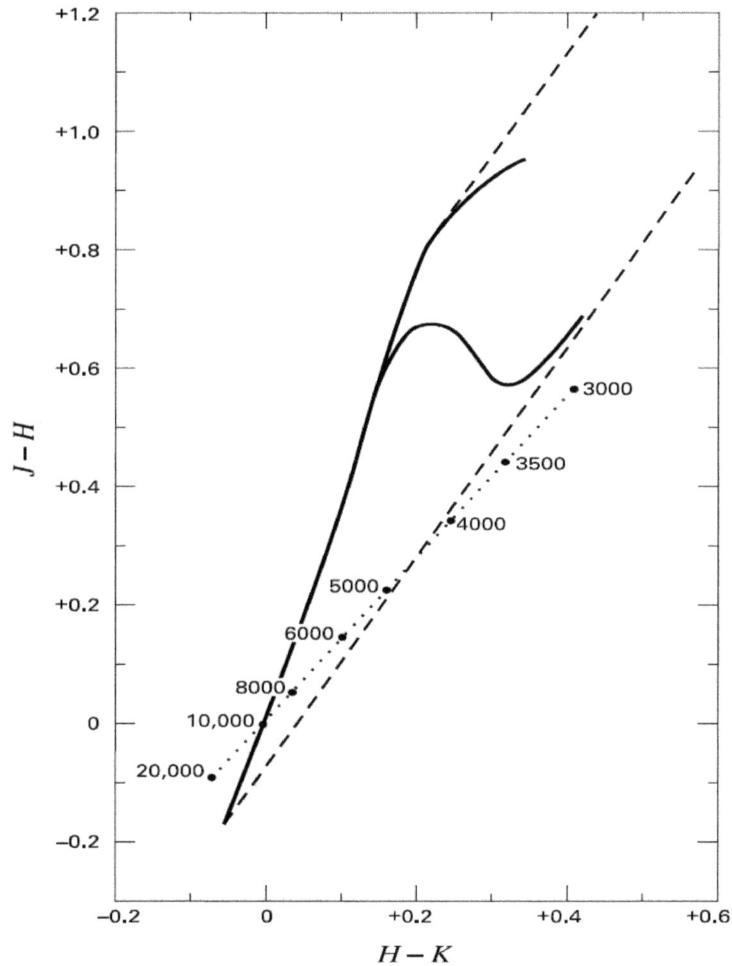

Figure 4.13: Color-color diagram in JHK colors in near-infrared. The curved line indicates the position of main sequence (dwarf) stars from B type to M type. The finger-like curve extending from the dwarfs is for giants. The dotted line indicates the Blackbody at different temperatures. The two dashed lines show the extinction vectors – the loci of stars with interstellar extinction (from Stahler & Palla 2004).

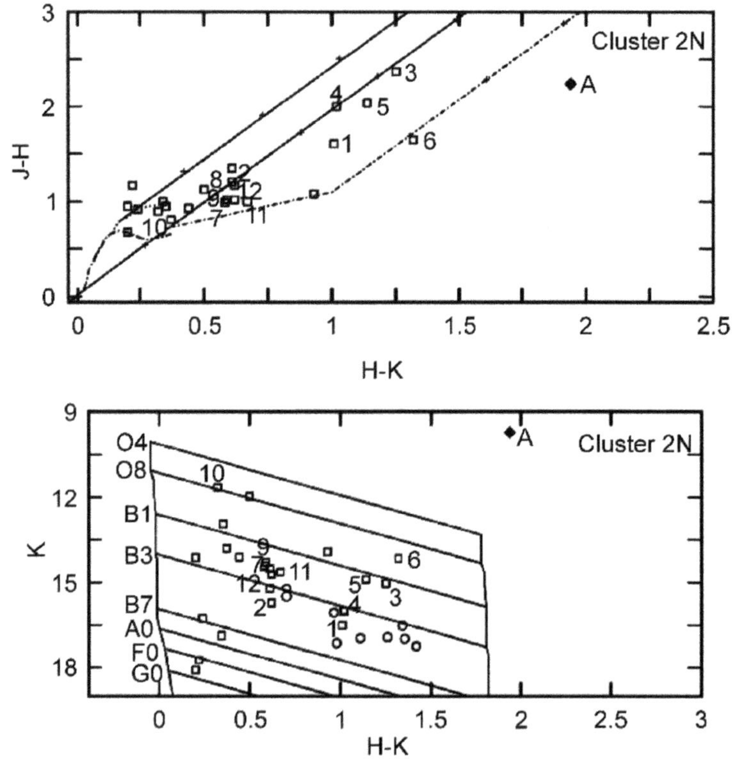

Figure 4.14: Color-color (top) and color-magnitude (bottom panel) diagrams in JHK bands for the region of young star clusters, IRAS 21413+5442 (from Anandarao et al 2008). See text for details.

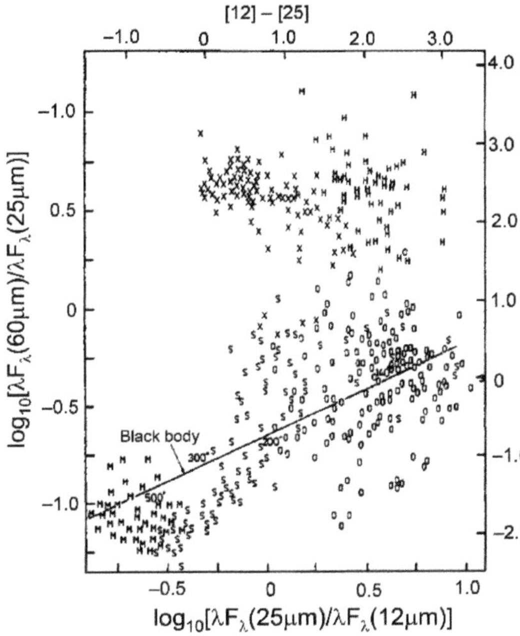

Figure 4.15: IRAS colour-colour diagram *(60-25)* vs *(25-12)* showing distinct regions occupied by different types of objects: M denotes M type and Mira variable stars; S the OH-IR stars; O denotes Planetary Nebulae; H shows HII regions and X indicates galaxies (from Pottasch 1993). The straight line across the figure is the blackbody of temperature from 140 K (right side) to 500 K (to the left).

4.6 Atmospheric Effect due to Zenith Angle

Astronomical observations, made from the ground-based telescopes, need to be corrected for the effect caused by the atmospheric air column along the line-of-sight (LOS). Due to this effect, depending upon the elevation or zenith angle of the object at a given local time, the light from the object undergoes absorption resulting in the reddening of light. This effect is clearly a function of wavelength of the light being observed. Since we are dealing here with visible and near-infrared wavelength regions, it is customary to use magnitude system for defining the power output from an astronomical source. The observed magnitude therefore needs to be corrected for each wavelength (filter) using the formula known as Bouguer's law as follows,

$$m_0(\lambda) = m_\chi(\lambda) - a_\lambda \chi \quad (4.14)$$

Here $\chi = (sec\ z)$, m_χ is the magnitude of the object observed at a zenith angle of z and m_0 corresponds to un-reddened magnitude (or the one with zero air mass) at a wavelength λ (see Figure 4.16 (A)). The factor a_λ is the extinction coefficient that depends on wavelength, and it varies with wavelength, site of observation and the local time.

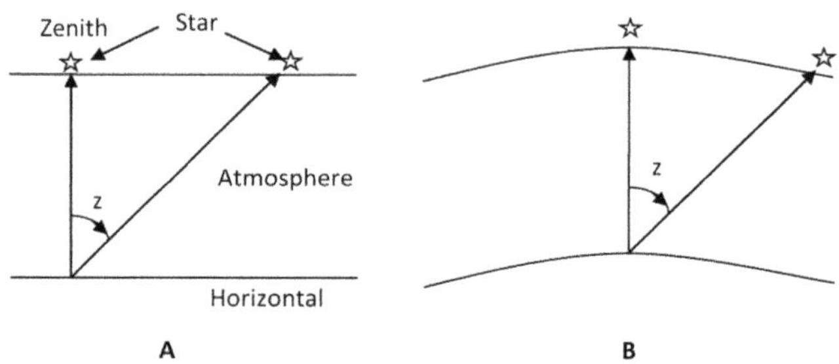

Figure 4.16: Shows star observed at the zenith and at a zenith angle z for plane parallel atmosphere case (A), and for the case where curvature of atmosphere is considered (B).

As shown in the Figure 4.16 (B), for larger zenith angles we need to consider the curvature of the Earth (and hence the atmosphere). Thus, in general we can write down the air mass for curvature effect χ_c as a function of zenith angle as

$$\chi_c = \chi - 0.0018167\ (\chi - 1) - 0.00287\ (\chi - 1)^2 \ - 0.0008083\ (\chi - 1)^3 \quad (4.15)$$

With $\chi = sec\ z$, as defined earlier, we may notice that for zenith angles approaching 0^0 ($sec\ z$ tends to 1) χ_c tends to χ.

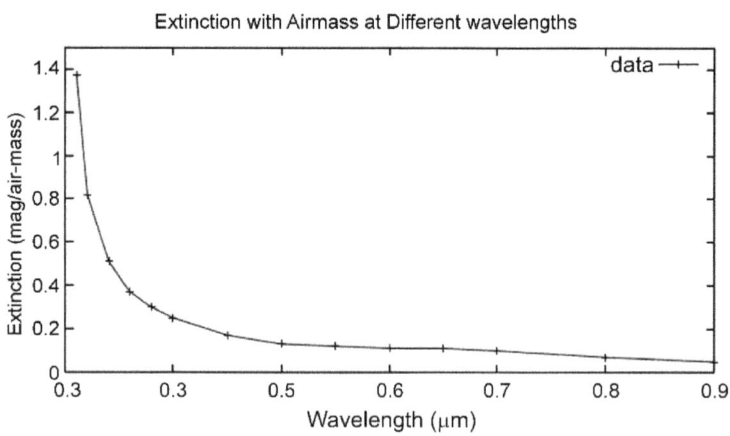

Figure 4.17: Typical plot of the extinction coefficient (mag/air-mass) as a function of wavelength for visible wavelength region (data from Krisciunas et al 1987; CFHT, Mauna Kea web site).

Figure 4.17 shows the extinction coefficient as a function of wavelength (data from CFHT, Mauna Kea website). As the wavelength increases the effect of air mass reduces. For small zenith angles (up to 60 degrees) we may approximately write down the effect of air mass for different photometric filters as: U ~ 0.50 mag/χ; B ~ 0.25 mag/χ; V ~ 0.20 mag/χ; R ~ 0.10 mag/χ; and I ~ 0.05 mag/χ (note that these values are obtained from Majewski, 2015).

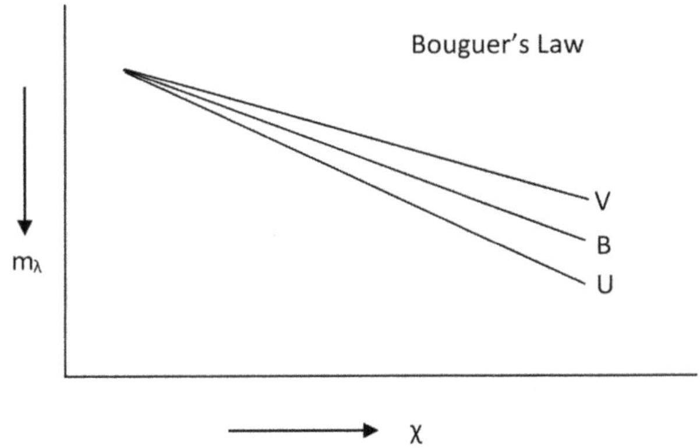

Figure 4.18: Shows schematically (not to scale and not real observations) the effect of air mass on different photometric magnitudes (in U, B, and V) for a star.

Figure 4.18 shows plot of apparent magnitude of a star as a function of air mass (m_λ vs χ) for three photometric pass-band filters (U, B, V) centred at three different wavelengths. We can see the effect of wavelength – as the wavelength increases the effect of air mass decreases. The slope of the straight line fit to the data points gives the constant (for a given wavelength) that may be called extinction coefficient (a_λ). An average value may be obtained for a given site, but we need to determine it for every night of observation for precision photometry.

The procedure to find this constant would be to observe several standard stars, of well-determined colour and apparent/absolute magnitudes, at different zenith angles and find an average value for the extinction coefficient (as per the Bouguer's law) - since the relationship is linear - the slope of the plot of magnitude with air mass gives the coefficient. For U, B, V magnitudes for instance we can write, with a_U, a_B and a_V as the extinction coefficients for U, B and V bands respectively,

$$m_{0U} = m_U - a_U\,\chi;\ \ m_{0B} = m_B - a_B\chi;\ \ m_{0V} = m_V - a_V\chi$$

We can also write down the corresponding equations for colour indices as follows,

$$(U_0 - B_0) = (U_\chi - B_\chi) - (a_U - a_B)\,\chi$$
$$(B_0 - V_0) = (B_\chi - V_\chi) - (a_B - a_V)\,\chi \tag{4.16}$$

Since the air mass effect might vary within the passband of a photometric filter, being broad, it is necessary to use a standard filter (see section 4.3). It is clear that a star appears redder as the air mass increases. We see this effect in the case of the Sun during the dawn and dusk.

Due to the differential effect of the air mass on the colour indices, it is required for precise determination of colours to eliminate this effect through multiple observations using standard stars. From a large volume of diligent observations by several earlier observers, colours for these standard stars are well established to the accuracy of a milli-magnitude (for details on photometric accuracies, see Bessell 2005). To account for the differential colour effect, we may write the extinction coefficient as a sum of two parts – one that is average part and the second the differential part. We can now redefine the extinction coefficient as (see Majewski, 2015),

$$C_0 = C - a_C\chi - a'_C C\chi \tag{4.17}$$

Here a_C is the average value of colour extinction coefficient ($= a_{\lambda 1} - a_{\lambda 2}$) and a'_C the differential colour extinction coefficient ($= a'_{\lambda 1}-a'_{\lambda 2}$); C_0 the colour index at the zenith observation = ($m_{\lambda 1}-m_{\lambda 2}$) and C_χ for any zenith angle χ. This method requires observations on (i) stars of same spectral type at different zenith angles (or air masses) to determine the coefficient a_C and (ii) usage of blue/red pairs of filters at the same zenith angle to determine the coefficient a'_C. Usually ignoring the colour effects, we can simply take care of the air mass effects for a single filter for given site for a given night. For a few well-known observatories the photometric standards for different filters for stars of different colours have been obtained by several workers (see the review by Bessell 2005). In order that observations taken at any other site (having telescope, instrument, and filters different from those used in standards) be in conformity with these standardised values we need to make a series of observations to solve the equation below for the constants (or coefficients)

$$m_{true} = m_{inst} + a_0 + a_1 C + a_2 C^2 + \cdots \tag{4.18}$$

Here the subscript "*true*" stands for standards and "*inst*" for a new site; *C* is colour index and the coefficients "*a*" represent effects due to simple

Bouguer's law and the differential colour effects. The coefficients (a's) may be determined by observing several stars.

4.7 Relative, Absolute and Differential Photometry

Apart from determining the continuum source from SED modelling, photometric observations are also made in general for various other scientific purposes: in some cases just knowing the relative flux of a source (with respect to a standard source) would be all that is needed; in certain cases finding the absolute value would be needed; and in some other cases we need to find if a source is undergoing variability in which case we need to do differential photometry using two or more sources that are stable in their luminosity or flux.

All the above-mentioned methods, nonetheless, require calibration and correction for contribution from several extraneous factors. We discuss these in the following paragraphs. Differential photometry is further dealt with in a later section.

4.7.1 Differential Photometry: As mentioned in Chapter 1, starlight varies temporally for variable stars with periodic or aperiodic changes on time scales from a few days or longer to hours or shorter. When we observe a variable star in order to monitor and analyse its variability with time (temporal variability), then a technique called differential photometry is employed (see for latest exposition Hartley & Wilson 2023). In this technique we observe two or more stars along with the programme star. One of the two (or more) stars that are known to be non-variable is chosen as the primary star and all the other stars including the programme star are then analysed for variability with respect to the primary. The small random fluctuations with time for the standard stars provide a template of standard deviation due to observational errors and environmental changes. The programme star then needs to be assessed against this template for variability. A difference in magnitudes of the primary and programme star and other standard stars is plotted as a function of time of observation.

$$m_\lambda(p) - m_\lambda(v)\ vs\ t; \quad m_\lambda(p) - m_\lambda(s_1)\ vs\ t; \quad m_\lambda(p) - m_\lambda(s_2)\ vs\ t$$

Here p denotes primary star; v the (variable) programme star; and s_1 and s_2 the other standard stars for comparison. All the magnitudes need to be corrected for atmospheric effects before taking the differences although the differencing would have taken care of the air mass effect provided the stars are of similar colours.

4.8 Point and Extended Sources

A source is said to be resolved by the telescope when the seeing disc is smaller than the angular size of the object. If on the other hand the size of the seeing disc is larger than the angular size of the object, then the object is considered spatially unresolved. Figure 4.19 shows schematically this difference. If a telescope is working at its diffraction-limited resolution using adaptive optics, then the seeing-disc angular resolution needs to be replaced by that prescribed by the Airy disc. All the stars (other than the Sun of course) are unresolved as their radii subtended angles at the observer much less than a milli arcsec. Even stars in giant phases barring those very close by are unresolved by telescopes using the conventional techniques.

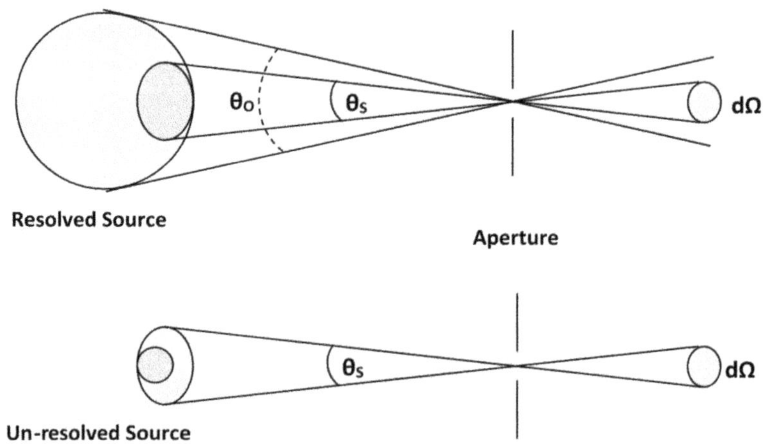

Figure 4.19: Geometry of an un-resolved or point source (such as a star) and a resolved or extended source (such as a nebula) as seen by a telescope. The solid angle $d\Omega$ is defined by the seeing-limited angular resolution θ_S. An un-resolved source subtends a smaller angle than the limiting angle of resolution while a source that is said to be resolved makes a larger angle. Even if the telescope is working at its diffraction limit there will be sources that it cannot resolve. The sketch is a general one and not to scale (adopted from Taylor 2022).

In general, stars of all types including those in their giant phase are considered unresolved by telescopes and are called point sources having angular sizes much less than one arcsec. All the other sources having angular diameters of a few seconds of arc are extended sources. Except those in the local group and nearby, clearly, extragalactic objects (external galaxies) even if intrinsically are extended sources due to their large distances have sizes of a few arcmins to arcsecs in deep sky.

4.8.1 Surface Brightness: For extended objects we use a parameter called surface brightness defined as the brightness of the object (I) divided by its area or solid angle subtended at the observer or the area on the sky (A). The surface brightness thus depends on the magnitude m of the object and its area A (in units of square arcsec) by,

$$S = -2.5\log_{10}(I) + \log_{10}(A) + ZP = m + 2.5\log_{10}(A) \qquad (4.19)$$

Care must be exercised not to simply take it as magnitude of object divided by the area, since magnitude is in logarithmic scale. ZP is the zero-point magnitude.

4.9 Photometric Observational Procedures and Data Processing in the Visible and Infrared Regions

While making photometric observations on a point source such as a star or an extended source such as a star cluster, we need to apply corrections for (i) the sky background radiation in the same filter bandwidth as for the program object and (ii) extinction due to atmospheric column in the line of sight using one or more photometric standard stars preferably at nearly (within about 10 degrees of) the same zenith angle as the object. While using a single element detector, a physical aperture of suitable diameter (e.g., field stop) is to be used. Aperture photometry is generally carried out on point sources such as stars or very distant extended objects such as galaxies or active galactic nuclei. The aperture is usually chosen to consider the seeing conditions and is around a few arcsec. For array detectors such as CCDs, however, this task of choosing the right aperture is done in the analysis software such as the Image Reduction and Analysis Facility (IRAF). In the case of visible region, the sky is usually sampled within the same image frame of the object (which contains incidentally the sky background also). In the case of near-infrared wavelength region (see, e.g., Heidt 2022), a sky frame is generated using a method called "dithering" that involves taking images all around the source object frame. The set of dithered frames so obtained is median-averaged and used as a sky frame F_S to be subtracted from the object frame. The sequence of observations is therefore (i) dark frames of suitable exposure times of same duration as for object frames; (ii) a set of object frames of suitable exposure times (iii) a set of sky frames of suitable exposure times, by using the dithering method and (iv) a set of standard star frames of suitable exposure times. It must be noted here that the standard star observations also need dithering frames (as in step (iii)). Normally, we use a minimum of 4-8 frames for dithering by off-setting appropriately around the programme object. To avoid ghost images, it is normal practice to reject initial couple of frames in every set of observations. We repeat this sequence (except for dark frames) over a required number of times for achieving higher *S/N* ratio. The sky-corrected object frame F_O may then be obtained by using the following equation in which F_{O+S} stands for the uncorrected object frame (object + sky), F_S the sky, F_F the flat and F_D the dark frames,

$$F_O = (F_{O+S} - F_S)/(F_F - F_D) \tag{4.20}$$

In the above equation the dark component is automatically removed in the numerator due to the subtraction. The flat frame represents the image obtained by exposing a uniformly lit background – usually a screen inside the telescope dome or sky background in the twilight time. The flat frame division removes the inter-pixel sensitivity differences of an image detector (such as CCD or an infrared array detector). As explained above, the dithered frames are usually median-averaged to obtain a single background frame (or sky frame).

A similar analysis exercise must be done for the standard star frames also. Using the photometric magnitudes available in the literature for standard stars as calibration, we then find the object magnitudes in different bands. As mentioned already the data processing is a bit more involved in the case of near-infrared bands than the visible bands. As shown earlier, the emission from the atmosphere as well as the telescope environment is high for infrared bands. Infrared bands also suffer from absorption from atmospheric constituent gases making the definition of bands subjective to the site of observation. Hence, we need to choose the set of standard filters in such a way that they match with the standards available in the literature (for a comprehensive study on the standardization we refer Bessell & Brett, 1988).

Standard analysis packages (working mostly in the Linux OS) are available; for instance, the Image Reduction and Analysis Facility (IRAF) package developed by National Optical Astronomy Observatory, USA (http://iraf.noao.edu/) or the GNU Astronomical Utilities 1.0 by GNUASTRO (Akhlaghi & Ichikawa 2015; Akhlaghi 2021).

For flux/absolute calibration, we use the well-studied and documented A0V star α Lyrae (Vega) assuming that its magnitude is 0 (zero) for all optical and near-/mid-infrared bands (U band through N band in the mid-infrared). The flux tables for Vega are available in literature (see e.g., Glass 1999 and Tokunaga 2000). Appendix A3 lists fluxes of Vega in different bands.

Photometry on a single isolated point source (i.e., a star) is a straightforward task while that in a crowded cluster of stars (such as a star forming cluster region) requires proper care and judgement to choose a sampling aperture while analysing the 'cleaned' image data. Such clusters also have gaseous matter around its constituent stars and present a very difficult task on hand.

Many an extended source including a galaxy, or a gaseous nebula requires surface photometry information. Here the sky sampling requires larger off-sets to avoid contribution of the matter surrounding the source itself. We note that usually for such an extended source when photometry is required, we measure/estimate the brightness of the gaseous matter in terms of "magnitude/flux density per square arcsec" as opposed to simply "magnitude" in the case of a point source.

IRAF (or GNUAU) code has tasks that help do various tasks required for processing and analysing the image data for both visible and infrared regions. We can also develop routines within the framework of IRAF for tasks not included in the software. The data required for gainfully performing IRAF tasks are dark frames; bias frames; flat fields, dithered frames (for infrared region) in addition to the raw object (that include the sky contribution) frames. Further, as mentioned earlier, we need data on a standard star for flux calibration. The standard star should be chosen in such a way that its position in the sky is not too far away from the programme object – typically within 10^0 of the objects is good enough. Farther than this requires corrections for zenith angle differences between the object and the standard star. In order to get object image frames with a reasonable S/N ($\geq$ 3), we need to give sufficient exposure and if the exposure required is too long (more than a few minutes) then we need to take several exposures and then add them after processing each set of data separately. Co-adding of frames of same exposure would lead to improvement in S/N. Care need to be taken in matching the position of the object in all the co-added frames to the accuracy of one pixel in the centroids of a few chosen stars within the image frame.

Typical analysis steps are given below while using the IRAF software.

Step 1: Find an average value of standard deviation (σ) of sky background counts from several randomly chosen locations on a given image frame to sample sky background for a dark sky.

Step 2: Find the average fullwidth at half maximum (FWHM) from the images of a good number of isolated (not crowded in a cluster) stars in the image frame. Usually, we should get the FWHMs the same within a few pixels. Avoid saturated images of stars and those that are very faint.

Step 3: Define an annulus around a given star for sampling the sky background immediately in front of the star. The annulus diameter is usually chosen as the same as the FWHM of the star image. The width of the annulus may be chosen to be a few pixels.

Step 4: Subtract the star peak count from the average count within the annulus and then divide the result with the standard deviation of dark sky background (σ). The resultant value is then multiplied by a factor of 4-6 to give a detection threshold count. If the peak count of any star is above this threshold, then the star is considered to be detected. Therefore, the detection threshold should be chosen carefully to detect but not miss any real star in the image frame.

Step 5: Repeat steps 1-4 for standard star. Calibrate the object frame with reference to the standard star. This will also set the zero-point magnitude at the pass band being considered. We may note that zero-point varies from night to night and from site to site.

Step 6: Using Vega standard find absolute flux values of all the stars in the image frame.

Step 7: Repeat steps 1-6 for images of object in other pass bands.

Step 8: Now we are in a position to construct SEDs and Colour-Colour and Colour-Magnitude diagrams for the programme stars. For this purpose, we may even use observations in other pass bands from other facilities.

4.9.1 Photometric Errors: Both systemic and systematic errors in the determination of photometric fluxes or magnitudes of astronomical sources need to be estimated (usually, the analysis packages such as IRAF do calculate photometric error, but it refers to analysis error). Sky-background, stray light (significant contributor in optical bands), instrumental emission (significant contributor in infrared bands), detector noise, fluctuations in the sky background, seeing errors, and zenith angle related extinction errors (especially when the programme source and the standard have large differences in zenith angle and seeing in their respective lines of sight) are some of the important contributors to the errors in photometric accuracy. As mentioned earlier, the signal-to-noise ratio (SNR) for an observation is usually approximated for two cases – namely, the back-ground-limited case in which the sky-background is the most predominant source of noise and the second, the detector-limited case in which the detector (or photon) noise is overwhelming.

4.9.2 Completeness Limit for Observation on a Star Cluster: As the brightness of the objects decreases, so does the detection ability for an observational set-up (comprising of front/back optics and detector and site/sky conditions) at sufficiently high S/N ratio. An important statistical parameter may then be defined as the fraction of objects of similar brightness that can be detected using the analysis criteria on an image obtained from an instrument. This parameter is known as the "Completeness Limit". This is a useful statistical criterion for detecting large number of stars in a crowded field. For sufficiently bright objects this parameter would be 1 – since there is a high probability that all such objects would be detected. As we approach fainter limits this fraction would deviate from 1. Usually for a given exposure the magnitude of objects where the completeness drops below 0.9 (or 90%) is considered as the limit.

As an example, we may refer to the Figure 2.25 (Ch. 2), an H band image of the star forming region S235 obtained by using HAWAII-1 imaging array detector at Mt Abu 1.2 m telescope after analysis following the procedures mentioned above.

4.9.3 Standard Stars in Visible and Near-infrared Regions: Stars with stable brightness (or flux or magnitude) may be qualified as photometric standard stars and may be used for calibrating unknown stars while making photometric observations. Lists of such photometric standards had been prepared by astronomers and are available in the literature (for UBVRI photometric standards see Landolt 1992 and Graham 1982; and for near-infrared JHK photometric standards see Hunt et al 1998).

4.10 Spectral Energy Distribution

We have already mentioned that a plot of flux received from an astronomical source as a function of energy interval (or equivalently frequency / wavelength interval) determines the nature of the principal energy source mechanism. It is quite conceivable that an object may have more than one source of energy responsible to account for the observed SED. Several objects reveal multiple sources of energy from the analysis of their SEDs. The SED in principle should encompass a major portion of the electromagnetic spectrum but in practice since the nature of observational techniques is different for different wavelength or frequency regimes, we rarely get an opportunity to obtain a complete spectrum from one end to the other. However, the wavelength region that is most important is from UV to Infrared within which both very hot sources as well as cold sources normally are covered. While a SED in the visible/infrared regime is often modelled by invoking one or two blackbody distributions, the high energy and low energy ends may need additionally non-thermal sources for reconciling with observations. Sources such as main sequence stars may follow a single blackbody distribution with an equilibrium temperature (the surface or effective temperature), stellar sources that have evolved beyond the main sequence phase and those that are in the pre-main sequence phase would require multiple blackbody sources – one accounting for the stellar surface and the other the circumstellar matter. The SEDs of diffuse or gaseous nebulae are usually modelled using a blackbody representing a central source that powers the nebula and another one representing thermal emission due to dust envelope(s) as well as other non-thermal energy sources such as 2-photon continuum, free-free emission, and synchrotron emission. It is therefore clear that photometric observations of good precision are needed to model SEDs. Often, just by combining UV, visible and IR photometric observations we would be able to discern the physical nature of the underlying energy sources for an object. The spectrophotometric modelling software codes used for interpreting the SEDs include DUSTY (Ivezic et al 1999) among others.

A few examples of SEDs for selected post-asymptotic giant branch stars and their modelling are shown in Figure 4.20. As may be seen from the figure, the data are modelled using two sources – one for the star (a blackbody) and a second one for the circumstellar shell (see Venkataraman et al 2017).

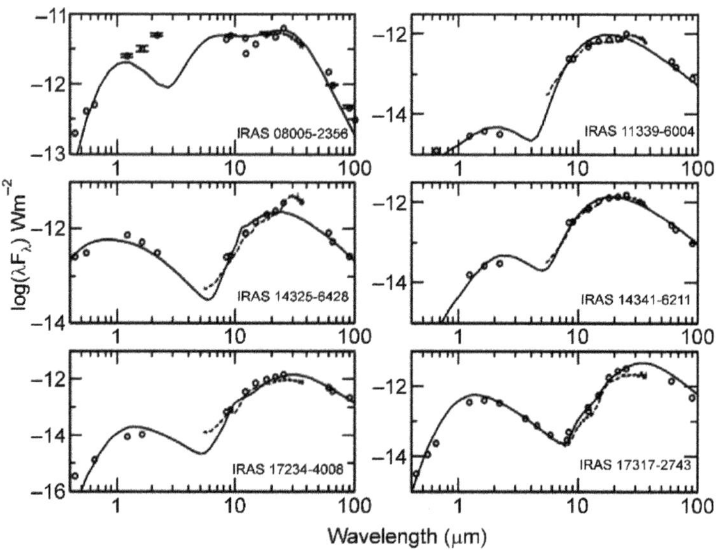

Figure 4.20: Model SED of the photometric observations from visible to mid-infrared wavelengths for a sample of evolved post-asymptotic giant branch stars (Venkataraman et al 2017). The solid curves show model fits using DUSTY (see text). The sources of photometric data are given in Venkatraman et al (2017).

4.11 A Brief Note on Polarimetry

By virtue of its wave nature light radiation (photons) demonstrate polarization property. From the electromagnetic theory in the framework of Maxwell's equations, the electric vector (and magnetic vector) vibrates in the plane perpendicular to the direction of propagation. Normally for an unpolarized radiation the electric vector vibrates in all directions but because of interaction with matter the vibrating electric vector rotates in an ellipse. This is termed as elliptical polarization. Two special cases of elliptical polarization are linear polarization that has vibrating electric field in a single direction and circular polarization in which the vector rotates in a circle with the amplitudes in two orthogonal directions being the same with phase difference of 90^0 (see, for the theory of polarization of light, Jenkins and White 1995; Born & Wolf 2000). Some of the physical mechanisms (most relevant to astrophysical point of view) which produce polarized radiation are: (1) reflection of light on plane surfaces; (2) scattering of light by particles such as electrons (Thomson scattering), molecules (Rayleigh scattering) and by dust grains (Mie scattering); and (3) Synchrotron emission for relativistic electrons in presence of magnetic fields among other processes (see Serkowski 1974; Serkowski et al 1975; Kitchin 2000; Tinbergen 2009; Trippe 2014). These physical processes are wavelength dependent and in principle polarization is an important parameter at all wavelength regimes for understanding the nature of the sources of polarization. Further, polarization caused by scattering of dust grains also depends on the size and shape of the dust grains. Stokes and later Chandrasekhar mathematically defined the parameters (called Stokes parameters) that characterize polarization, namely *I*, *Q*, *U* and *V*. These are defined as follows:

$$I = E_{x_0}^2 + E_{y_0}^2$$

$$Q = E_{x_0}^2 - E_{y_0}^2$$

$$U = 2E_{x_0}E_{y_0}\cos\delta$$

$$V = 2E_{x_0}E_{y_o}\sin\delta$$

Here E_{x0} and E_{y0} are the maxima of the electric field components in an orthogonal plane (x, y) for a general case of elliptical polarization as shown in Figure 4.21. The angle δ represents the phase difference between the electric field components. We can see that the total energy flux or intensity I is $(Q^2 + U^2 + V^2)^{1/2}$. The parameters Q and U represent the linear polarization for which the phase angle $\delta = 0$; while V represents the circular polarization when the phase δ becomes $\pm 90^0$ and E_{x0} equals E_{y0}.

The degree of polarization, defined as the fraction of incident light beam that is polarized, is given by

$$P_l = \frac{(Q^2 + U^2)^{1/2}}{I}$$

for linear polarization.

And for circular polarization, it is,

$$P_c = \frac{V}{I}$$

The angle of polarization θ_{pol}, defined as the angle between the x-axis (aligned with the north-south direction) and the direction of the semi-major axis of the polarization ellipse (see Kitchin 2000), can be found as

$$\theta_{pol} = \frac{1}{2}\tan^{-1}\frac{U}{Q}$$

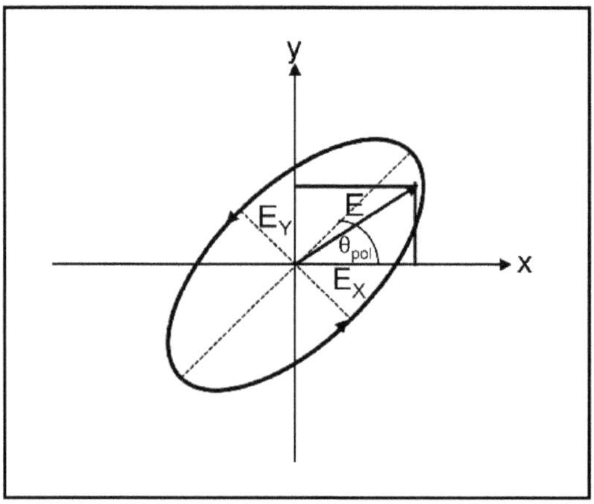

Figure 4.21: Polarization Ellipse. E_x and E_y are components of the electric vector E. The angle θ_{pol} represents the degree of polarization.

A photopolarimeter is essentially a photometer into which appropriate optical elements are introduced for the measurement of degree of polarization and the angle of polarization. Polarizers are usually employed (a combination of waveplates and Wollaston prisms) for producing two (or four) orthogonal beams (at position angles of combinations of 0^0 & 90^0 and/or 45^0 & 135^0) of linearly polarized light from the input of unknown light beam and then a rotating retarder such as a waveplate for analysing the polarized beams for finding the degree of polarization and the phase angle (for details, see Serkowski et al 1975). Traditional broad/narrow band photometric filters are used for determining wavelength dependency of polarization of light from a program celestial object. Calibration of the instrument and telescope for instrumental polarization is made by observing standard (nearby) stars with supposedly unpolarized light. Similarly, we can modify a spectrometer as well into a spectropolarimeter. For details of polarimetric instrumentation, in optical and near-infrared regions, we may refer to Serkowski et al (1975), Kulshrestha et al (1987),

Aitken (1988), Tinbergen (1996), and Trippe (2014). Advancements in polarimetry include imaging (photo- and spectro-) polarimetry for studying extended sources such as the star forming regions. For details of PRL photo-polarimeter for the near-infrared region, we refer to Esakkiappan et al (2019).

CHAPTER 5

Spectroscopic Techniques

It is the job of Physics to analyze why each atom wants what it wants

– Richard Feynman

The techniques, and the purpose of spectroscopic observations will be discussed in this chapter. Basic parameters or merit factors of spectroscopy like resolving power, dispersion, spatial and spectral multiplexing will be discussed. Analysis techniques for spectra and spectral lines will be introduced. Calibration methods will be discussed. In Chapter 6 we deal with the relevant decomposition and deconvolving techniques applied to spectroscopic analysis.

We may recall here one of the most important theorems in geometrical optics called the principle of Fermat or the principle of least time (see Born & Wolf (1999) and Jenkins & White (1981) for a thorough exposition). In geometrical optics, the laws of reflection and refraction are based on Fermat's principle that states the path of a light ray traversing from one point to another is the one that takes the least time. Ray tracing analysis in geometrical optics involving lenses, mirrors, and various dispersive elements uses this principle.

Essentially, there are three methods to obtain spectrum from a source. The first one uses the principle of refraction of light through materials for achieving dispersion. The second method uses diffraction of light, while the third one exploits the principle of interference between two or multiple light rays. Dispersion itself, the basis of spectroscopy, is the spread of constituent colours of a spectrum over angular or linear spatial elements without overlapping. It may be defined as,

$$D(\lambda, \theta) = \frac{d\theta}{d\lambda} \tag{5.1}$$

In the above form it is called angular dispersion which may also be expressed in terms of linear dispersion. Herein what follows, we discuss the basic optical layout of a spectrometer, its figures of merit and some of the most extensively used spectrographs in astronomy, based on the three principles enunciated above.

5.1 Optical Layout and Figures of Merit of a Spectrometer

Figure 5.1 shows, in general, the schematic optical layout for a spectrograph. The light beam from a telescope (A) having an aperture size 'D' is passed through a field stop and a collimator (a) with an aperture size 'd' that makes the converging rays parallel. The parallel beam is then passed through a dispersing device such as a prism or grating. The dispersed light is then focused on a detector using an imaging lens (imager). The solid angle subtended by a celestial object at the telescope is marked as $\Omega(\varphi)$ and that by the image at the detector is $\Omega(\theta)$.

5.1.1 Design Parameters of a Spectrometer: The most fundamental characteristic features of a spectrometer are: (i) dispersive element (prism; grating; interferometer); (ii) resolving power (R); (iii) throughput or light collecting capacity (acceptance solid angle times its area); (iv) Luminosity (or throughput) - Resolving Power product (L x R); (v) spectral multiplexing or the number of spectral elements sampled simultaneously (M_λ); (vi) spatial multiplexing or the number of spatial elements sampled simultaneously (M_{xy}); (vii) figure of merit (M_λ x M_{xy}); (viii) telescope field angle or focal-ratio matching - for most spectrometers $f_T/D = f_{col}/d$ with f the effective focal length (refer Figure 5.1); (ix) detector pixel matching; (x) sensitivity.

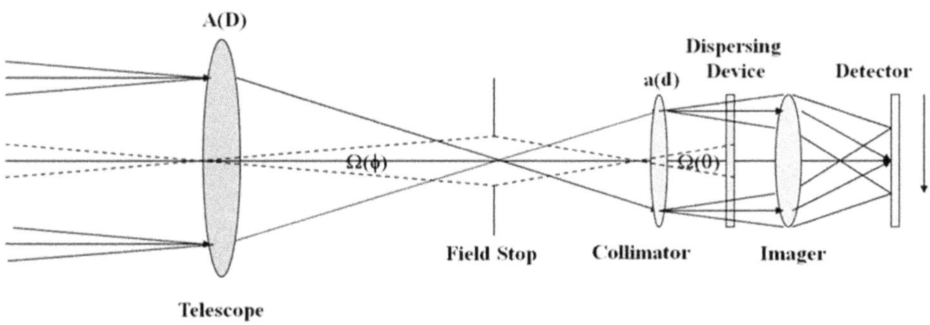

Figure 5.1: Schematic optical layout of a typical spectrometer. Telescope A, field-stop, collimator with a dispersive device, imager and detector are shown. The solid angles subtended by a celestial object at the telescope and at the detector are also shown.

5.1.2 Rayleigh Criterion for resolving two spectral lines: It is instructive to include here the criterion for detecting two components of a spectral line or two closely spaced emission lines to be resolved. Usually, the spectral element of a high dispersion spectrometer is narrow enough to separate two close-by spectral lines, but it is sometimes possible that two or more lines get partially overlapped. It is also possible that a single line emitted by an object is structured having components representing two or more kinematically different flows or two objects emitting the same line overlapped in the line of sight. But classically the criterion proposed by Rayleigh deals with two emission lines represented by two diffraction patterns superposed on each other. The criterion clearly defines the lines as resolved when the principal maximum of one profile overlaps on the first minimum of the other. As shown in Figure 5.2, this typically happens when the dip between the two maxima is 81% of the maxima (assuming the maxima of the same height). Usually the spectral line (as discussed earlier) may be assumed to be a Gaussian or a Lagrangian (or the combined Voigt profile); hence any significant deviation from these shapes may be interpreted as due to one or more of the effects mentioned above. Such line profiles need to be decomposed using standard methods (see Chapter 6).

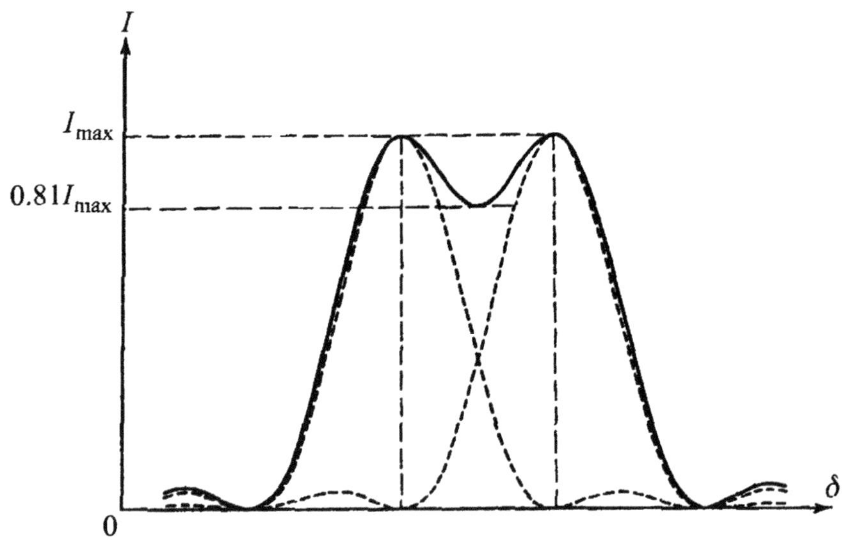

Figure 5.2: Rayleigh's criterion for resolving two spectral lines (from Born and Wolf 1999).

5.2 Prism Spectrograph

Historically this is the oldest used spectrometer. Prisms are rarely in use in today's observatories as principal dispersive elements. We discuss here briefly though for the sake of completeness. When we pass white light through a prism, dispersion occurs splitting the light into the constituent colours of different wavelengths. This was the famous Newton's experiment with sunlight dispersing it into the constituent VIBGYOR spectrum – violet, indigo, blue, green, yellow, orange, and red colours. From Snell's law of refraction, we have for light ray passing from one medium to another,

$$\mu_1 \sin i \;=\; \mu_2 \sin r \tag{5.2}$$

where "*i*" and "*r*" are respectively the incident and refraction angles with respect to the normal at the point of incidence, and μ_1 and μ_2 are the refractive indices of the two media in which the process of refraction is occurring. If the incident beam is travelling in air, we can assume μ_1 to be 1 so that we have,

$$\frac{\sin i}{\sin r} = \mu_2 \tag{5.3}$$

Now the refractive index of a material such as glass varies with wavelength and hence the angle of refraction would be different for different wavelengths leading to dispersion. For a limited range of wavelengths, the variation of refractive index with wavelength is given by the Hartmann's formula,

$$\mu_\lambda = A + \frac{B}{\lambda - C} \tag{5.4}$$

where *A*, *B* and *C* are constants that may be determined knowing the refractive index in three wavelengths for a given material. These constants had been determined by earlier scientists to a great degree of accuracy for various glasses that are used in making optical components (refer Born & Wolf 1999).

Figure 5.3 shows the basic ray diagram of refraction and dispersion in a prism with an apex angle α. A collimated beam from the telescope is passed through the narrow slit and the dispersed beam is imaged on a detector on the righthand side.

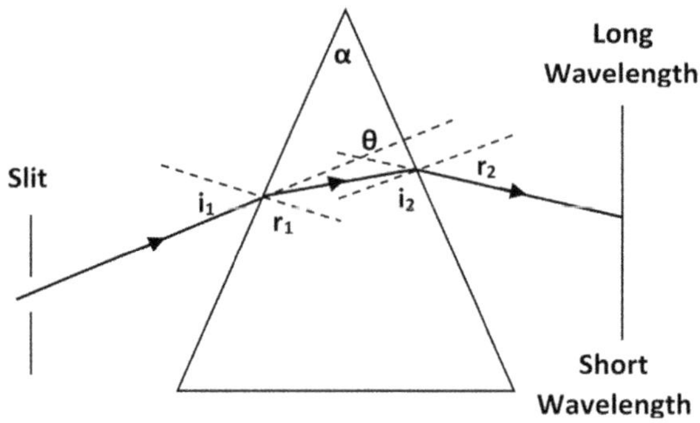

Figure 5.3: Dispersion process in a prism. The prism angle α, is by definition, at the apex of the prism. The angles of incidence are i_1 and i_2 at the outside and inside of the prism respectively, while r_1 and r_2 are the corresponding angles of refraction. The angle θ is called the deviation angle.

5.2.1 Dispersion in a prism: The angle of deviation or dispersion can be shown to be,

$$\theta = i_1 + r_2 - \alpha$$

and

$$\alpha = r_1 + i_2 \qquad (5.5)$$

Substituting the equations (5.3), (5.4) in (5.5), the angle of dispersion can be derived as,

$$\theta = i_1 - \alpha + \sin^{-1}\left\{\left(A + \frac{B}{\lambda - C}\right) \sin\left[\alpha - \sin^{-1}\left(\frac{\sin i_1}{A + \frac{B}{\lambda - C}}\right)\right]\right\} \qquad (5.6)$$

The angular dispersion may then be written as

$$\frac{\delta\theta}{\delta\lambda} = \frac{\theta_{\lambda_1} - \theta_{\lambda_1}}{\lambda_1 - \lambda_2} \qquad (5.7)$$

The suffixes 1 and 2 denote two wavelengths being dispersed. From this equation the maximum dispersion may be calculated by differentiation, and it occurs for an apex angle of $73^0.88$ that corresponds to the minimum deviation angle. For this apex angle, the ray passes through the prism nearly parallel to its base. In this case, the incident and emerging (refracted) rays would be nearly at 90^0 to the faces of the prism. This is practically a difficult condition to achieve and hence usually the apex angles are chosen close to 60^0. For this apex angle the dispersion for a flint glass prism can be found to be 1.39 x 10^7 degrees/metre for minimum deviation condition with angle of incidence at $55^0.9$. For details we may refer to Kitchin (2003).

Writing the deviation angle (eqn. 5.6) in a different way we have

$$\theta = \sin^{-1}(\mu_\lambda \sin r_1) + \sin^{-1}[\mu_\lambda \sin(\alpha - r_1)] - \alpha$$

Using this equation, we can write down the dispersion angle in terms of the Hartmann's constants *A*, *B*, and *C*. For the optimal apex angle of 60^0 we can show that the dispersion equation reduces to,

$$\frac{d\theta}{d\lambda} \approx -\frac{180\,AB}{\pi(\lambda - C)^2} \propto \frac{1}{(\lambda - C)^2} \tag{5.8}$$

The above simplified equation shows that at shorter wavelengths the dispersion increases. The linear dispersion which depends upon the focal length (say, f_{im}) of the imaging lens used to focus the spectrum on a detector (on the right side in Figure 5.1) is given by,

$$\frac{dx}{d\lambda} = f_{im}\frac{d\theta}{d\lambda} \tag{5.9}$$

5.2.2 Resolving Power: In practice, a prism-based spectrograph would have a slit at the focal plane of the telescope (see Figures 5.1 and 5.3). The resolving power of a prism spectrograph also depends on the width of the slit. For a monochromatic beam incident through a slit width of w the spread of the spectral image W is given as (with f_{col} as focal length of the collimating lens as shown in Figure 5.1),

$$W = w\frac{f_{im}}{f_{col}} \tag{5.10}$$

When the prism with an apex angle α is fully illuminated by the incoming beam of light then the width of the beam (D_B) and the length of the face of the prism (L_P) are related by

$$D_B = L_P\left[1 - \mu_\lambda^2 \sin^2\frac{\alpha}{2}\right]^{1/2} \tag{5.11}$$

This is analogous to diffraction by a rectangular slit (Kitchin 2003) and the limit of resolution may be given by the Rayleigh's criterion as

$$W' = \frac{f_{im}\lambda}{D_B} = \frac{f_{im}\lambda}{L_P\left[1 - \mu_\lambda^2 \sin^2\frac{\alpha}{2}\right]^{1/2}} \tag{5.12}$$

When W equals S (the incident beam width) the resolution is optimal. Then the slit width may be given by

$$w = \frac{f_{col}\lambda}{D_B} = \frac{f_{col}\lambda}{L_P\left[1 - \mu_\lambda^2 \sin^2\frac{\alpha}{2}\right]^{1/2}} \tag{5.13}$$

The resolving power of a prism spectrograph may be shown to be (see Kitchin 2003),

$$R = \frac{\lambda}{d\lambda} \approx \frac{ABL_P\left[1 - \left(A + \frac{B}{\lambda - C}\right)^2 \sin^2\frac{\alpha}{2}\right]^{1/2}}{(\lambda - C)^2} \tag{5.14}$$

From the above formula, clearly, for a given wavelength the prism resolving power depends directly on its size (L_P): the larger the size the higher its resolving power. It may be noted that the product of the third and fourth factors in the numerator of the equation for resolving power represents the emergent beam width, D_B. R depends also on the apex angle α and the constants A, B, and C which depend upon the type of material used for making the prism. Figure 5.4 shows the resolving power R as a function of the prism angle α. We notice that R depends inversely on the square of wavelength. For a dense flint glass prism, the resolving power achievable is about 10^4 for prism of a size of 10 cm.

However, it should be noted that the peripheral rays entering the spectrograph undergo optical aberrations. For spectroscopy of wider field angles prism spectrographs become unwieldy with complex optical and spectral distortions. It also lacks the throughput advantage that other spectrographs enjoy as we shall show later in this chapter.

Finally, the slit width needs to be chosen to match the seeing disc size so that the star light is not partly lost. Then this slit width would decide the resolving power achievable by the prism spectrograph. In some applications the slit is removed so as to allow a larger sky angle so that simultaneously several stellar spectra can be recorded though at a compromised resolving power. Star clusters are a good target for such an application. This is called slit-less spectroscopy. This technique is further discussed in the upcoming section on grating spectrograph.

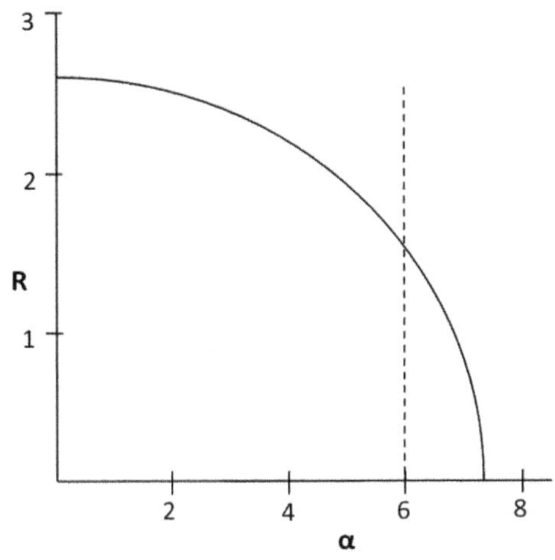

Figure 5.4: Variation of resolving power with prism angle (adapted from Kitchin 2003).

5.3 Diffraction Grating Spectrograph

The phenomenon of diffraction occurs when light waves are passed by an obstruction or passed through a narrow opening. If the source of light and the observing screen are at infinity with reference to the obstruction or orifice the diffraction is called Fraunhofer diffraction. Alternatively, if the light source and the screen are at finite distance the phenomenon is termed as Fresnel diffraction.

Grating spectrograph works by the principle of Fraunhofer diffraction and interference of the diffracted beams. Diffraction gratings made with narrow slits (for transmission) or grooves (for reflection) have been in use for obtaining stellar spectra for a very long time since the beginning of the 20th century. If light is passed through a narrow-slit diffraction occurs. If we pass light through two narrow slits with separation larger than the slit width, then the two diffraction patterns emerging from the slits interfere to produce narrow fringes. By extension, if the number of slits is increased, then the fringes become not only narrower but also better resolved.

In principle gratings represent diffraction phenomenon in equally spaced multiple narrow rectangular slits that can cause periodic amplitude and/or phase variations in the incident light rays. As already mentioned, it is an extension of the classical two (narrow rectangular) slit diffraction phenomenon. The emerging diffracted rays interfere to produce fringe pattern with maxima and minima depending up on their phases. Gratings can also be used in reflection mode wherein the light rays are reflected on a surface made of equally spaced grooves (defined by groove density ρ_g

measured in number of grooves per inch). Astronomical spectrographs in both near-UV, optical and infrared wavelength regimes utilise reflection gratings. A schematic diagram of a diffraction grating is shown in Figure 5.5 in which the two diffracted rays would interfere constructively to produce maxima according to the relationship called the grating equation for maxima,

$$m\,\lambda = d\,(\sin\alpha \pm \sin\beta) \tag{5.15}$$

where d is the spacing of the grating rules (or grooves $= 1/\rho_g$), m is the order of interference (= 0, +/- 1, +/- 2, ...) and α and β are angles of incidence and diffraction respectively as shown in the figure in the two panels. Conventionally, the angles of incidence and diffraction are considered positive or negative according to whether they are measured anti-clockwise (left) or clockwise (right) with respect to the grating normal. The positive sign in the equation refers to reflection grating and the negative for transmission: here onwards we discuss only the reflection grating and hence use positive sign. The right-hand side of the above equation represents the optical path difference between the rays A and B in the figure. For a monochromatic light beam the grating equation means different orders would dilate the spectral line progressively, while for a polychromatic beam different orders represent regions of spectra, for example, with the beginning of the second order coinciding with the end of the first order. This is further discussed later in subsections on spectral overlap and free-spectral range. For a comprehensive treatment of grating spectrographs, we may refer to Schroeder (2000) and Kitchin (2003).

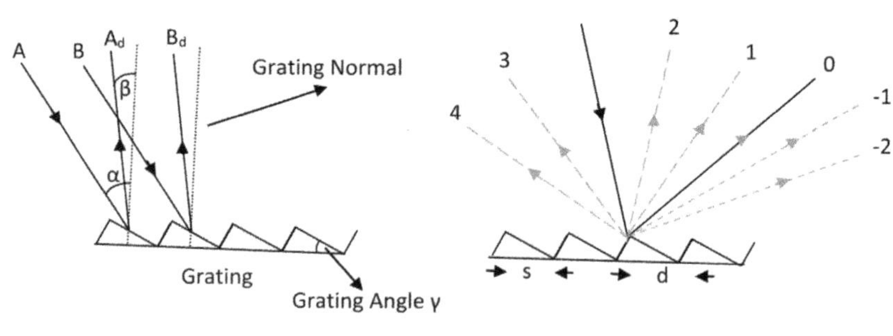

Figure 5.5: Diffraction grating principle. Left: A and B denote two rays diffracted in the directions of A_d and B_d. The angle of incidence is α. β is the diffraction angle. The dotted vertical line is the grating normal. If the diffraction angle is opposite to the incidence angle with reference to the normal, then it is considered negative. Grating angle that defines the groove is also shown. Right: Orders of grating (m) on either side of the 0th order. Also shown are the grating groove facet size *s* and the separation of consecutive grating grooves *d*.

For an incident intensity *I(0)*, the resultant dispersed intensity pattern *I(θ)* for a grating is given by, with *s* the groove facet size (cf. slit width) and *d* the groove spacing (see Fig 5.5),

$$I\,(\theta) = I(0)\left[\frac{\sin^2(\Delta)}{\Delta^2}\right]\left[\frac{\sin^2(N\delta)}{\sin^2(\delta)}\right] \tag{5.16}$$

$$\Delta = \frac{\pi\, s \sin\beta}{\lambda}$$

$$\delta = \frac{\pi\, d \sin\beta}{\lambda}$$

In the above equations, *N* is the number of grooves, and β is the angle with respect to the optic axis at which the pattern is viewed (the angle of diffraction). The factor in the first brackets in the eqn. 5.16 represents the diffraction pattern as modulated by the intensity structure of a single aperture. The second factor (in the brackets) arises due to the interference between patterns from *N* apertures or grooves. Figure 5.6 (from Born & Wolf 1999) shows the two patterns in detail – the interference fringe patterns and the diffraction envelope. A mirror image of the same occurs on the left side of the *y*-axis. It should be noted that in the figure, *p* represents the phase varying in multiples of λ/s or λ/d. The figure shows the resultant fringe pattern (see eqn. 5.16 above). The central diffraction envelope (zero order; $m = 0$) has the highest amplitude but the interference fringes are not separated leading to zero resolution ("white light"). In comparison, the first order ($m = 1$) shows increased resolution but with decreased amplitude.

The principal fringe maxima for the interference pattern occur at values of δ which are integer multiples of π (as shown in the top panel of the figure) at angular positions of

$$\beta = \sin^{-1}\frac{m\lambda}{d} \tag{5.17}$$

(This is grating equation at normal incidence, i.e., $\alpha = 0$ in eqn. 5.15).

The angular width of the principal maximum of the diffraction pattern is given by,

$$\Omega = \frac{2\lambda}{Nd\cos\beta} \tag{5.18}$$

It is inversely proportional to the number of grooves *N* and its peak intensity is inversely proportional to N^2. For a polychromatic input beam the principal maximum being independent of the order, represents bright white fringe flanked by spectra. Thus, as the diffraction angle β is a function of wavelength, the grating would decompose a monochromatic light beam into spectral orders except at $m = 0$.

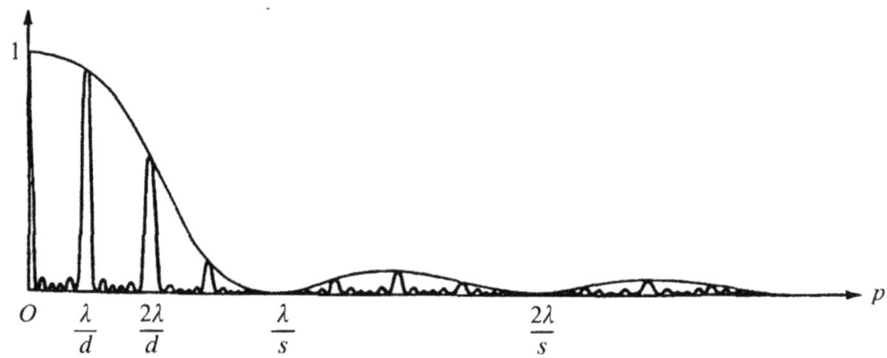

Figure 5.6: Diffraction Grating fringe patterns: shows the normalised resultant intensity pattern - the narrow interference pattern with phase *p* given by $m\lambda/d$ for maxima, m taking values of +/- 1, +/-2 and so on, enveloped by the diffraction pattern with phase p given by $n\lambda/s$ for minima, n taking values of +/- 1, +/-2 and so on (from Born and Wolf 1999). Since both signs are possible for phase, the figure is extended on the negative side also (a mirror image of the positive side).

Thus, as the order increases, the resolution (fringe separation in the interference pattern) increases while the amplitude decreases. This means that the diffracted light into higher orders decreases thus decreasing the grating efficiency. This is the reason why the gratings are carefully tuned for a desired wavelength in order to increase the amplitude in that region by working on the groove facet angle or shape and depth. This process of tuning is called "blazing". The wavelength at which the efficiency of a grating is highest is called the blaze wavelength and the groove angle (see Figure 5.5) is called the blaze angle. Further it may be noted that if a wavelength λ_1 satisfies the grating equation for order $m = +/- 1$; then $\lambda_2 = \lambda_1/2$ satisfies for $m = +/- 2$ and $\lambda_3 = \lambda_1/3$ satisfies for $m = +/- 3$ and so on.

It should be noted that for a reflection grating the phase difference between the centre of the groove or facet and its edge accounts for diffraction pattern while in the case of transmission grating it is across the line width *s* (see right-side diagram in Fig 5.5). The phase difference between the centres of two adjacent grooves or facets *d* (in the case of transmission grating, the line spacing *d*) accounts for interference pattern.

Figure 5.7a shows schematic optical arrangement for a grating spectrograph. The grating is a reflection type. We need to mention a few optical arrangements that are used in grating (or prism-based) spectrometers. Firstly, in a configuration called Littrow, the optics is such that the angle of deviation, i.e., the angle between the incident and dispersed light is zero, i.e., $\alpha = \beta$, or in other words the deviation of the incident light ray is zero. Figure 5.7a shows a non-Littrow configuration. Two other configurations need to be mentioned here, namely, Czerny-Turner and Ebert-Fastie – in the former the collimation and imaging are achieved by using two separate elements (lenses or mirrors) while in the latter a large single element (usually a mirror) does both the tasks.

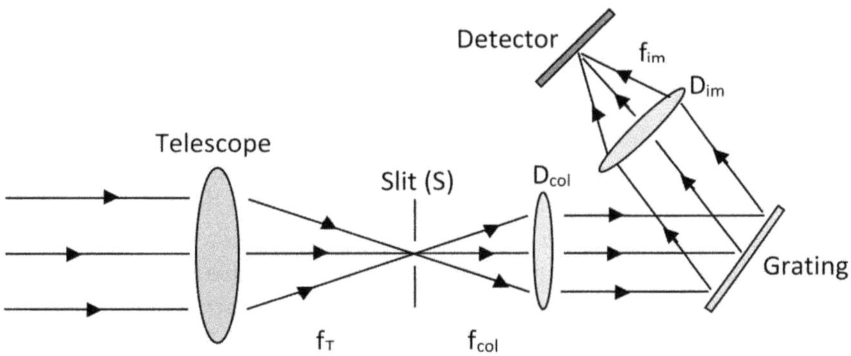

Figure 5.7a: Schematic optics for a grating spectrograph. The telescope, collimator lens and imaging lens foci are designated by f_T, f_{col} and f_{im} while their apertures are D, D_{col} and D_{im} respectively. The slit has a width of S which is normally matched with the seeing-limited angular size φ (see text and figure 5.7b).

In Figure 5.7b, the telescope beam is focussed on a slit of width *S*. (Like in the case of a prism spectrograph, the grating is also used in conjunction with a slit – essentially to define the sky angle). The beam then passes through a collimating lens of focal ratio $N_{col} = f_{col}/D_{col}$ (usually the same as the telescope focal ratio) and the diffraction pattern of the reflection grating G is then imaged by the imaging lens of focal ratio $N_{im} = f_{im}/D_{im}$. D_{col} and D_{im} are the apertures or beam- sizes of the collimator and imaging lenses respectively. It may be noted here that Figure 5.7b shows a general non-Littrow optical configuration.

5.3.1 Dispersion and Resolving Power: The angular dispersion may be obtained by differentiating the grating equation, holding the angle of incidence constant. We get,

$$\frac{d\lambda}{d\beta} = \frac{d}{m}\cos\beta \tag{5.19}$$

The linear dispersion depends on the focal length of the imaging lens f_{im}. If dx is the incremental element on the detector, then it is equal to $f_{im}\, d\beta$ for small increments in β. Thus, the linear dispersion is given by,

$$\frac{d\lambda}{dx} = \frac{d}{m f_{im}}\cos\beta \tag{5.20}$$

The slit width S is decided depending upon the angular size of the seeing (Airy) disc φ (see figure 5.7b) – usually it is considered as twice the size of the seeing disc so that we do not lose any starlight and at the same time not allow avoidable sky light. Having decided the slit width, we need to see the corresponding size of the slit S' as imaged on the detector as shown in Figure 5.7b. The slit width S' as projected on the detector is usually sampled using 2 pixels. More than 2 pixels makes the image over-sampled while less than 2 pixels under-sampled. This depends on the ratio of the focal lengths of the collimator and imaging optics (lenses) (f_{col}/f_{im}). We can show that,

$$S'\frac{D_{im}}{f_{im}} = S\frac{D_{col}}{f_{col}} \tag{5.21}$$

or

$$S' = S\,\frac{N_{im}}{N_{col}} \tag{5.22}$$

The ratio of the focal ratios of imaging and collimating lenses. The spectral resolution $\delta\lambda$ may now be calculated as

$$\delta\lambda = \frac{d\lambda}{dx}S' = S\,\frac{d}{m f_{im}}\cos\beta\,\frac{N_{im}}{N_{col}} = \frac{S\,d}{m\,N_{col}}\,\frac{\cos\beta}{D_{im}} \tag{5.23}$$

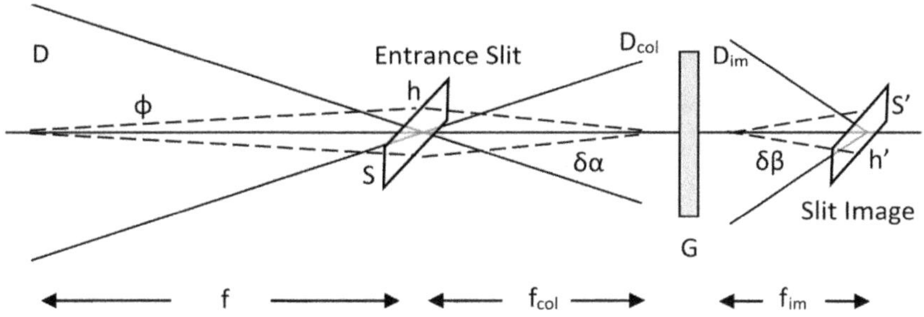

Figure 5.7b: Schematic optics of a grating spectrometer showing the slit of width S (and height h) as imaged by the optics (to form an image of width S' and height h'). The angle φ corresponds to the seeing-limited Airy disk. The angles δα and δβ are the angles subtended by the collimating and imaging lenses at the grating (dispersing) optics G (D_{col} and D_{im} with the reflection grating in between, as shown clearly in Figure 5.7a). (Adapted from Schroeder 2012).

As shown in the Figures 5.7 a & b, when the grating is fully illuminated or irradiated by the incoming light rays it makes a rectangular beam at the imaging optics – the height "*h*" of which has no bearing on the spectral resolution, as it represents the spatial extent of the sky, but its width *S'* has. If the length of the grating reflected by the imaging lens is given by $W \cos\beta$, whence, $W = (D_{im} / \cos\beta)$. Thus, the spectral resolution becomes,

$$\delta\lambda = \frac{S\,d}{WmN_{col}} \tag{5.24}$$

The resolving power of grating spectrograph can now be written as

$$R = \frac{\lambda}{\delta\lambda} = \frac{\lambda\, M_g\, m\, W\, N_{col}}{S\, d} \tag{5.25}$$

where we have considered M_g grooves for the grating. We can see that the resolving power of a grating spectrograph is essentially proportional to the product of the order of the spectrum and the number of grooves (i.e., $M_g\, m$). It also depends inversely on the slit width, the constraint being that the detector should sample the slit width adequately (following Nyquist criterion, two pixels), preserving the theoretical resolving power as much as possible.

5.3.2 Blazing: The diffraction factor in the intensity equation (eqn. 5.16) is also termed as the blaze function. Its maximum occurs when $\Delta = 0$ where the order $m = 0$. At this point the resolution is also zero. This therefore makes the grating inefficient. Ideally the maximum should occur at higher orders so that the resolution is large as well as the light intensity is sufficient. Blazing is the process of engineering the grooves of a grating to maximize the efficiency (intensity) at a chosen wavelength at higher orders. Let us examine closely the blaze condition for a grating. The manufacturers of gratings shape the grating groove in such a way that the groove angle γ satisfies the condition (see Figure 5.8),

$$\Delta = 0 = \frac{\pi d \cos\gamma}{M_g \lambda} (\sin i - \sin r) \tag{5.26}$$

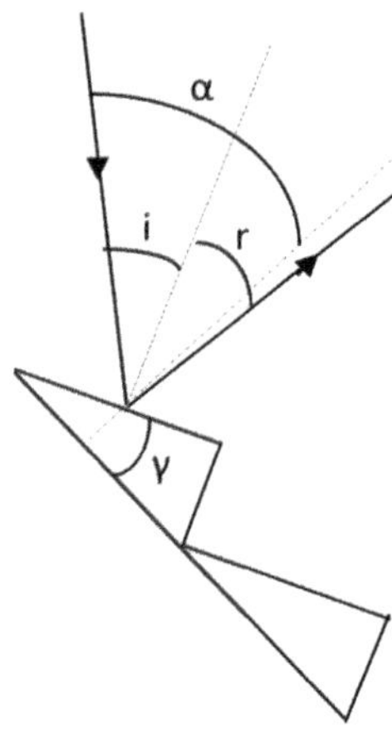

Figure 5.8: Blazing condition in a grating. The dashed line shows normal to the groove face and the dash-dot line the normal to the grating. The angles i and r are the angles of incidence and reflection with respect to the grating groove face and the angles α and β are the corresponding angles with reference to the grating normal. The angle between the incident and diffracted ray (not shown) is ψ. The angle γ is the blaze angle. Figure is adapted from Allington-Smith (2000).

From the above figure we can see that, in general, the grating/groove angle γ is related to the angles of incidence i and reflection r measured with respect to the normal to the grating facet by the following relationships (with α and β the incidence and reflection measured with respect to the normal to the grating),

$$i = \alpha - \gamma \;\; and \;\; r = \gamma - \beta$$

The negative is as per the adopted convention of angles with respect to the grating normal. We may note here that for Littrow configuration, $\alpha = \beta = \gamma$. The condition for the maximum of the diffraction pattern occurs when angles of incidence and reflection become equal. This means that the grating angle which is now called the blaze angle γ_B is given by,

$$\gamma_B = \frac{1}{2}(\alpha + \beta) \tag{5.27}$$

And the grating equation now becomes,

$$\lambda_B = \frac{2d}{mM_g} \sin\gamma \ \cos\psi/2 \tag{5.28}$$

In the above equation ψ is the angle of deviation between the collimator and imager beam directions and λ_B is the blaze wavelength at which the grating would be most efficient. We can see that the blaze wavelength is dependent upon the order as well. Figure 5.9 below illustrates this.

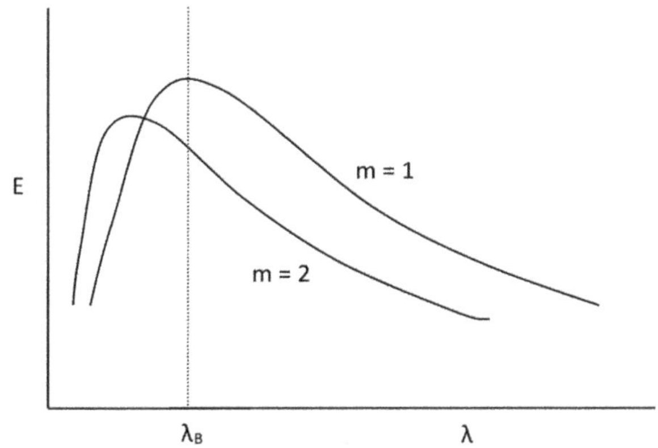

Figure 5.9: Efficiency of grating E as a function of wavelength for orders $m = 1$ and 2. Blaze wavelength is shown by the dotted vertical line. Adapted from Allington-Smith (2000) and Schroeder (2000). Not to scale.

The efficiency drops to about 40% on either side of the blaze wavelength and these wavelengths are given approximately by (Schroeder 2000),

$$\frac{2m}{m-1}\lambda_B \ \ and \ \ \frac{2m}{m+1}\lambda_B$$

The difference between the two wavelengths as given above would then decide the range of wavelengths for which a grating may be considered reasonably efficient given by λ_B/m.

It can be seen from eqn. 5.28 that the ratio of the blaze angles for general and Littrow conditions is cos ψ. Thus, Littrow condition usually renders a slightly better efficiency. The data on these parameters are normally given for Littrow condition.

5.3.3 Anamorphism: Another important aspect of the grating spectrometer which needs attention is what is known as anamorphism, a condition that arises because the collimated and imaging beams are not collinear, as shown in Figure 5.10. Consequently, for the reflection grating, except for order zero, the magnification in the two orthogonal directions is not same. The magnification of the slit given by S'/S applies to the dispersion axis in the image. The space axis perpendicular to the dispersion axis would have a different magnification leading to the anamorphism. The magnification along the dispersion axis is given by the ratio of the slit image S' and the actual slit width S, thus,

$$\frac{S'}{S} = \frac{W\ \cos\beta}{W\ \cos\alpha} = \frac{\cos\beta}{\cos\alpha} = \frac{f_{im}D_{col}}{D_{im}f_{col}} = \frac{N_{im}}{N_{col}} \tag{5.29}$$

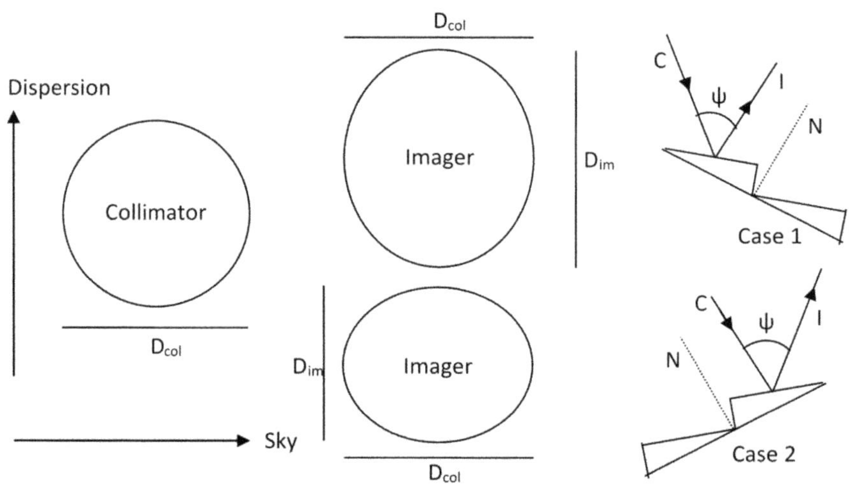

Figure 5.10: Illustration of anamorphism. The circular collimator beam can be distorted in two ways depending upon whether β is less than or greater than α as shown in the top and bottom patterns on the right side. Case 1 shows when the grating normal is towards the imaging lens while Case 2 corresponds to the configuration when the normal is towards the collimator. Notice the dilation of spectral image in case 1 and compression in case 2.

And the magnification in the spatial axis is simply the ratio of the focal lengths of the imaging and collimator optics. Quantitatively, the anamorphism may be determined by the so-called anamorphic factor given by the ratio of these magnifications and turns out to be,

$$A = \frac{D_{im}}{D_{col}} \tag{5.30}$$

As shown in the figure, the collimator beam is circular, but the spatio-spectral image pattern as imaged on the detector depends on whether the grating normal is towards the imager or towards the collimator. In the first case, the diffraction angle β is less than the incidence angle α resulting in a higher spectral resolution and larger range of wavelengths. In the second case, where the grating normal is towards the collimator $\beta > \alpha$, both the spectral resolution as well as the coverage of wavelengths are smaller. However, in this case the sampling of wavelength is better (over-sampling) than in the first case. Therefore, the designing of the spectrograph must make a proper choice. The blaze conditions would remain the same in both the configurations. The two cases represent positive and negative orders when grating is simply rotated 180^0 on its face and mounted – with grating normal towards imager taking positive order. Usually the normal-towards-imager design is preferred.

5.3.4 Band-pass: This refers to the range of wavelengths that is imaged on the detector – so the size of the detector determines this parameter along with the imaging lens aperture.

5.3.5 Overlapping Orders and Free-spectral Range: As may be seen, with increasing order number the spectra spread out (i.e., resolution increases) leading to one order overlapping on the next order and so on. This is called the spectral contamination. The overlapping orders would make the grating unusable in those orders unless a pre-filter is used to block the unwanted range of wavelengths from higher orders. The first and second orders are usually well separated while the second and third orders are partly overlapped. If the spectral lines of wavelengths λ and $\lambda + \Delta\lambda$ just coincide in two successive orders m and m+1 then we have (see Born and Wolf 1999)

$$(m + 1)\lambda = m(\lambda + \Delta\lambda)$$
$$\Delta\lambda = \frac{\lambda}{m} \tag{5.31}$$

Here $\Delta\lambda$ is called the free-spectral range (FSR), the range of wavelengths in which there is no overlapping.

5.3.6 Stray light: Gratings often suffer from stray light from the sharp edges in the groove surroundings, reflections of the diffracted beam by surfaces including that of the detector mount and optical components. This can be alleviated by using anti-reflection coated optical components. Stray light can also enter from sky. The slit can be made as narrow as possible – limited to seeing-limited or diffraction limited disc in case of large telescopes.

5.3.7 PRL Grating Spectrometers for Optical and Near-Infrared Region: At PRL, two grating spectrometers had been built one each for optical and near-infrared regions. Details of the optical spectrometer may be found in Banerjee et al (1999). A grating based spectrometer was acquired by PRL, called NICMOS3 (Nanda Kumar et al 1999) having a HgCdTe 256x256 array as the imager. Here we describe in detail the spectrometer for near-infrared region that has spectroscopic and wide-field imaging capabilities (Anandarao et al 2008a). The wide-field near-infrared camera and spectrograph (NICS) for Mt Abu 1.2 m, f/13 Cassegrain telescope of Physical Research Laboratory is shown in Figure 5.11. The design goals were to achieve seeing-limited imaging with angular resolution in an un-vignetted field of view of ~ 8'x8' with 0.5″ per pixel (of 18.5 μm) on a HgCdTe 1024x1024 infrared array, requiring a two-fold Cassegrain focal reduction. In addition to the imaging, the instrument has spectroscopic capability with a resolving power of 1000 in the 0.85 - 2.5 μm region, needing a dispersion of 1 nm per pixel.

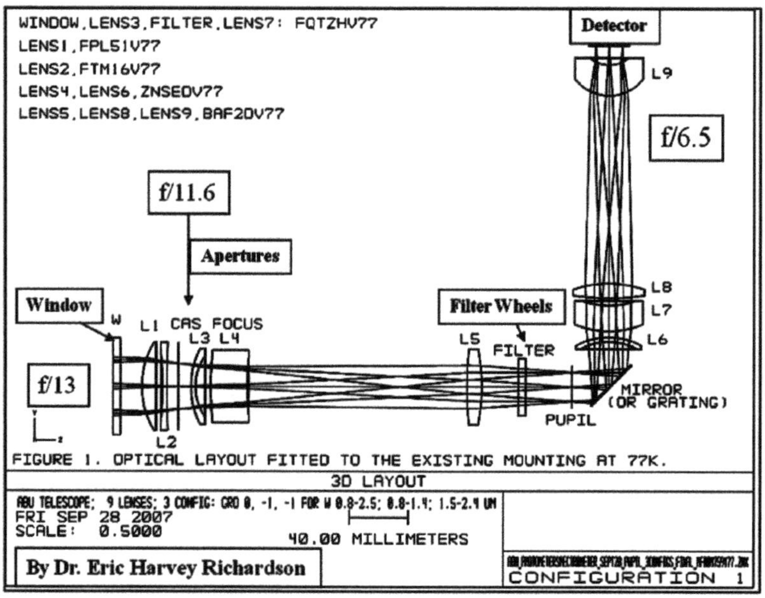

Figure 5.11: Optical Layout of a Near-infrared Spectrograph and Camera (NICS). (From Anandarao et al 2008a).

A typical first order spectrum is shown in Figure 5.12 in the H band (near-infrared). The space coordinate axis is in the horizontal direction and the wavelength is represented by the vertical axis. The bright narrow vertical patch shows the trace of spectrum from a star; the sharp horizontal lines are OH vibrational-rotational lines from the atmosphere at the height of ~ 80 km. The details of the grating spectrograph, from which the spectrum is shown here, are given in Anandarao et al (2008a).

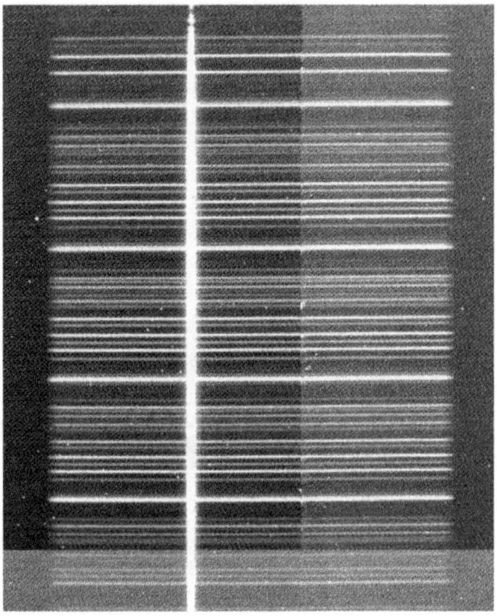

Figure 5.12: Typical trace of spectrum from a grating spectrograph – the horizontal axis shows the spatial extent, and the vertical axis is the wavelength (from Mt Abu 1.2 m IR telescope and NICS). The bright vertical trace is the spectrum from a single star; the sharp horizontal lines are OH spectra from extended atmosphere. The spectrum show is in H band (centres around 1.65 μm) of near-infrared region.

5.3.8 Echelle Grating Spectrograph: Echelle refers to diffraction gratings in which the groove spacing is made larger than in conventional ones and works at higher orders so that high resolving powers could be attained (Schroeder 2012). The groove density, ρ_g, for echelle is typically 30-300/mm while for a normal grating working in 1st or 2nd order it is 300-1200/mm in the visible range (see Schroeder 2012). The angles of incidence and diffraction (and the blaze angle) are typically made large in echelle so that the angular dispersion is large producing a large linear dispersion when the spectra are imaged on an array detector. For echelle gratings the FSR is small and hence the spectrum consists of several overlapping high orders. For small wavelength ranges we may use defining filters to separate the orders but if we allow a large range, we need a cross disperser to separate the overlapping orders. Usually, a prism is employed as a cross-disperser. At lower resolution compared to the echelle grating, the cross-disperser produces separated orders spread in a perpendicular direction to the original echelle spectrum.

As an example of an echelle spectrograph, we present here the one used for extrasolar planet detection working at a resolution of 60000 – 70000, the design of which is shown in Figure 5.13 (called PARAS, for details see Chakraborty et al 2014).

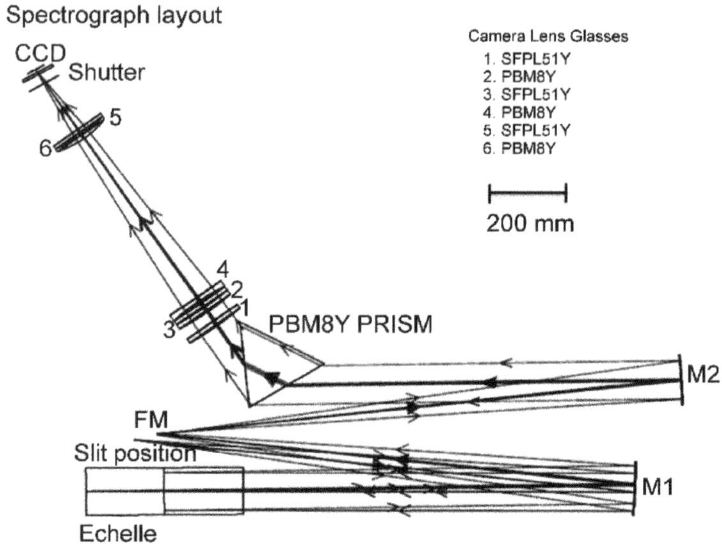

Figure 5.13: Schematic optical diagram of an echelle spectrograph (Chakraborty et al 2014). Telescope beam enters the echelle through an optical fibre and the prism is used as a cross-disperser.

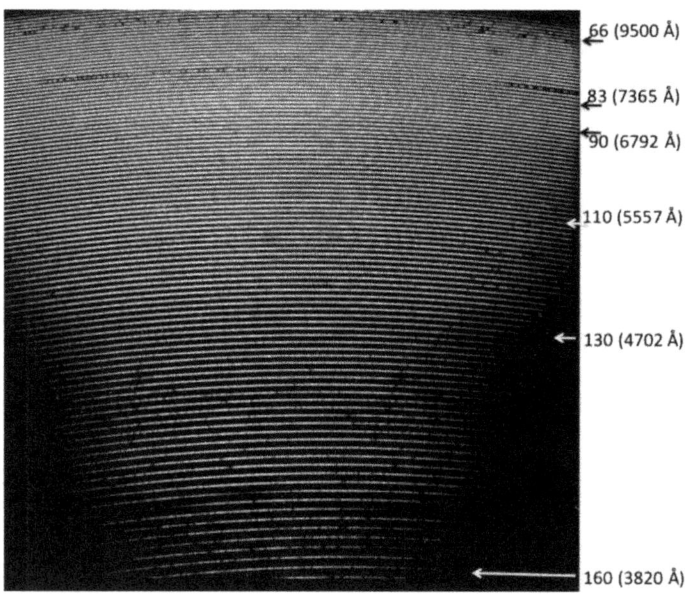

Figure 5.14a: Complete spectrum of the echelle spectrograph (PARAS)– several orders cleanly shown from lower wavelength limit (3800 Å or 380.0 nm) to higher at 9500 Å (from Chakraborty et al 2014). The spectrum is from the star 47 UMa.

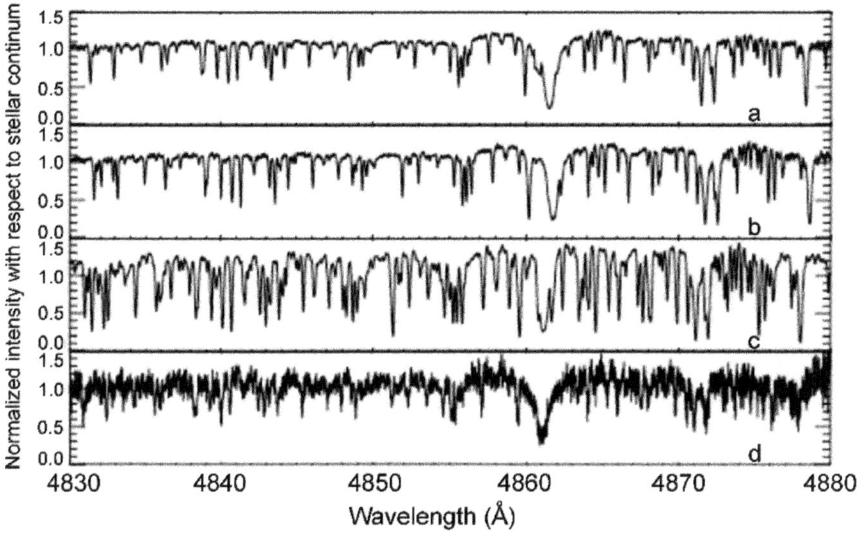

Figure 5.14b: A portion of the echelle spectrum on the stars (from top to bottom) 47 UMa (GI Va), σ Dra (K0 Vb), HD 137759 (K2 IIIc) and HD 185374 (G0 IVd). We may see the strong Hβ absorption line at 4861 Å, in all the stars. From Chakraborty et al (2014).

In PARAS echelle spectrograph shown in Figure 5.13, the star light is brought to the grating slit through an optical fibre (and a spectral reference beam from a standard source such as Argon spectral lamp by another fibre). The emerging spectrum from the grating is cross-dispersed using a glass prism as shown in the figure such that each order representing a wavelength range within the visible region 350.0-900.0 nm is separated and spread across the CCD detector as shown in the Figure 5.14a. A scan across one of the orders (after making the necessary data analysis) is shown in Figure 5.14b in a small region of 483.0-488.0 nm – we may notice the prominent absorption in Hβ line at 486.1 nm in all the stellar spectra.

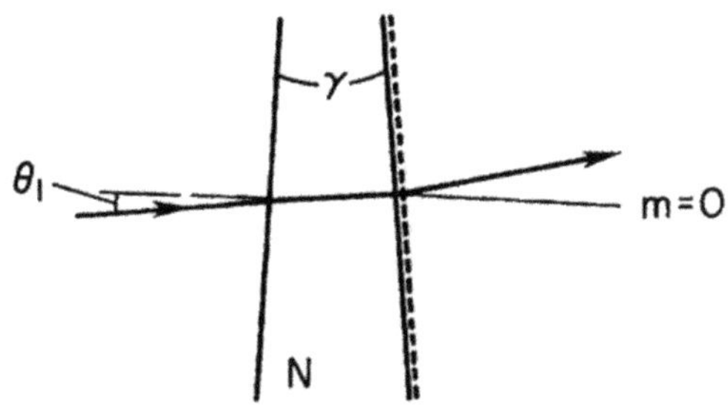

Figure 5.15: Schematic of a Grism: a combination of a prism with apex angle γ and a grating shown as a dashed line on the right (adopted from Schroeder 2000).

5.3.9 Grism Spectrograph: A grism basically contains a prism on the side of which is cemented a transmission grating as shown in the Figure 5.15; the main dispersing component being the grating. The figure shows part of a prism having an apex angle of γ and a grating shown as a dashed line parallel to the prism side on the right side.

In the case of a grism clearly the grating works in transmission mode. Assuming that the refractive indices of the prism and grating are the same, namely *n*, the grating equation gets modified as

$$m\rho\lambda = n\,\sin\alpha - \sin\beta \tag{5.32}$$

When the deviation of the incident optical beam by the prism is counteracted by that from the transmission grating the grism becomes very handy for both spectroscopy and imaging without changing the optics of the instrument. The resolving power that may be achieved is in the range of 2000-3000.

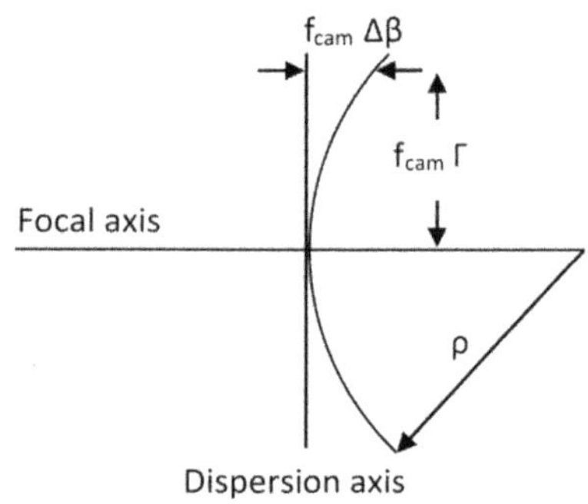

Figure 5.16: Schematic showing curvature (ρ) of grating spectral image (see text for details).

5.3.10 Image Curvature and Tilt in Grating Spectrograph: Often, in spectrographs described above namely, the grating and the prism, we find the image of the slit is not straight but curved due to the dependence of the angle of dispersion β on the angle Γ subtended by the incident ray on the plane perpendicular to the grating slit (see Figure 5.16). The tilt may be estimated by (see for details Schroeder 2000),

$$\frac{d\beta}{d\Gamma} = \frac{\sin\Gamma\,(\sin\alpha + \sin\beta)}{\cos\Gamma\cos\beta} = \lambda\,A_{ang}\tan\Gamma \tag{5.33}$$

Here A_{ang} is the angular dispersion for $\Gamma = 0$. One may show that the difference in the deviation angle between the extreme cases of zero and maximum possible $\Gamma\,(\Gamma_0)$ may derived from the above equation as,

$$f_{cam}\,\Delta\beta = \frac{(f_{cam}\Gamma_0)^2}{2r_{cur}}\,;$$

$$with\ r_{cur} = \frac{f_{cam}}{\lambda\,A_{ang}},\ as\ the\ radius\ of image\ curvature.$$

The curvature effect would be more pronounced in echelle than in normal gratings due to its dependence on the angular dispersion (being larger in the former).

5.3.11 Slit-less Spectrograph: In the foregoing paragraphs we discussed the role of the slit in the prism and grating (or grism) spectrographs in deciding the resolving power. One important variant in these spectrographs may be realised when the slit is removed from the optical arrangement of the three spectrographs, they become an imaging spectrograph at low spectral resolution. This arrangement gives such spectrographs the advantage of spectral multiplexing. This compromise is well-suited for spectroscopic survey of large areas of sky containing clusters of stars or clusters of galaxies. It not only gives spatial and spectral coverage but also a faster method to record spectra over large areas of sky. We must, however, be wary of confusing spectra that may result from overcrowded fields. This technique is also used for obtaining time-resolved solar coronagraphs. This technique is being used in many space-borne spectrographs in infrared region. James Webb Space Telescope employed this technique for integral spectroscopy of large areas of the sky using two orthogonally placed grisms for cross dispersion (Greene et al 2017; Willott et al 2022). The JWST near-infrared camera slit-less grism-based wide-field spectrograph offers a low–medium resolving power in 0.8-2.3 and 2.5-5.0 μm regions (Greene et al 2017). In addition, JWST uses single-object slit-less spectrograph for taking time series observations on exoplanet transit spectroscopy in the near-infrared region (Albert et al 2023).

A typical image of a slit less spectrograph is shown in the Figure 5.17 (from Pasqualis et al 2006). Figure shows several orders around the zeroth order of the slit-less grism spectra from the advanced camera of Hubble Space Telescope (see for details Pasqualis et al 2006). We discuss techniques for ground-based integral field spectroscopy at high resolution later in this chapter.

5.4 Two-Beam Interferometry: Michelson Interferometer

When two quasi-monochromatic light beams are made to interfere, a bright and dark fringe pattern appears based on the path difference and hence the phase difference between the two interfering beams. In the two-beam interferometry, two possibilities need to be considered – interference by the division of wavefront and by the division of amplitude. First let us consider classical interference by division of wavefronts.

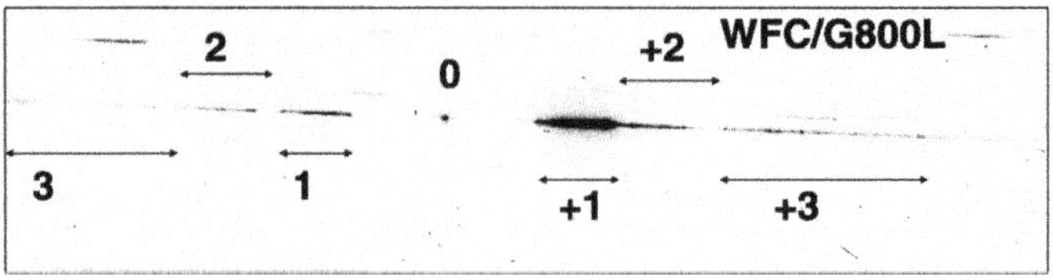

Figure 5.17: Image (1k x 1k pixels) of low-resolution slit-less spectra from the grism with the wide-field advanced camera of HST (from Pasquali et al 2006). The numbers indicate orders of the grism – from negative 3, 2, and 1 via 0th order to positive +1, +2 and +3 orders. The object imaged is the Wolf-Rayet star WR 96. The dispersion in the first order is 2.4 nm per pixel. The colinear streaks in the middle of the image are from the program star; the other streaks from other objects in the FoV.

Young was among the first to demonstrate the two-beam interference by his famous experiment in which a monochromatic beam of light was passed through two pin holes equidistant from the source of light as illustrated in Figure 5.18 below. As shown in the figure, monochromatic light rays from source S are passed through two pin holes S_1 and S_2 separated by a distance $d$whereby the two beams emerging from the holes interfere as shown in the figure. In this arrangement, two coherent light beams are created, the phase difference between which remains constant (Jenkins & White 1989; Born & Wolf 1999). This is the process of interference by the division of wavefront. The phase difference between the interfering beams generated by division of wavefront at the point P is given by,

$$\delta = \frac{2\pi\mu x d}{\lambda_0 a} \tag{5.34}$$

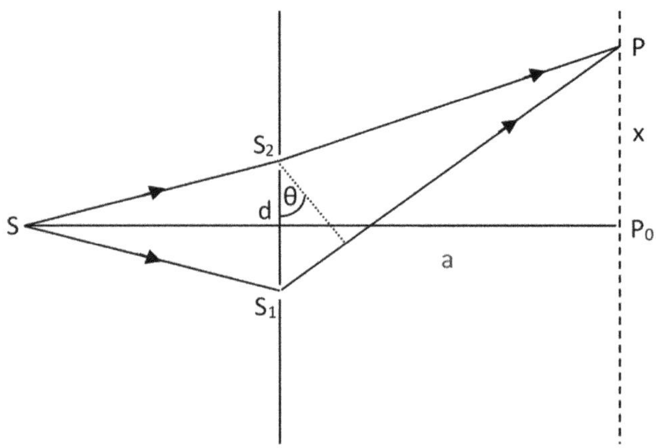

Figure 5.18: Young's interference experiment with two apertures S_1 and S_2 separated by d. The monochromatic light source is S. The beams shown on the right interfere to produce bright fringes if the phase difference is 2π.

Figure 5.19: Interference pattern in Young's double-slit experiment. (from Jenkins & White 1989).

The refractive index of the medium in which the interference is occurring is μ; x is the coordinate (in the orthogonal coordinate system $[x, y]$) on the screen and a is the distance between the pinhole apertures S_1 and S_2, and the screen. The monochromatic wavelength is λ_0. The order of interference m is given by,

$$m = \frac{\delta}{2\pi} = \frac{\Delta S}{\lambda_0} \tag{5.35}$$

The optical path difference is denoted as ΔS (μ times the geometrical path difference). All integer values of m correspond to maxima and half-integral values to minima. The resulting straight-line fringes run in the direction perpendicular to the line joining the two apertures (S_1S_2) at the image plane, as shown in Figure 5.19. The separation between any two consecutive bright fringes may be shown to be equal to $(a\lambda_0/\mu d)$.

Let us now discuss the interference by the division of amplitude. In Young's experiment we have considered monochromatic light; while we consider here what is called a quasi-monochromatic light that satisfies the condition (see Born and Wolf 1999 for details)

$$\frac{\Delta\lambda_0}{\lambda_0} \ll 1 \tag{5.36}$$

Here $\Delta\lambda_0$ is the wavelength range around the central wavelength λ_0 considered for the light source.

Now let us consider a plane parallel plate of thickness d and refractive index μ as shown in Figure 5.20 below. A quasi-monochromatic beam of light from a point source falls on the parallel plate at an angle of incidence θ'. A part of the beam is reflected at the point A and another part is refracted into the plate and gets reflected at the inner side of the plate and comes out of the plate at the point B. The two emerging beams undergo interference, and the interference pattern is seen at infinity (or localised at infinity). We note here that in the case of Young's experiment the two holes create division of wavefront, while here in the case of the parallel plates, division of amplitude occurs. The interference pattern is brought to focus by using a converging lens as shown in the figure. The optical path difference between the two beams is given by the paths AD for the reflected beam and ABCE for the refracted beam: $2\mu d \cos\theta$. Then the phase difference is given by

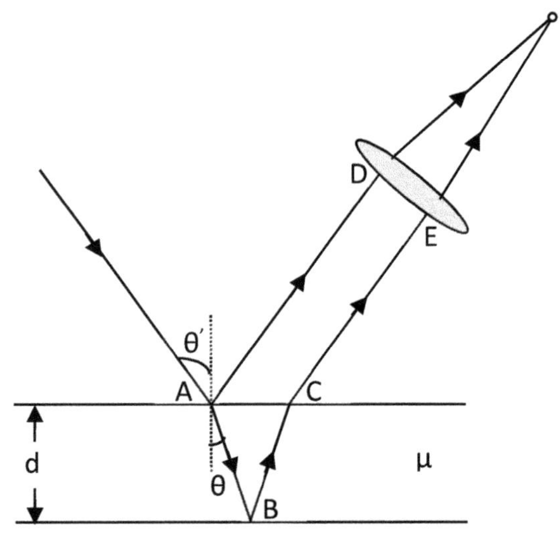

Figure 5.20: Two beam interference in a parallel plate of thickness d and refractive index μ.

$$\delta = \frac{2\pi}{\lambda_0} 2\mu d \cos\theta \pm \pi = \frac{4\pi}{\lambda_0} \mu d \cos\theta \pm \pi \quad (5.37)$$

The central wavelength of the beam is given by λ_0 and term $+/-\pi$ accounts for phase change at the reflection within the plate. Now for bright fringes the phase difference must be integer multiples of 2π, thus we get the condition for the order of interference m,

$$2\mu d \cos\theta \pm \frac{\lambda_0}{2} = m\lambda_0 \,; m = 0, \pm1, \pm2 \,.. \quad (5.38)$$

For dark fringes the phase must be half-integer multiples of 2π and hence m takes values of 1/2, 3/2, 5/2, and so on in the equation above. The concentric circular fringe pattern that we obtain at the focal plane is called fringes of equal inclination – each circular fringe corresponds to one particular or specific angle of inclination i or refraction θ. The central fringe corresponding to θ = 0, satisfies the equation,

$$2\mu d \pm \frac{\lambda_0}{2} = m_0\lambda_0 \ ; \ m_0 = m_1 + e \tag{5.39}$$

Here m_0 is often not an integer, and e is called the central fractional order. For the nth bright fringe from the centre at angle θ_n we have the equation,

$$2\mu d \cos\theta_n \pm \frac{\lambda_0}{2} = m_n \lambda_0 \tag{5.40}$$

Writing $m_n = [m_1 - (n-1)]$ we can write using the equation for central fringe

$$2\mu d(1 - \cos\theta_n) = (n - 1 + e)\lambda_0 \tag{5.41}$$

For small angles we can write down the equation for the angular radius of nth fringe as,

$$\theta_n = \theta_n' = \left[\frac{\mu\lambda_0}{d}(n - 1 + e)\right]^{1/2} \tag{5.42}$$

This shows that the angular radii of the circular fringes are proportional to $(1/d)^{1/2}$ and if $e = 0$ (central bright fringe) they are also proportional to square root of (positive) integers. Sometimes the gap between the two surfaces of the parallel plate is filled with air so that the refractive index taken as 1.

It may be noted that the classical Newton's rings are fringes of equal thickness in contrast to the fringes of equal inclination discussed here.

If the source of light is an extended source that can be imagined to be constituted by a large number of point sources so that it can be considered continuous, then at the focal plane of the imaging lens we find uniform light (see Born and Wolf 1999 for details).

5.4.1 Michelson Interferometer: Michelson interferometer is based on two-beam interferometry by the division of amplitude. The typical optical diagram is shown (see Born and Wolf 1999) in Figure 5.21.

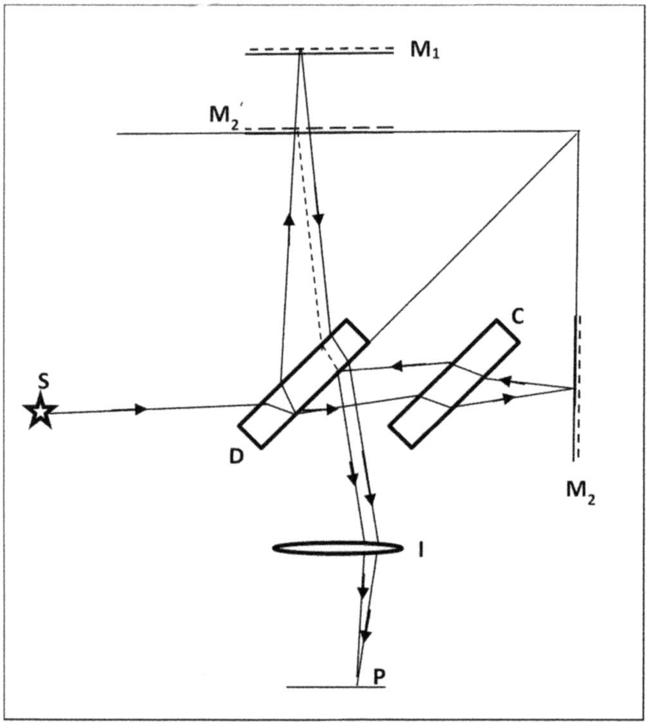

Figure 5.21: Michelson Interferometer optical arrangement (taken from Born and Wolf 1999).

Light rays from a quasi-monochromatic extended source S pass through the glass plate D coated on the backside for reflection (A). (In Young's experiment the source is a point source, and, in principle, an extended source may be considered as composed of many incoherent point sources. Thus, the treatment in the case of extended sources is quite complex.) Part of the beam is reflected into the plate and reaches a plane mirror M_1 while the transmitted part of the beam reaches another plane mirror M_2 kept perpendicular to the first mirror. The two beams are returned by the mirrors back to the plate D and then they interfere to give fringes localised at infinity and brought to focus P through the lens I. The reflected beam travels through D thrice while the transmitted beam only once. To compensate for this path-difference, another glass plate C, identical in material and thickness to D, is kept parallel to the latter as shown in the figure. This plate is called compensator. The path difference between the two parallel interfering beams can be varied by moving M_1 mechanically or by using an electronically controlled transducer. A change of path length by $\Delta m\lambda_0$ corresponds to a displacement of the fringe pattern by Δm orders. The back-silvered plate D forms a virtual image (M_2') of the mirror M_2 parallel to the movable mirror M_1 as shown in the figure. So, with an air gap d between M_1 and M_2' the arrangement resembles that of the two-beam interferometry.

In what follows, we discuss various conditions under which we may use the instrument.

5.4.1.1 Circular Fringes: These are fringes of equal inclination and may be seen at the focal plane (point) P, if the mirrors M_1 and M_2 are exactly perpendicular to each other in which case the mirror M_1 and the virtual image of M_2 (M'_2, see Figure 5.21) represent the parallel plate as discussed earlier.

5.4.1.2 Localised Fringes: If M_1 and M'_2 are not parallel but form a wedge, we obtain straight line fringes for a monochromatic light. These are called localised fringes. These are perfectly straight for separation $d = 0$.

5.4.1.3 White Light Fringes: If the source is white light (poly-chromatic) then we can see no fringes except when the path difference does not exceed a few wavelengths (see Jenkins and White 1981). If the separation $d = 0$ rays of all wavelengths coincide at the centre. As the path length changes rays of different wavelengths start separating from each other.

5.4.1.4 Visibility of Fringes: This defines the contrast or sharpness of fringes. If I_{ma} and I_{mi} are the maximum and minimum intensity of the fringes, then visibility is defined as,

$$V = \frac{(I_{ma} - I_{mi})}{(I_{ma} + I_{mi})} \tag{5.43}$$

By suitably adjusting path difference (depending upon wavelength) V can be made maximum so that the fringes become sharper. A plot of the intensity as a function of path-difference is called visibility curve. If the line has fine structure as we adjust path difference V shows fluctuation. For poly-chromatic light, the separation and intensity of the wavelength components need Fourier analysis of the visibility curve. In the early days of Michelson interferometry several variants of optical arrangement based on Michelson's original idea had been developed. The instrument had been used in the visible region for determining fine structure in a spectral line, lengths, and refractive indices of materials. With the advent of powerful panoramic infrared detectors, the interest in Michelson interferometry has been revived for astronomical high-resolution spectroscopy.

If the path difference between the two arms of Michelson interferometer (see Figure 5.21) is $\Delta Z (= \Delta\lambda/2\pi$, Δ being the phase difference) and taking the peak intensity of the light at wavelength λ as $I_0(\lambda)$, the intensity $I(\Delta Z, \lambda)$ at any point on the imaging lens focal plane may be written as,

$$I(\Delta Z, \lambda) = K\, I_0(\lambda)\left[1 + \cos\left(\frac{2\pi\Delta Z}{\lambda}\right)\right] \tag{5.44}$$

where K is a constant that accounts for reflection and transmission losses at different components. If the phase difference is 0 or integer multiples of 2π (or path differences of zero or integral multiples) then the intensity $I(\lambda)$ $(= K I_0(\lambda))$ would be maximum, while for phases of π and integral multiples of π, the intensity is zero or minimum (or path differences of half integer multiples). The first term on the right side of the equation represents a constant equal to the mean intensity of the light through the interferometer and may even be dropped from further treatment of the problem, and we may deal with the second term that depends on phase difference.

Typical intensity pattern for quasi-monochromatic light as a function of phase effected by moving the mirror M_1 is shown in Figure 5.22 below.

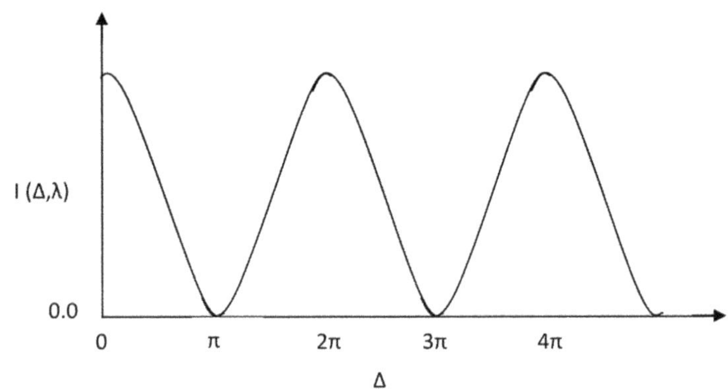

Figure 5.22: Intensity pattern in Michelson Interferometer as a function of phase difference Δ.

If the light is bi-chromatic then we find the superposition of the two patterns as shown in Figure 5.23, as a plot of the output intensity being I' with path difference ΔZ.

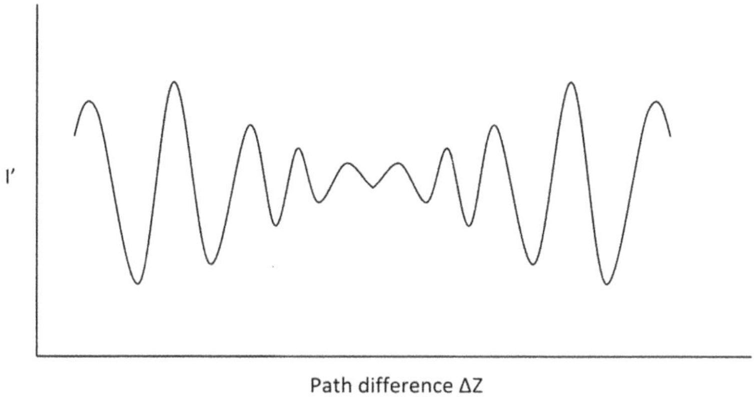

Figure 5.23: Schematic pattern for bi-chromatic light in a Michelson interferometer (not to scale; based on Kitchin 2003).

5.4.2 Fourier Transform Spectrometer: Now let us consider the poly-chromatic case and write down the total intensity of light at the image plane by integrating the equation 5.44 for all wavelengths for a given path length.

$$I(\Delta Z) = K\int_0^{\infty} I_0(\lambda)\cos\left(\frac{2\pi\Delta Z}{\lambda}\right)d\lambda \qquad (5.45)$$

This therefore represents the variations over the average light intensity that we chose to drop from the equation 5.44. The same equation may be written in terms of frequency of light ray instead of its wavelength as

$$\begin{aligned} I(\Delta Z) &= K'\int_0^{\infty} I_0(\nu)\cos\left(\frac{2\pi\Delta Z\nu}{c}\right)d\nu \\ &= \frac{K'}{2}\int_{-\infty}^{\infty} I_0(\nu)\cos\left(\frac{2\pi\Delta Z\nu}{c}\right)d\nu \end{aligned} \qquad (5.46)$$

The second equality of the above equation (with limits of integration from -∞ to +∞) is possible because we have taken $I(-\nu) = I(\nu)$. Now we can understand that the spectral information at the image plane of Michelson interferometer represents, and is embedded in, the real half of the Fourier transform of the spectral intensity distribution. If $F(\nu)$ is the Fourier transform of a function $f(t)$ (e.g., ν and t be frequency and time respectively) we can write,

$$F(\nu) = \int_{-\infty}^{\infty} f(t) e^{-2\pi i \nu t}\, dt \tag{5.47}$$

And the inverse Fourier transform as

$$f(t) = \int_{-\infty}^{\infty} F(\nu) e^{2\pi i \nu t}\, d\nu \tag{5.48}$$

Using this we may write the spectral energy distribution or simply the spectrum as

$$I(\nu) = 2K'' \int_{0}^{\infty} I\left(\frac{2\pi\Delta Z}{c}\right) \cos\left(\frac{2\pi\Delta Z\nu}{c}\right) d(\Delta Z) \tag{5.49}$$

The spectrum may be obtained, in practice, by scanning the interferometer in path difference ΔZ. The Fourier inverse transform needs to be obtained using discreet (with finite limits) fast Fourier transform algorithms. This scanning Michelson interferometer often called Fourier Transform Spectrometer (FTS) is used for infrared region quite effectively and efficiently (Connes & Michel 1975; Maillard & Michel 1982).

The resolution of FTS may be obtained using the Rayleigh criterion thus,

$$\delta\lambda = \frac{\lambda^2}{2\Delta Z} \tag{5.50}$$

If the maximum scanning length (the maximum path that the mirror M_1 can traverse) is Z, then we may write the resolving power as

$$R = \frac{\lambda}{\delta\lambda} = \frac{2Z}{\lambda} \tag{5.51}$$

Thus, the larger the scanning length the higher is the resolving power. For Z of the order of meters R can exceed 10^5 making FTS as one of the most powerful high-resolution spectrometers in the near-infrared region. In the visible region, FTS may exceed one million in resolving power for path lengths of a couple of meters.

The scanning intervals in wavelength, N, when the range is $\Delta\lambda$ $(\lambda_2 - \lambda_1)$ is given by,

$$N = \frac{\Delta\lambda}{\delta\lambda} \tag{5.52}$$

Since the FTS involves both positive and negative regions, we need to have two times this number. The sampling length interval δX itself is obtained by $\sim \lambda^2/\Delta\lambda$.

For high resolution spectroscopy with FTS in the visible wavelength range 500-550 nm requires 1 μm scan intervals while for near-infrared range of 2000-2050 nm the scan interval is 20 μm. Considering the demand on accuracy for a practical spectrometer FTS is certainly better suited for longer wavelength (infrared) region. But with the advent of large array detectors even in the infrared region the more conventional grating and echelle grating spectrographs are more commonly used rather than FTS that demands uncompromising accuracies. Further the instrumental function (the response of a spectrometer to a very narrow spectral line approximated by a delta function) is of the form $(\sin\psi)/\psi$ which has significant secondary maxima as shown in Figure 5.24. Weighting or more precisely, the apodizing functions are judiciously used to alleviate the problem at the cost of effective resolving power (see Naylor & Tahic 2007 and the references there in).

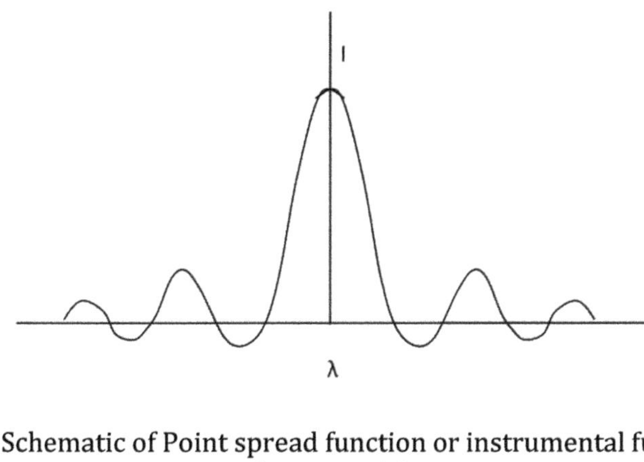

Figure 5.24: Schematic of Point spread function or instrumental function of FTS (not to scale).

5.5 Multiple-Beam Interferometer: Fabry-Perot Interferometer

By division of amplitude, we may create multiple beams from a single source. The interference fringes so produced are called the fringes of equal inclination. Multiple beam interferometry basically involves a parallel-sided plate of thickness d and of refractive index μ'. The refractive index of the medium around the plate is μ. The inner sides are coated for high reflectivity. Light beam from a (quasi-monochromatic) source S falls at angle θ (to the normal of the plate) on the topside of the plate and undergoes multiple reflections and refractions (at angle θ') as well as transmissions as shown in Figure 5.25 below. The transmitted beams ($S12T_1$, $S1234T_2$ etc) are parallel to each other and interfere to form circular fringes localised at infinity. Similarly, the reflected beams ($S1R_1$, $S123R_2$ etc) are also parallel and undergo interference to form fringes. These fringes are called fringes of equal inclination as in the case of two-beam interferometry with division of amplitude. The rays from point S on the extended source at an incident angle θ would be imaged at the corresponding point P on the circular fringe on the detector D (see Figure 5.25).

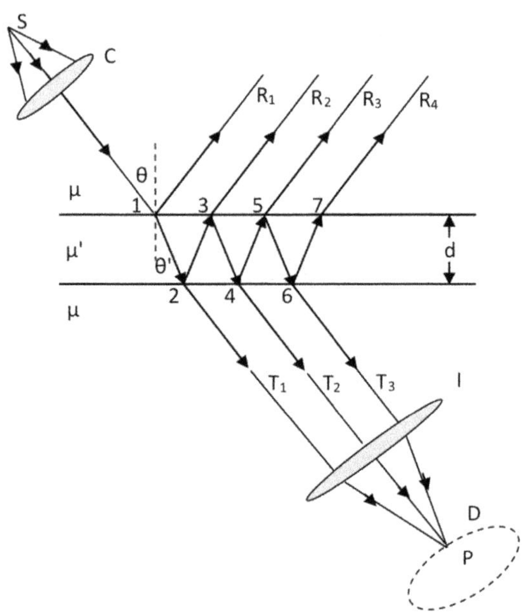

Figure 5.25: Principle of Multiple-beam interferometry: Fabry-Perot interferometer. D is the detector and point P on the fringe corresponds to point S at the extended source.

As shown in the case of two-beam interferometry the phase difference between any two consecutive reflected or transmitted beams (for a central wavelength λ_0) is given by,

$$\delta = \frac{4\pi}{\lambda_0}\,\mu' d\,\cos\theta' \tag{5.53}$$

Let us now compute the amplitude and intensity of reflected and transmitted beams. Let *r* and *t* represent the coefficients of reflection and transmission respectively (which are simply the ratios of reflected/transmitted amplitude to the incident amplitude) for the ray travelling from outer medium (usually air) on to the parallel plate. Let the corresponding coefficients be *r'* and *t'* for the ray that travels from the parallel plate on to the outer medium. These physical parameters are related thus,

$$tt' = \mathcal{T};$$

$$r = -r';$$

$$rr' = r^2 = r'^2 = \mathcal{R};$$

$$\mathcal{T} + \mathcal{R} + \mathcal{A} = 1 \tag{5.54}$$

Here $\mathcal{R}$, $\mathcal{T}$, $\mathcal{A}$ are reflectivity, transmissivity and absorptivity respectively. We assume that absorptivity is negligibly small for surfaces coated with highly reflecting materials. If the complex amplitude of the electric field associated with the incident wave is A^i, the complex amplitudes of the reflected waves are given by,

rA^i, $tt'r'A^ie^{i\delta}$, $tt'r'^3A^ie^{i2\delta}$ ..... $tt'r'^{2n-3}A^ie^{i(n-1)\delta}$,for rays $S1R_1$, $S123R_2$, $S12345R_3$ etc (See Figure 5.25).

Similarly for the transmitted complex amplitudes we have,

$tt'A^i$, $tt'r'^2A^ie^{i\delta}$, $tt'r'^4A^ie^{i2\delta}$ ..... $tt'r'^{2(n-1)}A^ie^{i(n-1)\delta}$ for rays $S12T_1$, $S1234T_2$, $S123456T_3$ etc.

Here *n* is the number of reflected (or transmitted) waves. If n reflected waves are superposed for interference, then the combined amplitude of the electric field of the reflected rays can be given by

$$A^r(n) = A^i[r + tt'r'e^{i\delta}\left(1 + r'^2e^{i\delta} + \ ... + r'^{2(n-2)}e^{i(n-2)\delta}\right)]$$

The terms in the parentheses in the above equation are a geometric (infinite) series. In the limit as *n* tends to infinity, we can write the above equation as

$$A^r(n) = A^r(\infty) = \ -A^i\,\frac{r'\left[1 - (r'^2 + tt')e^{i\delta}\right]}{1 - r'^2e^{i\delta}}$$

Using the relations for reflectivity and transmissivity we can write the following equation,

$$A^r = A^i \frac{(1 - e^{i\delta})\mathcal{R}^{1/2}}{1 - \mathcal{R}e^{i\delta}} \tag{5.55}$$

The intensity of the reflected light can then be written as

$$\begin{aligned} I^r &= A^r A^{r*} \\ &= I^i \frac{(2 - 2\cos\delta)\mathcal{R}}{1 + \mathcal{R}^2 - 2\mathcal{R}\cos\delta} \\ &= I^i \frac{4\mathcal{R}\sin^2\frac{\delta}{2}}{(1 - \mathcal{R}^2) + 4\mathcal{R}\sin^2\frac{\delta}{2}} \end{aligned}$$

Here the intensity of the incident light is given by $I^i = A^i A^{i*}$.

Similarly, we can write down the transmitted amplitude as

$$\begin{aligned} A^t(n) &= tt'A^i\left(1 + r'^2 e^{i\delta} + r'^4 e^{i2\delta} + \ldots + r'^{2(n-1)} e^{i(n-1)\delta}\right) \\ &= A^i tt' \left[\frac{1 - r'^{2n} e^{in\delta}}{1 - r'^2 e^{i\delta}}\right] \end{aligned}$$

In the limit as n tends to infinity this reduces to

$$A^t(n) = A^t(\infty) = A^i \frac{\mathcal{T}}{1 - \mathcal{R}e^{i\delta}}$$

Then the intensity of the transmitted light can be written as

$$I^t = I^i \frac{\mathcal{T}^2}{1 + \mathcal{R}^2 - 2\mathcal{R}\cos\delta} = I^i \frac{\mathcal{T}^2}{(1 - \mathcal{R})^2 + 4\mathcal{R}\sin^2\frac{\delta}{2}} \tag{5.56}$$

The expressions for the reflected and transmitted light are called the Airy function. It is a periodic function with phase δ. The maxima and minima are given by,

$$m\lambda_0 = \frac{\lambda_0\delta}{2\pi} = 2\,\mu' d\cos\theta' \tag{5.57}$$

For integral values of $m = 1, 2, 3, \ldots$, we get maxima and for half integral values of $m = 1/2, 3/2, 5/2, \ldots$, minima.

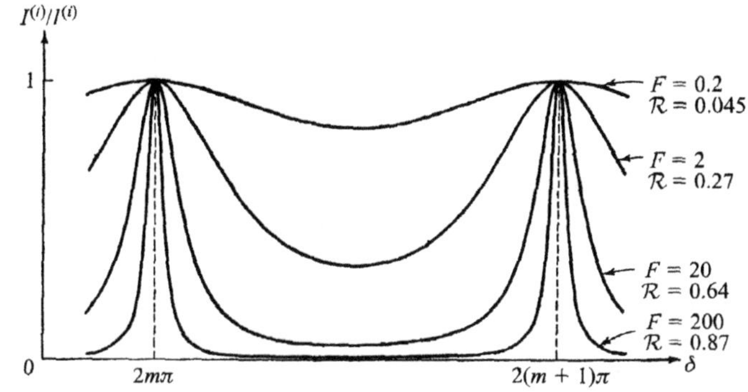

Figure 5.26: Transmitted intensity as a function of phase – the Airy function. It is also shown as a function of reflectivity (Born and Wolf 1999).

Figure 5.26 shows the Airy function for transmission normalised to the incident intensity. At the maxima we can see the transmission is equal to the incident intensity provided absorption loss is negligible. The Airy function depends also on the reflectivity as can be seen in Figure 5.26. As the reflectivity is increased, the peak of the Airy function becomes narrower and narrower. On the imaging device we see a system of circular fringes of equal inclination. Let us define,

$$\mathcal{F} = \frac{4\mathcal{R}}{(1-\mathcal{R})^2} \tag{5.58}$$

We can then write the Airy function as,

$$\frac{I^r}{I^i} = \frac{\mathcal{F}\sin^2\frac{\delta}{2}}{1+\mathcal{F}\sin^2\frac{\delta}{2}} \tag{5.59}$$

$$\frac{I^t}{I^i} = \frac{1}{1+\mathcal{F}\sin^2\frac{\delta}{2}} \tag{5.60}$$

It may be seen that since $\mathcal{R}$ is less than 1, so would be the second term in the denominator of eqn. 5.60, and hence, we may approximately write,

$$\frac{I^t}{I^i} \approx 1 - \mathcal{F}\,\sin^2\frac{\delta}{2} = 1 - \frac{\mathcal{F}}{2}\,(1 - \cos\delta) \tag{5.61}$$

The intensity of the transmitted light thus depends on $\mathcal{F}$ and as it increases to approach unity the second term in the above equation decreases and the fringes become very sharp against a dark background.

Since $I^r + I^t = I^i$, ignoring losses due to absorption and other factors, the transmitted and reflected patterns are complimentary (see equations 5.59 and 5.60). Thus, in the reflected light we see sharp dark fringes in a bright background.

5.5.1 Fabry-Perot Interferometer: Fabry-Perot Interferometer or spectrometer (FPI or FPS) is a product of one of the most innovative and apparently humble discovery by Charles Fabry and Alfred Perot at the end of 19th century. As discussed in detail in the preceding section, Fabry-Perot Interferometer is perhaps the most fascinating scientific instrument being applied to varied fields of research for over a century. Its applications are very wide-ranging from astrophysics to laser physics. The on-going technological improvements spread over several decades speak volumes of the potential of this instrument.

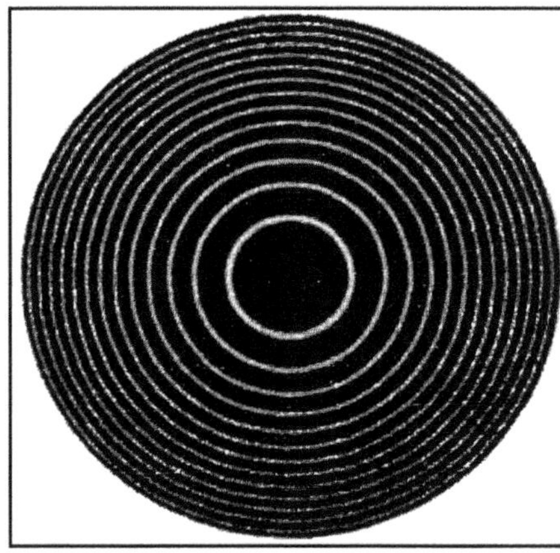

Figure 5.27: Transmitted circular fringe pattern in a Fabry-Perot interferometer (Born and Wolf 1999).

We discuss in the following sections, the actual instrumentation for spectroscopic measurements using a Fabry-Perot interferometer. As is the case for all the spectrometers in general, we use a collimator for making a parallel beam of the incident light and the transmitted beam through the system of parallel plates of Fabry-Perot interferometer is brought to focus on a detector by using an imaging lens. The system of parallel plates is known as an **etalon**. It is made of a glass substrate coated for very high reflectivity. The transmitted circular fringes of equal inclination are shown in Figure 5.27.

Let us now derive some merit factors relating to the FPS.

5.5.1.1 Reflection Finesse: The reflection finesse of an etalon is its defining parameter. It represents the ratio of the gap between two maxima and the half-width of a maximum. Let us now derive this parameter.

For a fringe of order m, the intensity would drop to half of the maximum on either side at

$$\delta = 2\pi m \pm \frac{\varepsilon}{2} \tag{5.62}$$

with ε as the phase required for the intensity to drop to half its maximum. We can see that this happens when the second term in the denominator of the equation for transmitted intensity becomes unity. Thus

$$\frac{1}{\left(1+\mathcal{F}\,\sin^2\frac{\varepsilon}{4}\right)}=\frac{1}{2} \tag{5.63}$$

Therefore, for small angles we have the condition,

$$\varepsilon=\frac{4}{\mathcal{F}^{1/2}} \tag{5.64}$$

The larger the $\mathcal{F}$, the smaller would be the fringe width. Since two adjacent fringes require a change of phase of 2π, expressing finesse, by definition, as the ratio of phase changes we have,

$$F=\frac{2\pi}{\varepsilon}=\frac{\pi\mathcal{F}^{1/2}}{2}=\frac{\pi\mathcal{R}^{1/2}}{(1-\mathcal{R})} \tag{5.65}$$

This parameter F is called finesse that characterises the interferometer. Physically finesse represents the number of interfering beams indicating that the larger the number the better will be the sharpness of the peaks. Finesse as defined above is a function of reflectivity and hence the parallel surfaces need to be coated for very high reflectivity and hence is called the reflective finesse. The actual or effective finesse is smaller than the reflective finesse due to various external factors which we shall discuss later.

5.5.1.2 Resolving Power: If suppose two wavelengths λ and $\lambda+\delta\lambda$ are to be resolved the two fringes formed by these wavelengths must cross each other at half-intensity point. This demands that in the transmitted intensity given above in the denominator the second term should be unity. As seen above in the finesse treatment this condition is satisfied if the phase change for full width at half maximum of a single peak is given by,

$$\varepsilon=\frac{2(1-\mathcal{R})}{\mathcal{R}^{1/2}} \tag{5.66}$$

Now from the phase difference equation we have by differentiation,

$$d\delta = -\frac{4\pi d}{\lambda}\sin\theta\, d\theta$$

Also, from the condition for maxima we have by differentiation,

$$m\, d\lambda = -2d\, \sin\theta\, d\theta$$

Combining the above three equations we get the resolving power R as,

$$R = \frac{\lambda}{\delta\lambda} = m\,\frac{\pi\mathcal{R}^{1/2}}{1-\mathcal{R}} = mF \tag{5.67}$$

Thus, resolving power turns out to be the product of order number and finesse. Alternatively, since $m = 2d/\lambda$ it also depends directly on the gap and inversely on wavelength. For a given gap shorter wavelengths have higher R than longer wavelengths.

5.5.1.3 Free-Spectral Range: The gap in wavelength between two consecutive orders is called free-spectral range (FSR). This can be derived as follows.

Suppose for a wavelength λ_1 the corresponding angle of formation of circular fringe is θ_1 so that we can write,

$$2\mu' d\cos\theta_1 = m\lambda_1$$

Now at the next order for the same wavelength satisfies the equation,

$$2\mu' d\cos\theta_2 = (m-1)\lambda_1$$

Suppose that another wavelength λ_2 separated from λ_1 by an amount $\Delta\lambda$ falls exactly at the order (m - 1) of the latter wavelength (for a multi-chromatic light source), this simply means that the wavelengths λ_1 and λ_2 are separated by one order. We can then write,

$$2\mu' d\cos\theta_2 = m(\lambda_1 - \Delta\lambda)$$

Now combining the above three equations we can get an equation for $\Delta\lambda$ as,

$$\Delta\lambda = \frac{\lambda_1^2}{2\mu' d\cos\theta'} \tag{5.68}$$

This quantity is called free-spectral range (FSR) and for small angles and air-gap etalons we have,

$$\mathrm{FSR} = \Delta\lambda = \frac{\lambda^2}{2d} \tag{5.69}$$

This is an important parameter for choosing an appropriate FP etalon for a specific purpose. As we can see the smaller the gap d the larger would be the free-spectral range. FSR is also a strong function of wavelength for a given gap. Figure 5.28 illustrates the two parameters of free spectral range and the resolution element.

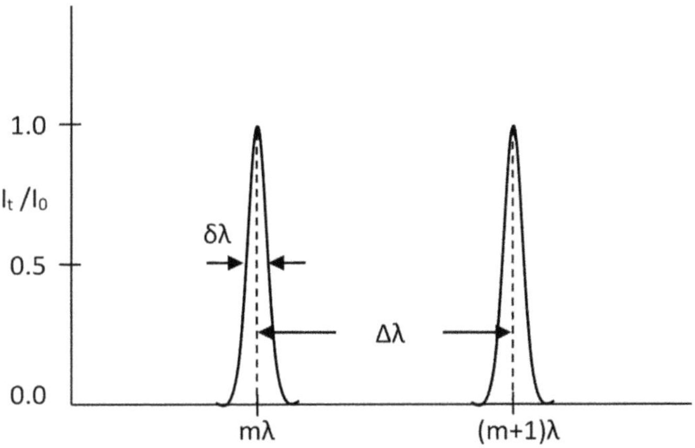

Figure 5.28: Normalised transmitted intensity with fringe orders. The free-spectral range (FSR) $\Delta\lambda$ and resolution $\delta\lambda$ are defined.

5.5.1.4 Fringe Contrast: Fringe contrast is the ratio of maximum to minimum transmittance. This is given by,

$$C = \frac{\mathcal{T}_{max}}{\mathcal{T}_{min}} = \frac{(1+\mathcal{R})^2}{(1-\mathcal{R})^2} \tag{5.70}$$

Thus, the fringe contrast increases as R increases.

5.5.1.5 Fractional Order: As mentioned earlier in the treatment of two-beam interferometry, in this case too there would be a phase change due to reflection and needs to be accounted for. Hence the condition for maxima becomes now,

$$m + \frac{\phi}{\pi} = \frac{\delta}{2\pi} = \frac{2\mu' d \cos\theta'}{\lambda_0} \tag{5.71}$$

Therefore, at the central fringe (θ′ = 0) the order is given by,

$$m_0 = \frac{2\mu' d}{\lambda_0} + \frac{\phi}{\pi} = m_1 + e \tag{5.72}$$

This shows that m_0 is generally not an integer but a fraction due to the presence of the phase change. The quantity e is then called the fractional order for the central fringe. We may estimate the value of the fractional order by measuring the fringe diameters as shown below.

5.5.1.6 Radii of Fringes: Again, as in the two-beam interferometry we can derive the angular radius of any circular fringe. For p^{th} fringe let us write down the angular radius (for small angles) as,

$$\theta_p = \left(\frac{\mu' \lambda_0}{d}\right)^{1/2} (p - 1 + e)^{\frac{1}{2}} \tag{5.73}$$

Here we have considered the refractive index for the environment as 1 (for air). For air-gap etalons μ'will also become 1. If the focal length of the imaging lens is f_{im} we can obtain the linear radius of the fringe as,

$$r_p = f_{im}\, \theta_p$$

Determining the diameters of two rings we can estimate the fractional order as follows. For the pth and q^{th} ring the ratio of their radii may be written as,

$$\frac{r_p^2}{r_q^2} = \frac{(p - 1 + e)}{(q - 1 + e)} \tag{5.74}$$

Therefore, the fractional order may be written as

$$e = \frac{(q - 1) r_p^2 - (p - 1) r_q^2}{(r_q^2 - r_p^2)} \tag{5.75}$$

We can obtain radii of several pairs of rings from a single interferogram and get statistically more accurate value for e by least square method.

5.5.1.7 Angular Width of Fringe and Resolving Power: The angular (or linear) width of the fringe, as illustrated in Figure 5.29, is also an important parameter that needs to be considered in several applications. The figure shows fringes as imaged by the imaging optics. The fringes subtend angles at the image plane as shown in the figure. Here we determine a relationship between the angular width of a fringe and the resolving power.

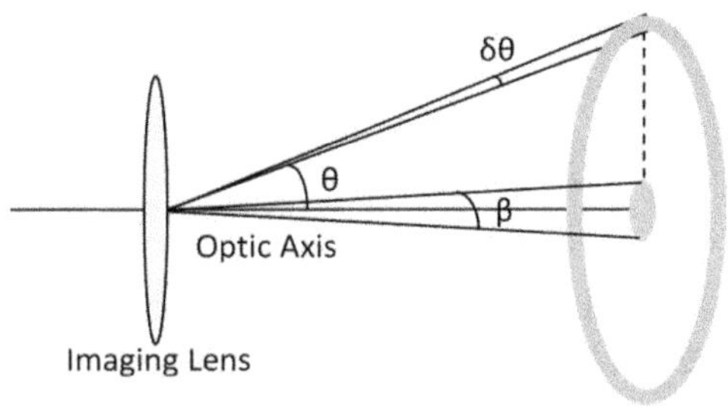

Figure 5.29: Angular widths of Fabry-Perot fringes formed by the imaging lens. The axial fringe width is β and that of the fringe with inclination angle θ is δθ. These widths are related to resolving power as shown in the text.

The equation for the fringe at inclination angle θ having a width of $\delta\theta$ at half intensity is given by,

$$2d\cos(\theta + \delta\theta) = m(\lambda + \delta\lambda)$$

For small angles the cosine function may be expanded and written as,

$$2d\left(1 - \frac{\theta^2 + 2\theta\delta\theta + \delta\theta^2}{2}\right) = m(\lambda + \delta\lambda)$$

Now the equation for fringe at θ is,

$$2d\left(1 - \frac{\theta^2}{2}\right) = m\lambda$$

Difference of these two equations and neglecting the square of the angular width $\delta\theta$, we get,

$$\theta\delta\theta = \frac{m}{2d}\,\delta\lambda = \frac{\delta\lambda}{\lambda} = \frac{1}{R}$$

In the above equation, R is the resolving power defined earlier.

The angular width of a fringe at half-intensity for an nth fringe is thus given by,

$$\theta_n(\delta\theta_n)_{1/2} = \frac{\Delta\lambda}{\lambda_0 F} = \frac{\delta\lambda}{\lambda_0} = \frac{1}{R} = \frac{1}{mF} \qquad (5.76)$$

The square of the angular diameter β subtended by an axial fringe (central bright spot shown in Figure 5.29) may be obtained as (here the angle subtended by the fringe varies from 0 to β/2 in the above equations) the solid angle Ω subtended by the cone of light passing through the FPS,

$$\Omega = \beta^2 = \frac{2\pi}{R} = \frac{2\pi}{mF} \tag{5.77}$$

This defines the aperture size for axial (central) fringe spectrometry. For area detectors, however, this consideration is not required. We have shown that all the fringes would have the same resolving power for any angle of inclination. For peripheral fringes, it is the matching with the detector elements (pixels) that would decide the effective resolving power as we will show later.

It may be mentioned that for a monochromatic light the fringes represent the same wavelength at different inclination angles in different orders; fringes are transmitted maximally all over the field. For a quasi-monochromatic light on the other hand the orders represent different wavelengths around the wavelength at normal incidence. Thus, it is evident that when the central or axial fringe peaks for a given setting (e.g., gap) the other fringes do not reach their maxima and the transmitted wavelength at an angle θ is given by,

$$\lambda_0 - \lambda_\theta = \delta\lambda_\theta = \lambda_0(1 - \cos\theta) \tag{5.78}$$

For a fringe at an angle θ to maximise in intensity, the etalon needs to be tuned to a slightly different gap setting. To determine the so-called phase change for maximally transmitting any wavelength at different spatial points requires a calibration phase map that may be generated by scanning the etalon using a standard laboratory source.

5.5.2 Types of Fabry-Perot Etalons: In its most basic design a Fabry-Perot "etalon" contains two parallel plates of substrates the inner sides of which are coated with highly reflecting material (reflectivity $> 95\%$), stuck together so that the gap d in between the plates forms a cavity. When a collimated beam of quasi-monochromatic light passes through the parallel plates, multiple reflections take place within the cavity. The substrate plates need to be flat and parallel to each other to a very high degree of accuracy (usually $\lambda/50$ to $\lambda/100$).

Fabry-Perot etalons are made in a variety of ways. The most common are optically contacted fixed air-gap etalons, fixed solid-spacer etalons, and air-gap etalons with piezo-electric transducers. In the case of the piezo transducers, a potential difference between two faces of a crystal is generated when subjected to a stress; and this phenomenon is known as the piezo-electric effect. Conversely, when a piezo-electric crystal is subjected to an electric field then it changes its shape, and this fact may be exploited to use such materials as position transducers. Among some of the most efficient transducers are the ceramic materials (e.g., Lead Zirconate Titanate) giving large linear change per volt of potential difference applied. The piezo-crystal actuated etalons are the most sophisticated and highly accurate types especially when combined with capacitance micrometry for monitoring/controlling parallelism and scanning (see Atherton 1995). In what is called close-loop etalon control, both piezo-electric transducers and capacitance sensors are used for real-time monitoring of parallelism and scanning by electronic feed-back mechanism; while in open-loop etalons, only piezo-electric transducers are used and are controlled by electronic circuitry for scanning and parallelism adjustments. The fixed airgap and solid etalons may be tuned in wavelength either by tilting or by changing pressure (for airgap) and temperature (for solid spacers). The substrates on which the reflection coating is done on inner sides are made of fused silica glass for visible (and UV) wavelengths and water-free fused silica for near-infrared region. The solid spacers are made of glasses with very low thermal expansion coefficient such as Zerodur or Ultra Low Expansion (ULE). The fixed airgap and piezo-electric actuated etalons are available commercially for gaps as small as a few micrometres while the solid spacers are available with thickness of spacers as small as 20-50 micrometres. The flatness of substrates is as highly accurate as $\lambda/100$ even for etalons of large usable apertures; while parallelism can be maintained to an accuracy of $\lambda/100$ using close-loop piezo-electric transducers coupled with capacitance micrometry. Similarly, the scanning step sizes are also very small to suit various astronomical applications. Wavelength ranges are available from UV to near-infrared with accuracies as specified above. Usually, the etalons are made very rugged and mechanically very robust (not to be dropped though!). Etalons are also available in hermetically sealed enclosures to isolate the plates, spacers, and actuators from environmental vagaries.

Nearly all the commercially available etalons – from those with fixed airgap to those with solid spacers – are made with **optical contacts** without using any glues (epoxy). Optically contacted etalons are made using highly polished, optically flat and near dust-free substrates. The optical contact is then made simply by the inter-molecular forces because the contact is made over considerably large enough surface area that the molecular forces are sufficient to hold the plates together. This process helps in reduction of absorption and scattering losses of light.

5.5.3 Effective Finesse: In the foregoing sections we have defined a merit parameter for an etalon called finesse, more precisely reflection finesse, a function of reflectivity of the etalon inner surfaces. However, there are factors such as plate flatness; plates not being exactly parallel and the focal plane aperture being large which affect and deteriorate the reflection finesse. After considering all these effects, the resultant or effective finesse is actually much smaller than the reflective finesse.

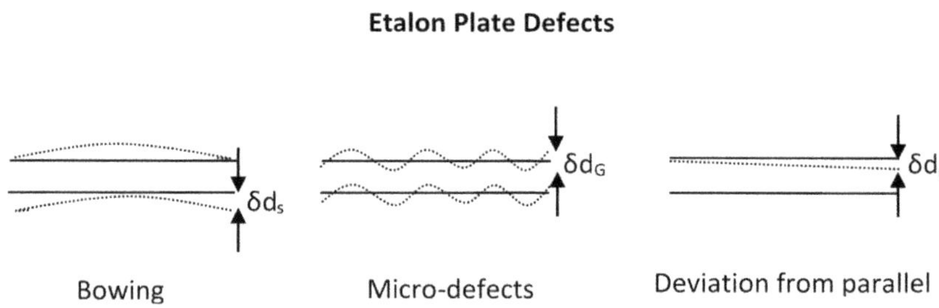

Figure 5.30: Fabry-Perot etalon plate defects.

Mainly three types of plate defects are important to be considered. These are illustrated in an exaggerated fashion in Figure 5.30 (see, e.g., Atherton et al 1981). (i) Bowing of surfaces of plates of etalon as shown in the first diagram on left in the figure is quite common occurrence in etalons either in the manufacturing process itself or by deterioration over time: the maximum deviation being δd_S. (ii) Micro-defects in the form of undulations that are inevitable for any surface: with a root mean-square value of $<\delta d_G^2>^{1/2}$. (iii) Deviation from parallelism. This can happen while scanning or inherently: the deviation being δd_p. The finesses associated with these defects are given below:

Finesse due to **curvature defect**:

$$F_{DS} = \frac{\lambda}{2\delta d_s} = \frac{K_s}{2}, where\ K_s = \frac{\lambda}{\delta d_s} \tag{5.79}$$

Finesse due to **surface defect**:

$$F_{DG} = \frac{\lambda}{4.7 < \delta d_g^2 >^{1/2}} = \frac{K_g}{4.7}, where\ K_g = \frac{\lambda}{< \delta d_g^2 >^{1/2}} \tag{5.80}$$

Finesse due to **parallelism defect:**

$$F_{DP} = \frac{\lambda}{\sqrt{3}\delta d_p} = \frac{K_p}{\sqrt{3}}, where\ K_p = \frac{\lambda}{\delta d_p} \tag{5.81}$$

The overall **defective finesse** is given by,

$$F_D = \left[\frac{1}{F_{DS}^2} + \frac{1}{F_{DG}^2} + \frac{1}{F_{DP}^2}\right]^{-\frac{1}{2}} \tag{5.82}$$

In addition to the defects mentioned above, the acceptance cone of rays would also affect the overall finesse and hence, is a factor that needs consideration. It arises from the fact that a finite solid angle of light beam enters the interferometer. Ideally the beam is collimated but if the beam deviates from collimation and has a finite f-number then the finesse and hence the resolving power deteriorates.

The **aperture finesse** can be derived from the acceptance solid angle for an axial fringe (equation 5.77),

$$F_A = \frac{2\pi}{m\,\Omega}$$

where Ω is the solid angle subtended by the sky acceptance cone into the FPS.

Then the overall effective finesse that accounts for reflectivity, etalon plate defects, and acceptance aperture, may be written as

$$F_{eff} = \left[\frac{1}{F_R^2} + \frac{1}{F_D^2} + \frac{1}{F_A^2}\right]^{-\frac{1}{2}} \tag{5.83}$$

where F_R is the reflection finesse; F_D is the defect finesse and F_A is the aperture finesse. The aperture finesse is an important consideration only in the case of central fringe scanning FPS. From the equation above, clearly the smallest finesse would be the deciding factor for the value of the overall finesse. It may be noted that the reflection finesse is the largest of all the three. However, if the defect and aperture effects are substantial, then the overall finesse gets deteriorated. In order to maintain the overall finesse as close as possible to the reflection finesse which can be very high for etalons having very high reflectivity, the optical quality of the parallel plates should be very high (better than $\lambda/100$), as well as the parallelism of the etalon plates must be maintained to a high degree of accuracy. In addition, the acceptance solid angle must be as small as possible.

The above expression follows the fact that the resolutions defined by various factors as shown above add up in quadrature (when assumed Gaussian distribution), thus,

$$(\delta\lambda)_{eff} = [(\delta\lambda)_R^2 + (\delta\lambda)_D^2 + (\delta\lambda)_A^2]^{1/2} \qquad (5.84)$$

5.5.4 Schematic of Optical Arrangement for FPS: Essentially, there are two ways of using the FPS: A. axial- or central-fringe scanning mode in which we select the central fringe of the interference pattern and scan in wavelength. The detector in this case would be a single element detector such as a PMT; and B. imaging mode in which the entire fringe pattern is imaged on an imaging 2-D detector such as an IPD or a CCD. In imaging mode again there can be two ways: (i) taking only a single interferogram at any setting of an etalon, and (ii) scanning in wavelength and obtaining a cube of data – 2-axes spatial coordinates and, the third, the spectral axis.

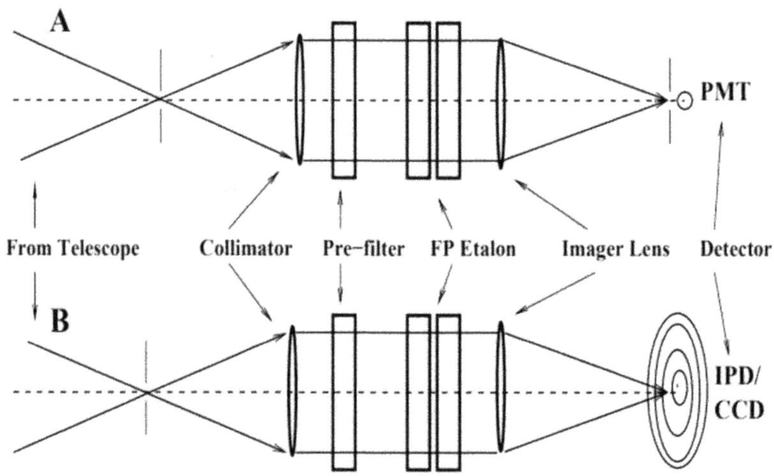

Figure 5.31: Schematic optical diagram for FPS. The upper panel (A) is for the central-fringe scanning mode and the lower panel (B) shows the imaging mode.

Figure 5.31 shows optical arrangement for a Fabry-Perot spectrometer used in central-fringe scanning mode (A) and interferogram imaging mode (B). The light beam from telescope is passed through a field-stop and a collimating lens. The collimated parallel beam goes through a pre-filter (to isolate a single order). The filtered light is passed through an FP etalon and the fringe system that emerges from it is brought to focus on a suitable detector by means of an imaging lens (imager). For the central fringe scanning (marked A in the figure) an aperture is used to isolate the central fringe and a PMT is used to detect the signal. For imaging mode (B in figure) the entire circular fringe pattern is imaged on an imaging detector such as an IPD or CCD.

Choosing of the bandwidth of the filter depends upon the FSR, for it should not exceed the latter. The choice of a suitable etalon (and FSR) depends up on the width of the spectral line from celestial object. Thus, FPS may be used in many ways depending upon the object, the available detector, and the type of etalon.

We describe now, various methods of scanning in wavelength.

5.5.5 Wavelength Scanning: As is evident from the equation for maxima, wavelength scanning may be done by one of the following three methods: by varying (i) the angle of incidence θ, (ii) the refractive index μ of the medium within the cavity or gap, and (iii) the gap "*d*" itself. Scanning by changing the incidence angle is usually very cumbersome. The refractive index is a function of environmental parameters, such as, the temperature and the pressure. Hence it is possible to vary, in a controlled way, the pressure or temperature of the gap medium for wavelength scanning. Let us examine each of these methods of scanning.

5.5.5.1 Tilt – scanning: For a fixed incident beam, the change of wavelength can be affected by tilting the etalon in steps of small angles from its normal-to-beam position. The angular dispersion can be derived by differentiating the condition for maxima,

$$\frac{d\lambda}{d\theta} = -\frac{2\mu' d \sin\theta'}{m} = -\lambda_0 \sin\theta' \qquad (5.85)$$

where λ_0 is the wavelength for normal incidence (θ' = 0).

This method used mostly in case of solid spacer etalons is applicable in situations where low resolution is acceptable.

5.5.5.2 Scanning by Refractive Index: This can be achieved by first mounting the etalon in an air-tight (hermetically sealed) chamber and then varying in steps either pressure or temperature within the chamber. The basic microscopic theory of variation of refractive index of dielectric materials with density was provided by Lorentz-Lorenz law and Gladstone-Dale approximation (details of this eponymous theory may be found in Born and Wolf 1999). Tables 2.1 and 2.2 in Born and Wolf (1999) give refractive index for some gases and liquids as a function of pressure.

The change in wavelength (from the central or reference wavelength λ_0) brought out by changing the refractive index of the air/gas inside the chamber may be written by differentiating the fringe maxima equation thus,

$$m\,d\lambda = 2d\,d\mu \;\; or\; \Delta\lambda = +\lambda_0\,\frac{\Delta\mu}{\mu}$$

$$\frac{\Delta p}{p} = \frac{\Delta\mu}{\mu - 1}; \quad \text{hence we get,}$$

$$\frac{\Delta\lambda}{\Delta p} = \frac{(\mu - 1)\lambda_0}{\mu p}\;;\; or\; \Delta\lambda = \frac{\Delta p}{p}\,\frac{(\mu - 1)\lambda_0}{\mu} \tag{5.86}$$

The above relation is for normal incidence and for gases whose refractive indices are nearly unity. Under conditions of normal temperature and pressure, for air (nitrogen: $\mu = 1.00029$) for instance the change for 1 atmosphere of pressure at 500 nm wavelength would give a shift of 0.145 nm while for the gas SF_6 ($\mu = 1.00078$) it is 0.39 nm per atmosphere of pressure (Roesler 1974; Sahu et al 1984 and citations therein). As long as the gas within the chamber is in ideal conditions, the change with pressure remains linear and hence for smaller shifts (for large gap etalons) the technique is very useful and serves the purpose well. It is to be noted that a full scan must cover more than one FSR of the etalon (at least 1.2 FSR), to facilitate phase correction. For normal air this condition requires high pressure and hence gases of higher refractive index are preferred over air: for instance, acetone that has a refractive index of 1.001090.

Pressure scanning method had been used highly successfully by many scientists for over a few decades. This technique gave useful results in gaseous nebulae such as PNe and HII regions (e.g., Arsenault & Roy 1984; Sahu et al 1984; Anandarao et al 1988; Banerjee et al 1990 and references therein). The only drawback being at reasonable pressures only small wavelength ranges can be covered. For larger coverage one requires high pressure systems especially in near-infrared. Since the 1980's more sophisticated gap-scanning etalons have been developed and the pressure scanning has taken a back seat but not disappeared.

Refractive index change can also be brought about by changing the temperature of the airgap. As much as a reasonably small temperature change may affect small ranges of wavelength shift, at higher temperatures the method may damage the coatings.

5.5.5.3 Gap Scanning: The most direct method to scan is to change the gap between the parallel plates of the etalon. This can be done by mechanically moving one plate with respect to a stationary plate using a micrometre screw as had been done a long time ago for longer wavelengths (than visible or near-infrared region). However, during the 1980's in England a revolution of sorts had started when scientists had successfully constructed an etalon with an air gap separated by small circular piezo-electric spacers stuck at the peripheral region of one of the two parallel plates. Now by applying suitable electric field externally to the piezo-electric studs they were able to change the gap suitably in small steps. In addition, the change in the gap as well as the parallelism was monitored to a high degree of accuracy by using what is known as capacitance micrometry. With this arrangement they were able achieve etalons of very large effective finesse. Figure 5.32 gives a schematic of the construction of an etalon with piezo-electric transducers for gap controlling and capacitance-micrometres monitoring parallelism and gap in real time (Atherton 1995).

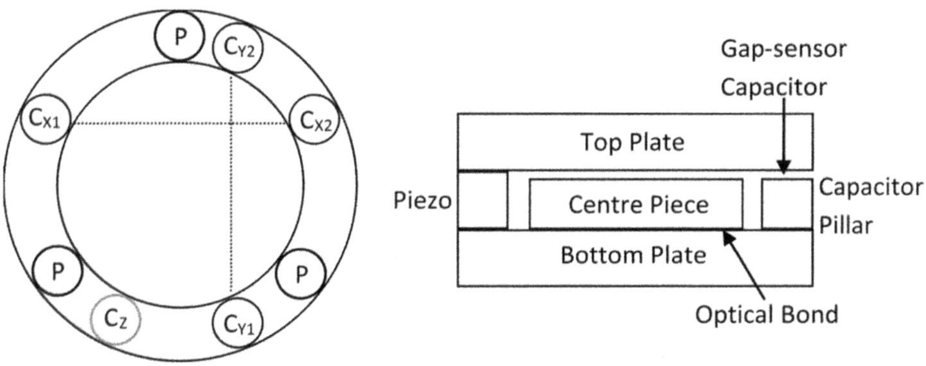

Figure 5.32: Construction of a Fabry-Perot etalon using Piezo-electric transducers and capacitance micrometry (adapted from Atherton 1995).

As shown in the figure the bottom plate is mounted on its periphery with three piezo-electric actuator studs (P) cylindrical in shape (positioned at 120^0 apart) which will act together in varying the gap between the top and bottom plates in a periodic fashion by using a ramp voltage variation. The gap sensing and error feed-back is done by two pairs of parallel plate capacitors mounted orthogonally as shown in the figure - C_{X1} and C_{X2} and C_{Y1} and C_{Y2}. A third capacitor C_Z is used as a reference for gap. The centre piece (see figure), made of thermally stable material such as fused silica or Zerodur is inserted between the two plates to account for the required manufacturer-set nominal gap d. The parallelism between plates is sensed by comparing between the two capacitor sets in X and Y directions. The error signal is fed back to the electronics so that required error voltages may be applied to the piezo actuators. Using this technique its developers were able to achieve unprecedented accuracies in parallelism and scanning, as given below.

Table 5.1 gives expected accuracies (from Queensgate Instruments, UK or currently IC Optical Systems Ltd., UK) from the performance of the etalon and its control unit. The capacitor wafers are separated by 50 μm so that we get a capacitance of 22 pico Farads. The capacitance micrometry can in principle detect a pico meter (10^{-12} m) in gap (see Atherton 1995 for details). Several scientists have used these etalons for imaging spectroscopy of astrophysical sources (TAURUS: Atherton et al 1982; HIFI: Bland and Tully 1989; Seema et al 1992).

Table 5.1: Piezo-electric transducer and capacitance micrometre accuracies.

Sensitivity of Capacitance Micrometry	10^{-12} m
Piezo-transducer Sensitivity	2.4 nm/V
Voltage Applied	+/- 750 V
Response Time	1 ms
Parallelism	+/- 0.5 μm
Parallelism Resolution	+/- 0.25 nm
Spacing	+/- 1.0 μm
Spacing Resolution	+/- 0.5 nm
Non-linearity	+/- 0.1% of FSR

The parallelism and spacing resolutions are given by 12-bit resolution (+/- 2048 or 4096 steps). This turns out to be +/- 1000 nm range of scanning at a resolution of 0.49 nm per step.

We can therefore fix the appropriate step size δd depending upon wavelength or the corresponding velocity resolution δv required for a specific application, keeping in mind that,

$$\frac{\delta d}{d} = \frac{\delta \lambda}{\lambda} = \frac{\delta v}{c} \tag{5.87}$$

Here d is the nominal gap as specified by manufacturer and c the velocity of light. Let us give an example: in one application the Hα line at 656.3 nm needs to be scanned using an etalon of gap $d = 100$ μm. We get for the wavelength chosen the FSR as 2.15 nm and for an effective finesse of 20 we get a resolution in λ of 0.11 nm. For scanning over one FSR we require to change the gap by $\lambda/2$, thus, the number of steps required to achieve this would be 656.3/(2x0.49) = 670 steps. The wavelength resolution we achieve with this scanning step would be 0.003 nm and the corresponding velocity resolution would be 1.5 km/s. Depending upon the precision required and time available for scanning, we can combine a few steps to make one big step – i.e., we can take 4 steps to make one single big step compromising on velocity resolution. The Nyquist criterion for sampling needs to be applied to choose the number of scanning elements to sample the required resolution element.

5.5.6 Environmental Factors Affecting Etalon Performance: Changes in the ambient environmental temperature and pressure as well as humidity during a scan would lead to change in the refractive index of the environment and hence adversely affect the precision that we are expecting from FPS. Let us examine these in some details.

Temperature variation: A change in temperature of ΔT would bring out a change of $\Delta\mu$ in refractive index and hence in the gap d. This is quantitatively given by,

$$\Delta\mu = \frac{(\mu - 1)\Delta T}{T} \tag{5.88}$$

T is average environmental temperature (= 273 K, for example). Then the corresponding gap change would be

$$\Delta d = \frac{d(\mu - 1)\Delta T}{\mu T} \tag{5.89}$$

We can see that for a change of 20 K in temperature for air (μ=1.00029) we get Δd = 2.1x10^{-5} μm for a gap of d = 50 μm.

Pressure variation: In a similar way as for temperature we can show that the change in gap for a change in pressure of Δp over the ambient average value of p can be given by (cf. eqn. 5.86),

$$\Delta d = \frac{d(\mu - 1)\Delta p}{\mu p} \tag{5.90}$$

For a change of pressure by ten percent of an atmosphere we get for an etalon of gap 50 μm a change in gap of 1.35x10^{-5} μm.

Relative Humidity variation: For one percentage change of relative humidity, the refractive index of air changes only by 1 part in 10^8.

5.5.7 Phase Correction: From the theory of interference, it is clear that in an etalon for a given value of pressure or temperature (for refractive index scanning), angle of incidence (for tilt-scanning), or gap between the etalon plates (gap scanning) only a particular wavelength gets maximised at the centre (on-axis) or at any other particular fringe on the focal plane of the imaging lens (that images the circular fringe pattern). There is a "phase" difference between one fringe to other for maximal transmission for a given setting of the etalon, for a given wavelength (the FP fringes being fringes of equal inclination the intensity is maximised for all points on a fringe). For this reason, scanning the etalon is necessary for covering the entire fringe system (by any of the three methods described) for maximal transmission, for a given λ.

For a monochromatic radiation the circular fringes with increasing order in radial direction (equal inclination θ) the spatial (angular) relationship with order is given by (for $\delta\lambda/\lambda << 1$),

$$\cos\theta = \frac{m}{m_0} \tag{5.91}$$

Here m_0 represents the order for the central fringe (including the fractional part of the order) and m is the order for the fringe at angle θ.

For a quasi-monochromatic or a multi-chromatic light beam passing through an etalon the fringe system involves phase difference; meaning that a wavelength maximum at the centre (not central fringe) does not ensure the maxima at other spatial locations for the same wavelength. By scanning the etalon (say by changing the gap d) we can attain maxima at different spatial locations. The relationship that gives the wavelength difference $\delta\lambda$ between the central and an off-axis location at an angle θ can be readily derived as

$$(\delta\lambda)_\theta = \lambda_0(1 - \cos\theta) \tag{5.92}$$

Expressing this with reference to the image plane

$$(\delta\lambda)_{xy} = \lambda_0(1 - \cos\theta_{xy}) \tag{5.93}$$

The subscript xy denotes the spatial coordinate (x, y) in the image plane (detector plane).

Thus, when we scan over a free-spectral range (FSR) of the etalon it is conceivable that the line profile for a given emission line (from the celestial object or for a laboratory standard source) may need a phase correction, as illustrated in Figure 5.33. The problem becomes more serious when we deal with interferograms covering an extended source (in scanning over one FSR or single interferogram).

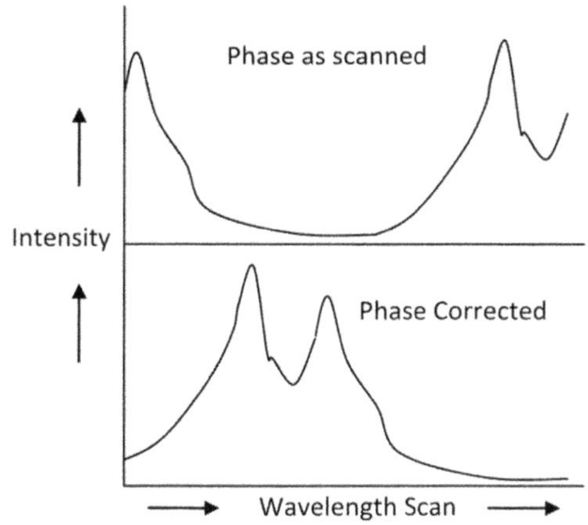

Figure 5.33: Schematic of a line profile at on off-axial pixel where phase correction is required. In the top panel is shown the profile without phase correction and the bottom the phase corrected one. We need to scan the FP etalon in gap for more than one FSR.

This problem can be dealt with by scanning a laboratory source giving the same line as for the celestial source. However, this is possible only for allowed recombination lines such as Hydrogen lines; but for forbidden lines such as [OIII] at 500.7 nm we need to calibrate using a line from a laboratory source as close as possible to the programme line. Still for extended sources we need to make the phase correction for the cube of data that is obtained. Usually this is done by obtaining the three-dimensional cube of data from a laboratory source which contains the spatial information (as given by the interferograms) and the wavelength scanning information – a function called phase function $\varphi(x,y,\lambda)$ (see Atherton et al 1982; Bland & Tully 1989, for details). This cube of data from calibration source is called a 'phase map'. For an etalon of a given gap and a given finesse, the phase map that may be used for phase correction would be the same for any wavelength.

5.5.8 Wavelength Calibration: Calibration of wavelength basically involves finding the exact order number for a given wavelength so that the etalon may suitably tuned to that wavelength. For this we should first know the fractional order we discussed in the foregoing pages. For a given etalon we can find the fractional order by imaging the circular fringe system on a digital detector array and from the measurement of the fringe radii we can obtain the fractional order accurately using the formulae derived earlier in this chapter.

We discuss here the methodology of calibration (see Atherton et al 1982 for details). Assuming that the scanning is made using a piezo-electric spacer etalon with gap d represented by an arbitrary parameter z, we can derive an expression for the exact gap (in terms of *z*) required for effecting a central maximum for a given wavelength. The actual gap *d* may be written as $d(z) = d(0) + a\ z$. Here $d(0)$ is the gap specified by the etalon manufacturer. Then we can show that the FSR in terms of *z* for a wavelength λ is $\Delta z = \lambda/2a$. Basically the parameter '*a*' represents the meter reading on the dial (voltage given to the piezoelectric spacer of the control unit in Queensgate instrument). By changing the voltage required to effect one FSR change for a given λ we can obtain Δz for different wavelengths.

Let us consider two wavelengths λ_1 and λ_2 and z_1 and z_2 the corresponding values of *z* required to obtain maxima on the (central) axis and Δz_1 and Δz_2 the corresponding FSRs. We can then write down the difference between the two orders corresponding to the two wavelengths as (see Atherton et al 1982),

$$m_2 - m_1 = \frac{m_1}{\lambda_2}(\lambda_1 - \lambda_2) + \frac{\lambda_1}{\lambda_2}\frac{(z_2 - z_1)}{\Delta z_1} \tag{5.94}$$

Here the refractive index of the environmental medium is taken as 1 (for air). It is noticed that the accuracy of determining the orders increases with increasing difference between the two chosen wavelengths. This also depends on how accurately we know the physical gap '*d*' of the etalon plates. Usually, the manufacturers of highly modern etalons quote t to the accuracy of a few micrometres. The above equation however assumes that $\lambda_1/\lambda_2 \sim 1$ for simplification. The value of *z* required for maximal transmission at the centre for any other wavelength (third wavelength) λ is then given by,

$$z = z_1 + \frac{m_2 \Delta z_2}{\lambda_2}(\lambda - \lambda_2) \tag{5.95}$$

Considering that at the central maxima, the quantity $m\lambda$ (= $2d$) is approximately constant (or invariant), the gap *z* may be determined, and hence the FP etalon may be tuned to any λ.

We have been able to tune reasonably accurately in all our observations with the Near-Infrared Fabry-Perot Spectrometer (NIFS) (see following sections), using Queensgate etalons with electronic controller for maintaining gap and parallelism to a very high degree of accuracy (see Table 5.1).

In the following we discuss various methods by which a FP etalon may be employed for spectroscopic observations of astrophysical objects. Apart from the type of object that we are interested in – whether it is a (quasi-)

point source or an extended source, the choice of the detector available also needs to be factored in while deciding on the method.

5.5.9 Central Fringe Scanning FPS: Axial fringe or central fringe scanning had been done in the past by several workers using pressure scanning technique in optically contacted air-gap etalons. In principle a suitable sky-area is matched with the entrance aperture of the FPS in such a way that the central fringe represents the object in part or in full. In case the object is extended beyond the aperture size then a few positions are sampled. A typical optical set-up for an axial fringe FPS for the 1.2 m Mt Abu telescope is shown in Figure 5.34 (from Banerjee et al 1987). This instrument used piezo-electric transducer etalon (manufactured earlier by Queensgate Instruments, UK; now by IC Optical Systems Ltd., UK). The *f*/13 Cassegrain beam from Mt Abu 1.2 telescope passes through a field aperture wheel (having apertures of different sizes) and a flip-mirror (or a beam-splitter with more than 90% transmittivity) to a collimator. The reflected beam from the mirror/beam-splitter (from a field star close by the programme object) is used for guiding the star/object to remain in the field throughout the observation. The collimated beam passes through order-selecting interference filter and then to a FP etalon. The central fringe of the FP fringe system is isolated by using a suitable aperture or exit pupil and then imaged by a Fabry lens on a single-element detector such a PMT. The purpose of the Fabry lens is to image the exit pupil of the telescope. The data, that consists of scanning steps versus corresponding signal counts, is then acquired by a data acquisition system for further analysis. Wavelength calibration is done by a standard spectral lamp at the mirror/beam-splitter as shown in the figure. Figure 5.35 shows instrumental profiles of the FPS in a few orders obtained by scanning.

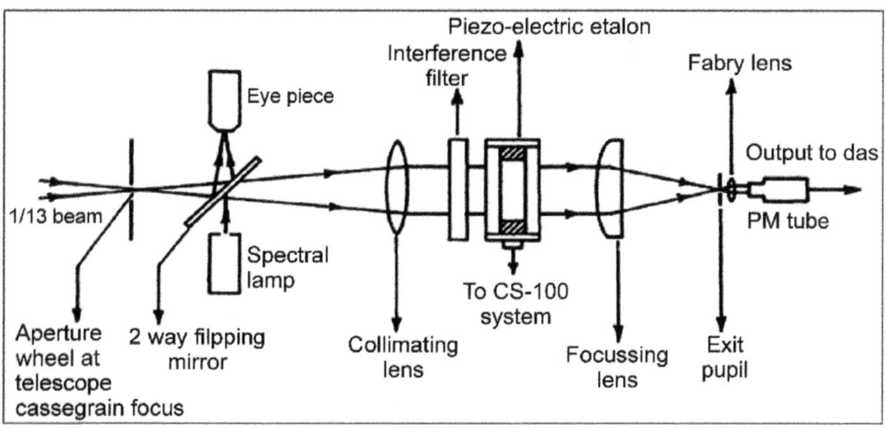

Figure 5.34: Schematic layout of a central-fringe scanning Fabry-Perot Interferometer (Banerjee et al 1999).

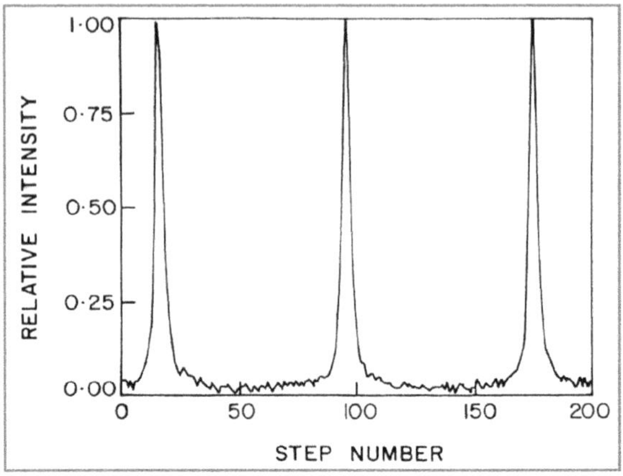

Figure 5.35: Orders of interference in a central-fringe scanning Fabry-Perot Interferometer (Banerjee et al 1987).

Figure 5.36 shows the optical image of a planetary nebula NGC 1535 along with line profiles of the optical emission line [OIII] 500.7 nm obtained by using the central-fringe scanning FPS at four positions marked by circles on the image (see Banerjee & Anandarao 1991). Evidently observations of such sparse samples (using central-fringe-scanning FPS) will not provide adequate coverage of kinematic information. It needs an imaging FPS (or integral field spectroscopy).

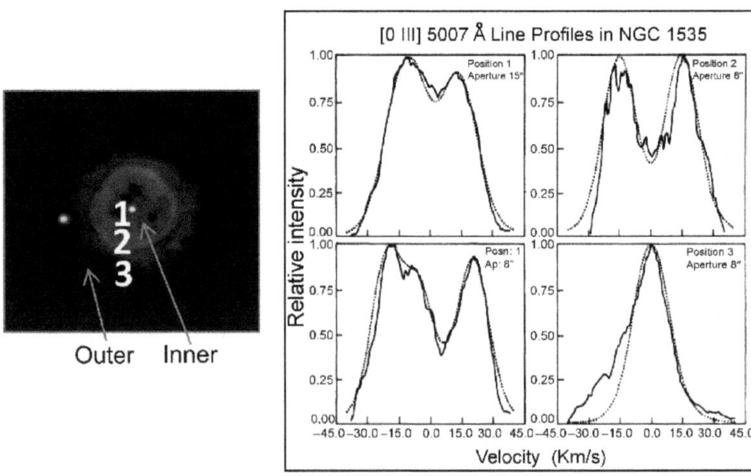

Figure 5.36: [OIII] line profiles at different positions in the double-shell PN NGC 1535 (marked as inner and outer shells in the image of the nebula). Left: Image of the PN with positions of observations marked in numbers; right: line profiles at 1-3 positions (Banerjee & Anandarao 1991).

Typically, we use this method of central-fringe scanning for point sources (such as stars) and also when we have a single element detector (such as a photomultiplier tube). Using a single-element detector for an extended source we will not be able to exploit the spatial multiplexing advantage that the FPS offers. For an extended source we need to point the instrument on several vantage/critical positions over the object and then make observations at each of these positions. This is not only time-consuming but also cumbersome and inaccurate in the sense that sharp changes in velocity field go undetected (say, in between the positions 1-4 shown in Figure 5.36). In addition, the fairly large aperture that is used (about 10 arcsec) averages out spatially the velocity differences within the aperture.

Wavelength calibration is done by scanning a lab source, ideally of the same emission line as the celestial object being studied. But in practice, it is not possible to have a lab source in the case of forbidden lines. Therefore, we normally use He, Na or Ne gas discharge lamp sources. This calibrates adequately the scanning step-size in terms of wavelength.

5.5.10 Imaging of a Single Interferogram: In the case of transient events involving certain celestial objects, due to time constraints only a single interferogram may be obtained to get kinematic information on the programme object. One of the well-known examples is the total solar eclipse which lasts typically for less than 5 minutes; sometimes even shorter than one minute. In such cases the available observation time permits only for a single or at best a few interferograms – as complete scanning over the entire FSR would take several minutes. Even with these constraints this method had in the past yielded several pioneering and significant results on galactic diffuse nebulae and external galaxies: velocity fields in spiral galaxies (de Vaucouleurs and Pence 1980; de Vaucouleurs et al 1983), planetary nebulae and HII regions (Georgelin & Georgelin 1976; Seema 1996; Chakraborty & Anandarao 1997, 1999 and references therein). Several events of total Solar Eclipse were also studied using a single (or a few) interferogram imaging method for solar coronal velocity fields (e.g., Desai 1998 and references therein). A typical interferogram is shown in Figure 5.37. Obviously on an extended source such as a galaxy, HII region or a planetary nebula, the fringes trace out only a few gaseous emitting regions in concentric circular patterns. Information in between the circular fringes would be missing because of the lack of inadequate coverage that involves wavelength scanning. The analysis of such a single interferogram deals with, at the outset, finding the exact centre of the fringe system (using a calibration interferogram prior to actual observation on program object). Then radial lines may be drawn outwards from the centre intersecting the fringes (see lines OA and OB in Figure 5.37) – a single radial line drawn in this way would yield line profiles along its path which represent the plasma at that particular position. In many an occasion, a single line would not be adequate due to poor signal to noise ratio; instead, a sector would be required (see arcs between the sector formed by the lines OA and OB in Figure 5.37). The angle of the sector is necessarily a compromise between adequate S/N and averaging over the extent of spatial element traced out by the sector which needs to be as small as possible (the sector arc length increases as it traces the outer most fringes).

Now the compromise on the spatial extent would mean that the line profile (obtained by integrating over the curved strip traced by the portion of the sector of the fringe) represents an averaged profile over the entire arc length being sampled. Further, as already mentioned, the sector arc length would be increasing as we extend to outer fringes. If within the sampled arc length there exist dynamically different regions or density inhomogeneities, then we would be unable to detect such features. The spatial resolution is thus compromised in this method. As the fringes narrow down radial direction outwards, the spectral resolution is also compromised due to sampling inadequacy.

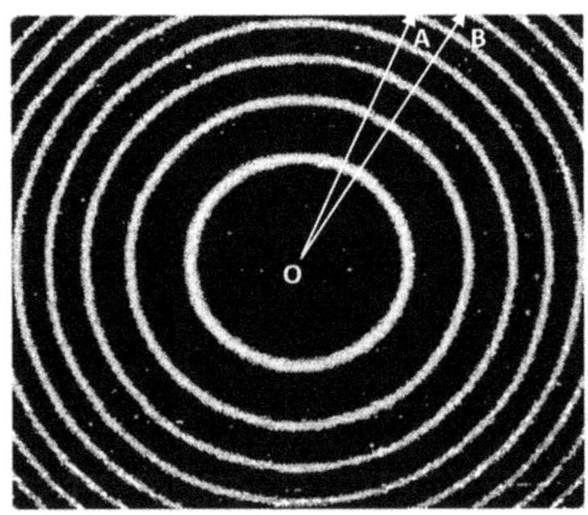

Figure 5.37: Single Interferogram showing the radial lines OA and OB; O being the centre of the fringe system.

5.5.11 Imaging FPS (IFPS): Imaging the full bloom of Fabry-Perot fringes in steps of wavelength covering a little more than one free spectral range is the most accurate method of realising its full capability. This can be achieved by using two-dimensional detectors or array detectors. Digital detector arrays (rather than analogue photographic film) are the best option for this purpose. As already mentioned, the scanning of the FP etalon a little over one FSR requires several interferograms to be taken at an appropriate step size. Further, with the help of a standard source, we need to generate a phase map with equal number of steps and correct the set of images on the object for phase. Figure 5.38 shows images of FP interferograms at different steps while scanning. We may notice that as the scanning progresses from channel 1 to channel 48, the central bright fringe starts expanding and covering the entire object field.

Figure 5.39 shows interferograms at different scan positions on the planetary nebula NGC 4361 and the line profile at a particular spatial position after applying phase correction. The spectral line scanned is the forbidden emission line [OIII] at 500.7 nm (Muthu & Anandarao 2001). These observations were made at Mt Abu 1.2 m telescope. The spatial resolution achieved here is a few arcsecs with velocity (spectral) resolution of about 10 km/s. Such an accurate mapping of velocity field is possible only with scanning IFPS. Even better accuracies, of course, may be obtained with larger telescopes. Using the velocity information so obtained, we may generate spatial velocity maps for an extended source. Such a map depicts the spatial distribution of intensity of an emission line at different velocity steps (i.e., iso-intensity map). An example of the velocity map is shown in Figure 5.40. This velocity map is generated from scanned interferograms on the planetary nebula NGC 1514 using the [OIII] emission line at 500.7 nm (Muthu & Anandarao 2003). In this map, the velocity difference between each panel is about 12 km/s. The central panel shows (referenced at 0 km/s) the bipolar structure of emission surrounding the central star of the planetary nebula.

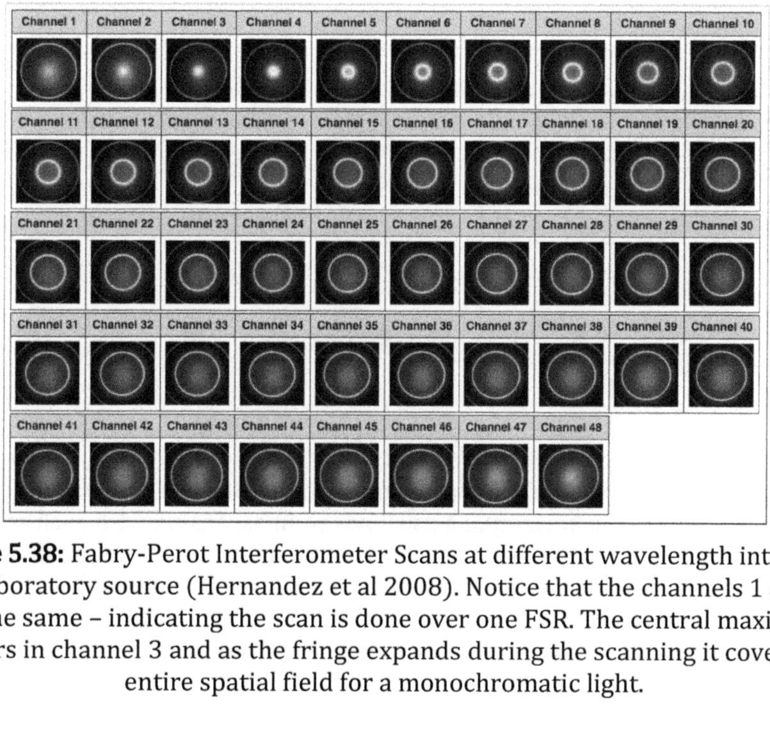

Figure 5.38: Fabry-Perot Interferometer Scans at different wavelength intervals on a laboratory source (Hernandez et al 2008). Notice that the channels 1 and 48 are the same – indicating the scan is done over one FSR. The central maximum appears in channel 3 and as the fringe expands during the scanning it covers the entire spatial field for a monochromatic light.

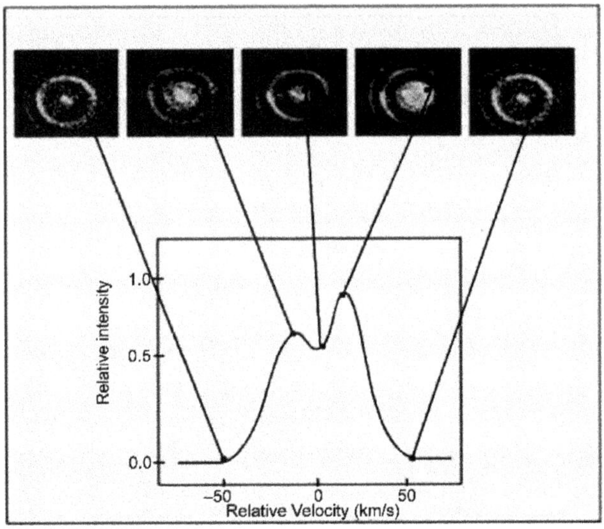

Figure 5.39: Scanning FP spectrometer images and the line profile at a spatial point on the PN NGC 4361. (Muthu & Anandarao 2001).

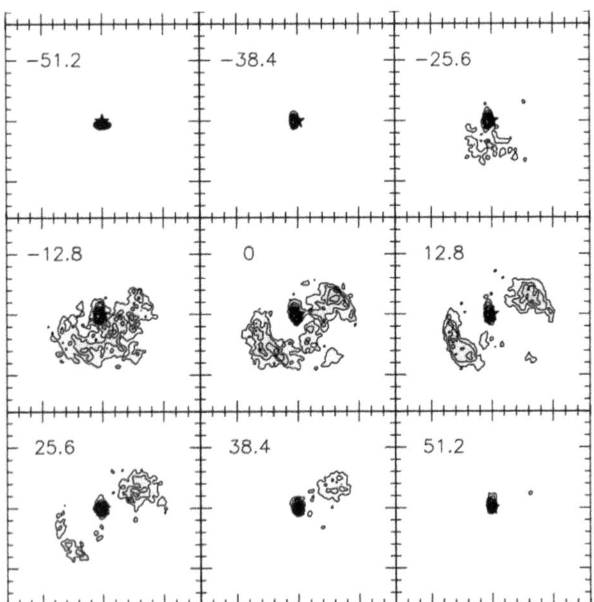

Figure 5.40: Velocity maps of iso-intensity for NGC 1514 (from Muthu & Anandarao 2003). The horizontal axis shows right ascension with each small tick corresponding to 10 arcsec. The vertical axis is declination at the same scale. The axes are centred with respect to the central star shown in the middle (see, e.g., top left panel). Velocity is shown in steps of 12.8 km/s with respect to the central panel where the intensity maximises.

5.5.12 Tele-Centric Mode - Infrared IFPS: In the case of infrared region obtaining a cube of interferogram data becomes very time consuming especially for smaller telescopes – having 1 or 2 m aperture. Also, an adequate velocity resolution (comparable with optical wavelengths) demands higher wavelength resolution at infrared region. For the same resolving power, the velocity resolution achievable in optical region (say 0.5 μm) is 4 times better than that in near-infrared region (say 2 μm); since the resolving power $R = c/\delta v = \lambda/\delta\lambda$. Increasing the resolving power would reduce the number of photons into the resolution element requiring thus a larger telescope aperture. For this reason, for smaller (1 m class) telescopes, we can use the tele-centric mode of operation for the set-up of the Near-Infrared Fabry-Perot Spectrometer (NIFS) as shown in Figure 5.41. The telescope beam directly images the object on to the detector passing through the etalon tuned to one specific wavelength (see eqn. 5.77). This then would produce a near-monochromatic image of the object on the detector. For 1 m -class telescopes, scanning may be done in this mode but only for coarse velocity resolution. Sufficiently narrow band samples may be taken on large telescopes on which the integration times for each frame may be reduced in comparison to smaller telescopes, and the scanning may be completed in reasonably short timescales during which the atmospheric changes may not be significant.

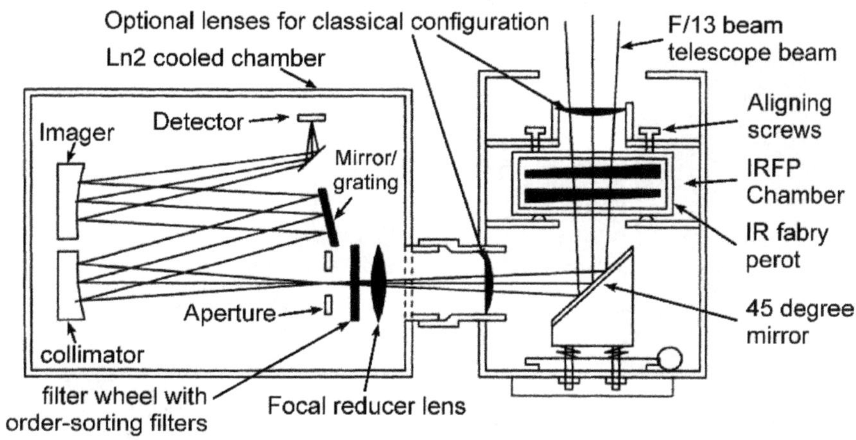

Figure 5.41: Near-Infrared FP Scanning Spectrometer (NIFS) working in "tele-centric" converging mode. (Anandarao et al 2000).

Using the $f/13$ Cassegrain beam of the Mt Abu 1.2 m telescope, we can obtain for NIFS an instrumental effective finesse of about 16 (largely governed by the aperture finesse) and hence a spectral resolution $\delta\lambda =$ 0.0014 μm. This corresponds to a resolving power of 1600 in the K band at interference order $m = 100$. This gives a velocity resolution of 215 km/s. If used in parallel/collimated beam instead of a converging beam, we may achieve an effective finesse of about 25 and a resolving power of 2500-3000. This corresponds to a velocity resolution of about 100 km/s (see Anandarao et al 2000).

Figure 5.42 shows images of Orion nebula in different gap positions of NIFS. The figure also shows the Orion nebula in K band (2.2 μm) on the top left; the box outlined is the region scanned by NIFS. The line profile for a specific position on the nebula read out from each of the 7 frames is also shown. The on-line and off-line frames 4 and 1 (or7) respectively, in Figure 5.42, may be used for tracing the emission regions very accurately.

This brings us to the use of Fabry-Perot Interferometer as a tuneable filter and this aspect is discussed further in the following section.

5.5.13 Fabry-Perot Etalon as a Tuneable Narrow-band Filter: Fabry-Perot spectrometer works as a narrow band filter when tuned to a particular wavelength by setting appropriately the gap between the etalons (Atherton et al 1981). By obtaining difference between two images - one on-line and the other off-line - we can delineate emission regions in ISM very effectively (refer Fig 5.42). This method is very accurate as compared to a conventional narrow band filter because of the fact that in the latter method we need to use a separate filter for background subtraction while in the former case the subtraction is done using the same set-up just by off-tuning.

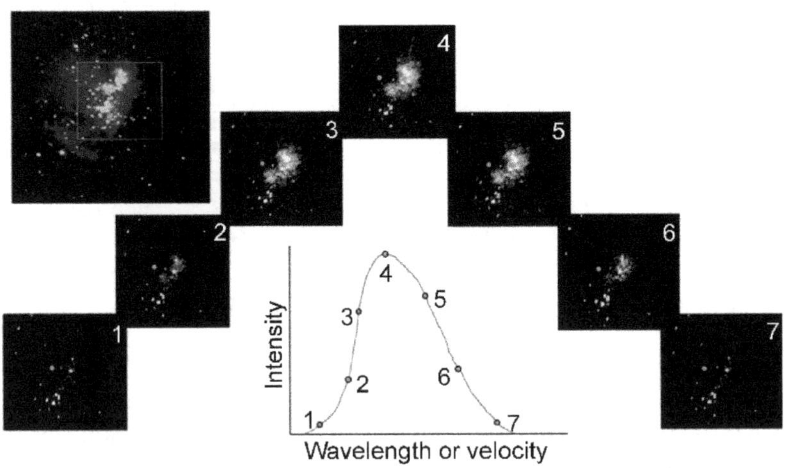

Figure 5.42: NIFS in scanning mode on Orion nebula for the 2.12 μm H_2 line. We may notice the off-line images in the 1st and 7th interferograms while the 4th one shows the maximum on-line image (Anandarao et al 2000).

We give here a few examples of the "filtergrams" (narrow-band images) obtained by a IFPS both in visible and near-infrared regions. Figure 5.43 shows a background-subtracted "filtergram" tracing the forbidden line [OIII] at 500.7 nm in the Planetary Nebula NGC 1514 observed using IFPS at Mt Abu 1.2 m telescope. Figure 5.44 shows filtergram images in two objects observed from Mt Abu observatory using IFPS in near-infrared K band – Planetary Nebula RAFGL 2688 in H_2 line at 2.12 μm and H Br γ line at 2.16 μm showing clearly equatorial torus (as two faint blobs seen perpendicular to the bright bipolar flows) in H_2 line but not in H Br γ, and the Orion Trapezium and Bar regions in H Br γ line (Venkata Raman, private communication 2018). Several dynamic features may be noticed in these filtergrams. Therefore, regions of various emission lines can be delineated accurately by using IFPS in tuneable filter mode.

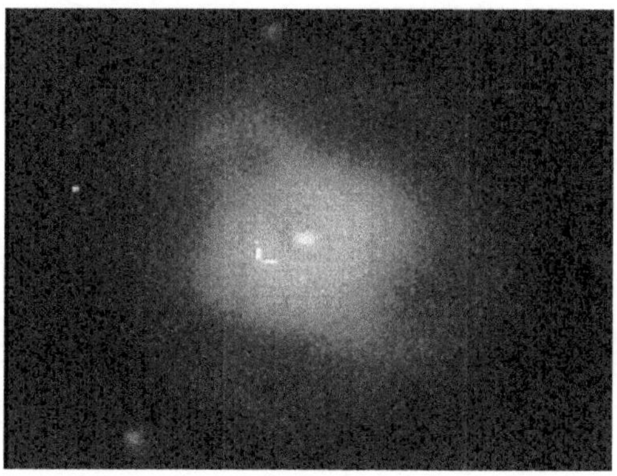

Figure 5.43: Filtergram of the PN NGC 1514 showing emission regions in [OIII] line at 500.7 nm.

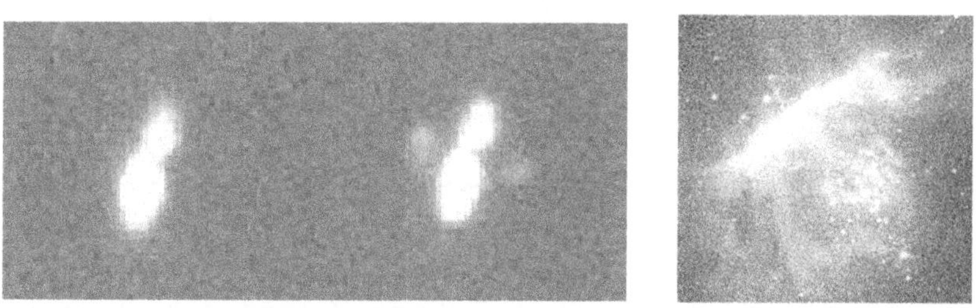

RAFGL 2688 Orion Trapezium & Bar

Figure 5.44: Left: Planetary Nebula RAFGL 2688 imaged in Molecular Hydrogen line at 2.12 μm (right) and Br γ (left) lines using FP etalon and NICS at Mt Abu: notice the faint blobs in the H_2 image indicating equatorial torus; Right: H Br γ line (2.16 μm) image of Orion Trapezium and Bar regions (Venkata Raman, private communication 2018).

5.5.14 Etalons in Tandem: As seen in the foregoing sections a single FP etalon is restricted by its free spectral range (FSR, $\Delta\lambda$; the spectral multiplexing) for its usability for spectral lines of very large widths (of a few thousand km/s), for example, those from galaxies or active galactic nuclei. This drawback also restricts its capability in the case of applications needing a tuneable filter over a large wavelength range. In order to improve its limited spectral multiplexing capability, Roesler (1974) and his collaborators suggested two or more FP etalons in tandem.

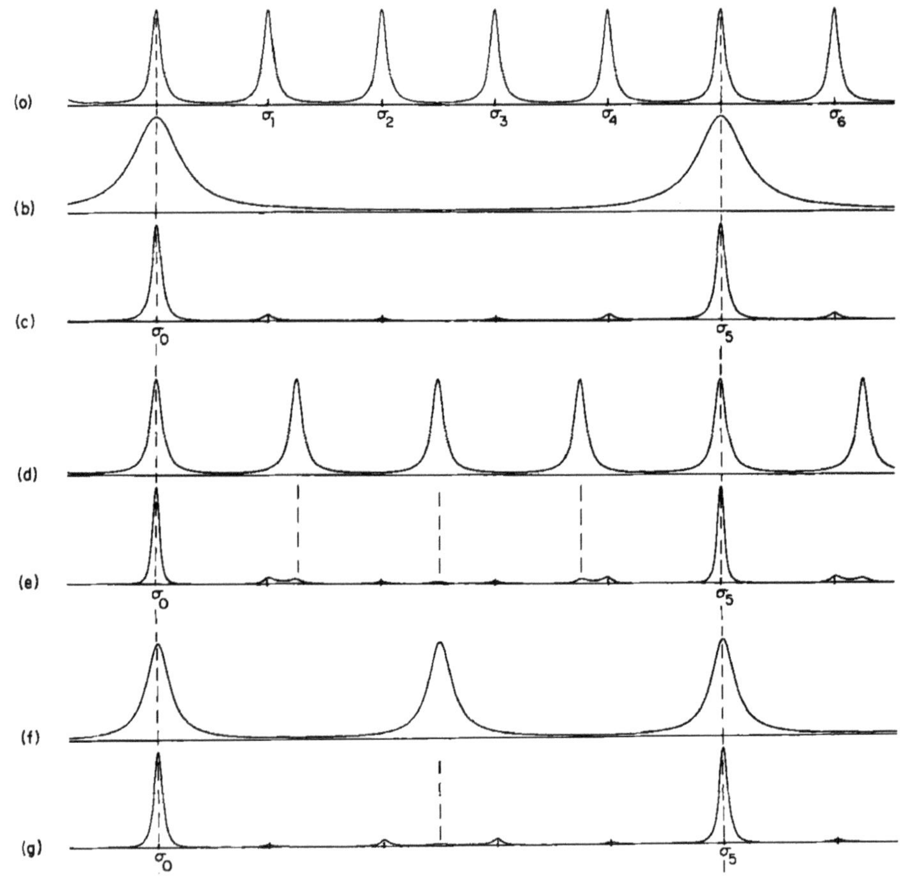

Figure 5.45: Tandem of two etalons producing larger FSR. The etalon pair (a) and (b) has the ratio of gaps equal to 5; the pair (a) and (d) has vernier ratio of 5/4 and the pair (a) and (f) has half-integral ratio of 5/2; their resultant patterns are shown in panels (c), (e) and (g) respectively. Note that wave number σ is used here ($\sigma = 1/\lambda$). The resultant FSR seen in (c), (e) and (g) is 5 times that of the first etalon (a) (taken from Roesler 1974).

Let us consider two etalons E_1 and E_2, with gaps d_1 and d_2. The first etalon will transmit at the normal incidence a wavelength of λ_0, as well as wavelengths that satisfy $\lambda_i = \lambda_0 \pm i\, \Delta\lambda_1$. Similarly, the second etalon tuned to the wavelength λ_0 will transmit the wavelengths given by $\lambda_j = \lambda_0 \pm j\, \Delta\lambda_2$. We can see that the transmitted wavelengths from these two etalons will not be the same except when i = j = 0, i.e., at λ_0. Now, a combination of pairs of etalons will give longer FSR than their individual values when their gaps are so chosen as to be in (i) multiple ratios (d_1/d_2= n); (ii) vernier ratios ($d_1/d_2 < 2$) or (iii) half-multiple ratios ($d_1/d_2 = n + 1/2$) where n is an integer. This is illustrated in Figure 5.45. As is shown in the figure, a combination of multiple etalons of shorter FSR not only increases the resultant FSR but also preserves the high resolution expected from a single etalon.

The figure shows the combination (tandem) of etalon 1 (whose maxima are seen in panel (a)) with etalon 2 (in panel (b)), etalon 3 (panel (d)), and etalon 4 (panel (f)). The resultant patterns of maxima from these three combinations are shown in panels (c), (e) and (g), respectively for combinations of etalons 2, 3 and 4 with etalon 1. The unavoidable small blips seen in between the two maxima in the resultant panels (c), (e) and (g) are due to the non-zero minima of the etalons (e.g., see panel (a) for etalon 1). We may notice the substantial increase in the FSR for FP tandems.

Small Gap and Large FSR: It may be noted however, that with the advent of etalons having piezoelectric transducers combined with capacitance micrometry as described earlier, gaps of a few micrometres are possible to be created. With the narrow spacers it is now possible to obtain very large FSR's in both visible and near-infrared regions.

5.5.15 FPS in Tandem with Other Spectrometers: FPS can also be used in tandem with other spectrometers such as a prism (to disperse the circular fringes formed by two or more wavelengths) and grating (as a pre-filter). Combination of FPS with Michelson interferometer is also possible (Anandarao et al 1983, SPIE). We may refer to Meaburn (1970) and Roesler (1974) for details of such interesting combinations.

5.6 Resolution-Luminosity Product for Various Spectrographs

As mentioned in the beginning of this chapter, one of the most important figures of merit of a spectrograph is the product of its spectral resolving power R and the luminosity or throughput. We find that the prism spectrometer can achieve resolving powers of about 10^4. Grating spectrometer can achieve better than the prism with resolving power of up to 10^5. The Michelson spectrometer (FTS) and FPS can achieve even higher resolving powers. Apart from the resolving power a spectrograph should also have a high light gathering power. Based on these arguments Jacquinot

(1954, 1960) defined a merit parameter for spectrographs given by the product of resolving power with the luminosity that the spectrograph can accept from the object. Jacquinot defined luminosity (different from the luminosity we defined earlier on photometry chapter) as the ratio of flux from an astronomical source received by the detector to the monochromatic luminance of the source,

$$L = \Phi_\lambda / B_\lambda \quad \text{with} \quad \Phi_\lambda = \varepsilon\, A\, \Omega\, B_\lambda$$

This gives the product,

$$L \times R = (\varepsilon A \Omega) \times R \tag{5.96}$$

The term in the bracket above, called the luminosity *L*, defines the fraction of light that enters the system (from a source). This quantity is also sometimes referred to as the 'throughput'. It may be mentioned here the term used by Jacquinot, the 'etendue' is luminosity without the transmittance factor ε. The area of the dispersive component is *A*, Ω is the solid angle subtended by the object at the spectrograph dispersive element, and ε represents the overall transmittance of telescope and other optics including the dispersive element.

This product *(L X R)*, called Jacquinot advantage, is compared for all the spectrographs discussed in this chapter for the same solid angle Ω:

$$(L \times R)_P < (L \times R)_G < (L \times R)_F \tag{5.97}$$

The subscript *P* stands for prism, *G* for grating and *F* for FPS or FTS. The difference between the prism/grating and the FTS/FPS arises because of the area of the dispersive element is larger in the latter (slit versus circular aperture). Although both need a slit (and their resolving power depends on it), grating scores over prism because the former can have high resolving powers at high orders (echelle mode). When working under detector noise-limited conditions the scanning speed of FTS scores over other similar spectrometers such as Fabry-Perot spectrometer (FPS) or grating spectrograph. It has Jacquinot advantage as well as wavelength-multiplexing advantage or Fellgett's advantage (see e.g., Meaburn 1970) and it scores over the FPS in the latter property. FPS requires observing time of *N x t* where *t* is time required for one sample and *N* the number of samples required for covering one free-spectral range. In comparison, FTS is faster by a factor of $(N/8)^{1/2}$ than FPS for same number of samples. Unlike the prism/grism/grating, FPS or FTS does not require an entrance slit to work at theoretically possible resolution.

5.7 Integral Field Spectroscopy Techniques

In order to obtain both spatial and spectral information on an extended object such a Planetary Nebula, HII region or a galaxy or a cluster of stars or galaxies, we need a spectrograph that has both spatial and spectral multiplexing advantage. A normal grating spectrograph does not possess both these advantages – it has spectral multiplexing or Fellgett's advantage, but its spatial advantage is limited to the slit length (see Figure 5.46). Similarly, an FPS has spatial advantage but does not have spectral advantage. Such an instrument that simultaneously provides both these advantages is termed as an Integral-Field Spectrograph (IFS). Natural choices for IFS are a scanning Fabry-Perot etalon-based spectrograph with large FSR or a Fourier Transform spectrometer (FFT) both of which have the advantage of spatial 2π symmetry. In comparison, the grating spectrograph that is more popular and versatile in using does not have this spatial symmetry. However, the instrumental function for the grating spectrometer is well-defined and less complex compared to the FPS or FFT. As already mentioned, the former can only provide the spatial information in a narrow strip over the object. It requires positioning of the grating over several locations on the object, significantly consuming precious telescope time, although the grating slit exposed to a crowded star field can in fact give spectra of two or more stars simultaneously, as shown in Figure 5.46. Therefore, we ask the question, "Can we somehow provide grating the spatial multiplexing also?"

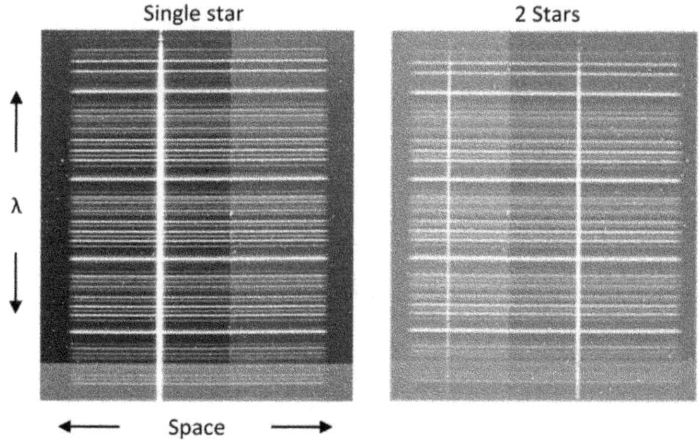

Figure 5.46: Grating spectra of a single star (left) and two stars (right) in the near-IR H band (Mt Abu 1.2m IR telescope and NICS). The horizontal axis represents the space and the vertical axis the wavelength. The intense lines in the horizontal plane are the atmospheric OH vibrational-rotational spectra. While the grating has spectral multiplexing advantage, it lacks spatial multiplexing.

Figure 5.47 illustrates conceptually the idea of integral field spectroscopy of an extended celestial object. While we need spatially resolved sampling as shown by the circles in the figure, the long-slit spectroscopy using a grating provides spectra only along the slit covering a few samples. From the figure we can imagine that a series of such spectral images, each one corresponding to a specific spatial section, needs to be obtained by a detector array. This then forms the concept for grating-based integral field spectroscopy (IFS): the entire length and breadth of the object in space needs to be imaged in spectra using some innovative technique - by a single grating in one shot. We must mention here that slit-less spectra can have spatial coverage for crowded star fields (star clusters for instance). For extended sources, such as gaseous nebulae or external galaxies, the slit-less spectra become useless. Further it cannot provide the required high spectral resolution.

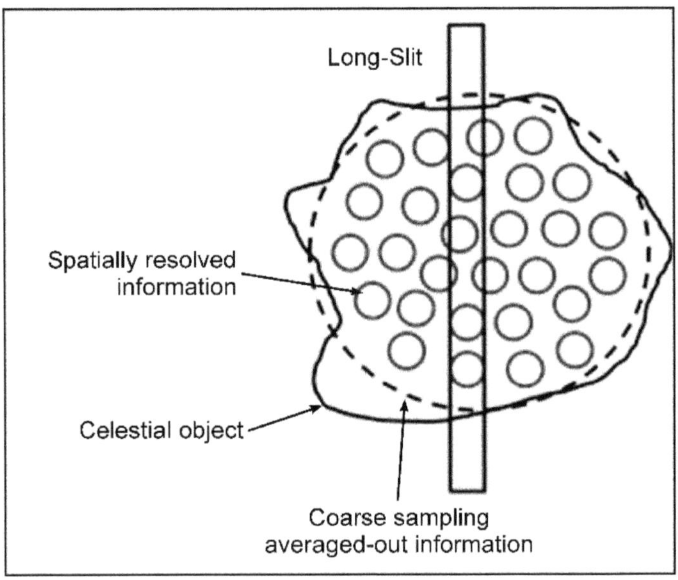

Figure 5.47: 3-D spectroscopy or Integral Field Spectroscopy of an extended celestial object: Concept.

Innovative techniques have been developed during the past two decades which make use of the reflection gratings for making IFS (see Bacon et al 1995; Bershady et al 2004, 2005; Allington-Smith 2006; Sanchez et al 2007a, b; Bacon &Monnet 2017). Essentially the two-dimensional object image from the telescope is divided into several portions and are fed in a linear sequence to the grating – by using lens-lets (pupil imaging), or a combination of lens-lets and fibres or image-slicing. Some of these techniques are shown in Figure 5.48.

As shown in the top panel of Figure 5.48, the lens-lets feed to the spectrograph, the image of an extended source as several spatially sampled sections. The grating spectrograph employed in the slit-less mode then gives the spectra of each section of the object as shown in the 3rd image in the figure as spectrograph output. These spectral images are then analysed to produce a 3-dimensional spectral output – two dimensions in spatial plane and one in the spectral element direction completes the 3-dimensional data cube, (x, y, λ), shown in the extreme right.

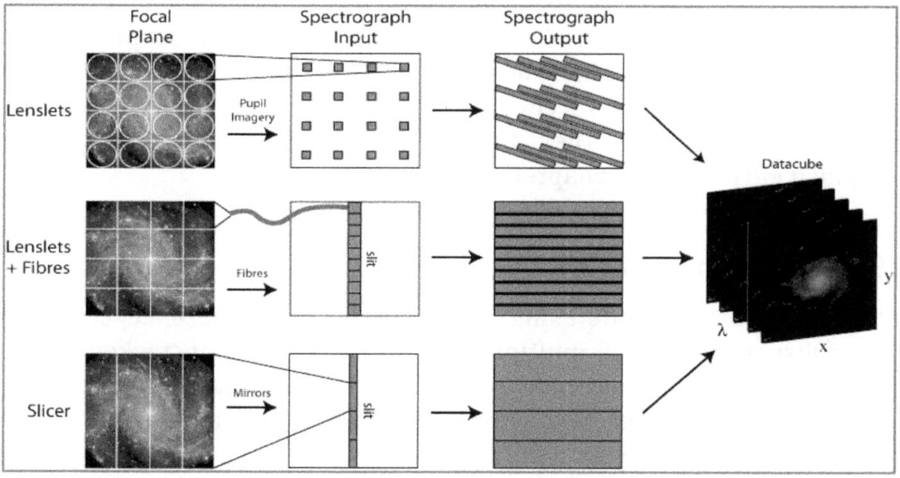

Figure 5.48: Integral Field spectra using a Multi-object grating spectrograph. Three methods are illustrated in the picture. Adapted from Bacon et al 1995; see also Bershady et al 2004, 2005; and Sanchez et al 2007a, b.

In the second panel of Figure 5.48, the slit is introduced in the grating spectrograph and the images of an extended source formed by the lens-lets are fed sequentially to the spectrograph slit by means of optical fibres. The spectral image now contains spectral information from each of the sections on the object thus completing the data cube.

In the third panel of the figure, the object is fed to the slit sequentially by slicing the image into several sections by plane mirrors. One rotating mirror reflects portions of the object onto a series of mirrors called pupil mirrors that in turn reflect the received portions sequentially onto the grating. The spectral image thus formed gives the data cube as shown. For details of these techniques, we may refer to the review by Allington-Smith (2006) and the references therein.

As already mentioned, the advantage of the grating spectrograph used as an integral field spectrograph over the imaging FPS (or Michelson-FTS) is that the instrumental function is well defined in the former than the later and the analysis is easier. Further in the case of IFPS scanning in wavelength is required to cover the spectral region, while the grating-IFS has this feature inherent in it (multiplexing). The Grating-IFS gives several lines at one time while IFPS gives only one line at one time (each with scanning over one FSR). It is, however, possible that a few lines may simultaneously be present in one FSR of an IFPS. It may be noted that the IFPS gives better spatial and spectral resolutions.

As we have described in the foregoing sections, many of the spectrographs require a spectral window-defining filter that not only restricts the wavelength to the desired range but also helps provide clean uncontaminated spectra. These filters usually have narrow bandwidth of less than a nanometre to several nanometres. These filters are called interference filters as they are based on two-beam interferometry described earlier in this chapter.

5.8 Interference Filters

Interference filters, made using the principles of interference, consist of a glass substrate on which multiple layers of alternatively stacked high and low refractive index materials are coated with each layer separated by a half-wave spacer layer. The resulting pass-band wavelength has high transmittance and narrow width. For a light ray of wavelength λ_0 incident at angle of θ, the transmitted wavelength is given by (assuming the surrounding medium is air for which the refractive index is 1.0),

$$\lambda_\theta = \lambda_0 \{(\mu_f^2 - \sin^2\theta)/\mu_f^2\} \tag{5.98}$$

where μ_f is the effective refractive index of the filter layers. This equation is valid for incident angles $< 10^0$. This shows thus that a filter may be tuned to a desired wavelength in a limited way. Tilting the filter will also alter the bandwidth. For a given bandwidth the maximum tilt angle that can be used is (see e.g., Petitdidier, 1983)

$$\theta_m^2 = 2\frac{\mu_f^2\,\Delta\lambda}{\lambda} \tag{5.99}$$

The tuning can also be achieved by changing the operating ambient temperature in which the filter is mounted. With increasing temperature, the wavelength shifts to longer values by ~ 0.02 nm/^{0}C (see Spectra-Film brochure for further details).

The transmission of the filter at the peak too changes as follows,

$$T_m(\theta) = T_m(0)\,(1 - B\theta^2) \qquad (5.100)$$

$$with\ B = 1.71 \pm 0.77$$

The critical parameters that we need to consider in choosing the right filters for a specific observation are:

(i) central or effective wavelength;

(ii) full-width at half-maximum (FWHM) or width of the pass band;

(iii) transmission at the maximum;

(iv) the shape of the transmission curve;

(v) the variation of the filter parameters with environment – change of temperature and pressure;

(vi) the usable aperture or size;

(vii) the optical quality or flatness of the surfaces;

(viii) the refractive index and the thickness of the filter.

5.8.1 Dichroic Filters: Dichroic filters (or mirrors) are thin film interference filters that transmit a certain range of wavelengths while the other range of wavelengths are reflected. Usually, the reflected and transmitted beams are at right angles to each other. Such filters are better suited for selectively passing a range of wavelengths into a photometric or spectroscopic instruments. These are also used as beamsplitters for guide-star optics.

5.9 Data Analysis of Spectroscopic Observations

In this section, we discuss basic principles in the analysis of data obtained with grating and Fabry-Perot spectrometers. There may be a few repetitions here, but we wish to summarise the critical points.

5.9.1 Grating Spectrometer Data Analysis: Data analysis steps are mostly common for visible and infrared wavelength regions. We give below a stepwise sequential procedure for analysing and extracting spectra of programme stars using a grating spectrograph.

Step 1: Observe sufficient 'raw' frames of spectra of a programme object and standard source for obtaining *S/N* not less than 3. If we have previous knowledge of the flux expected from the object, then we can estimate the exposure time required to get a desired *S/N* (from equations derived in earlier chapters).

Step 2: Obtain 'clean' object frame using the formula given in Chapter 4, $P = (O\text{-}S)/(F\text{-}D)$ where O is object(+sky), S is sky, F is flat-field frame and D is Dark frame. Dark gets subtracted automatically in the operation (O-S).

Step 3: Find the width and centre of spectral trace by taking a cut across the spectral trace. This typically looks like a sharp peak of width of a few pixels. Take sky or background sample of the same width and subtract from the object trace.

Step 4: Identify lines in standard spectra and calibrate the dispersion axis (nm/pixel). We may note here that typical night sky lines from the atmosphere may be found in Massey et al (1990). Standard spectral line lists for different elements are listed in literature, e.g., CRC Handbook for both visible and near-infrared spectra. In addition, we may use the atmospheric emissions as calibrators, especially in the near-infrared region. The near-infrared atmospheric OH lines are listed in Observatory web pages of Gemini/UKIRTF.

Step 5: Apply spectral calibration to the object spectra. Many applications need analysis up to this point.

Step 6: Flux calibration (relative or absolute) requires observations on one or more spectrophotometric standard stars. The flux calibration in spectroscopic observations is difficult and not as accurate as it is in the photometric observations. However, it is better to have at least a relative calibration to identify, understand or interpret the relative strengths of spectral lines arising from the same atom or molecule. Spectrophotometric standard stars for visible and near-infrared regions may be found in literature. We may refer to Massey et al (1988), Massey & Gronwall (1990), Oke (1990), and Stone (1996) for spectrophotometry of standard stars in the range 320-1000 nm and Lancon & Rocca-Volmerange (1992) for spectrophotometry of standard stars in the range 1.4 – 2.5 μm). After going through the steps 1-5 for the standard star, for flux calibration we need to identify a suitable spectrophotometric standard star. We need extinction and airmass corrections for the entire wavelength range of the spectrum. Once we specify the standard star the analysis package (such as IRAF) identifies the star and calculates what is known as sensitivity function which is a template for calibrated flux with wavelength. This template is then applied to the programme object. Usually, the star Vega (or any A0V type star with known extinction and airmass) is employed, for its magnitude in all the visible and infrared bands is considered as 0.0. These sequential procedures or tasks are built-in with several astronomical data analysis packages such as IRAF. We just need to feed the required inputs correctly.

Continuum Subtraction: Usually, the observed emission or absorption spectra ride over a continuum spectrum. It is important to present the emission/absorption spectra after removal of continuum. This is done by identifying a few critical nodal points in the continuum and then fitting a polynomial and then subtract from (or divide) the original spectra.

5.9.2 Analysis of Fabry-Perot Spectrometer Data: We outline here the data analytic techniques for the three modes of operation using FP etalon. We note here that the standard photometric and spectroscopic packages such as IRAF do not deal with the FPS data analysis. However, the initial cleaning of the data – dark-subtraction, sky-correction, flat-fielding and co-addition of frames - may be done using these packages when a digital array detector (such as a CCD for visible and an Infrared array detector) is used for recording data.

5.9.2.1 Central-Fringe Scanning FPS: The data analysis for this mode is rather straight forward (see section 5.5.9). At each scanning step we get a count corresponding to the number of photons detected in the narrow spectral window (for a PMT detector in photon-counting mode). Using a spectral lamp or even a bright source such as Orion nebula we can calibrate the scanning steps in terms of wavelength scale. The calibration lamp scans (over 1 FSR) give two maxima and hence the FP etalon may be calibrated at the wavelength being scanned. The manufacturer specified gap gives the FSR obtained by using the formula $FSR = \lambda_0^2/2d$ where λ_0 is the peak wavelength of the spectral emission line being scanned and '*d*' the gap of the etalon. The FSR as number of steps obtained from the scanned data is then calibrated using the FSR calculated from the formula. This gives the wavelength calibration.

For a piezo-electric scanned FP etalon (see section 5.5.5.3), we use a ramp voltage incremented by regular intervals for a desired range and traverse back to reach the starting position of the etalon gap. Each increment in voltage corresponds to a dial (of the control unit of the etalon) reading "*z*" in volts. The gap of the etalon varies as $d = d_0 + a\,z$, where "*a*" is the conversion factor between the gap "*d*" in micrometres and "*z*". We can determine the factor "*a*" by using a standard spectral lamp and observing the two consecutive maxima to get the FSR by using the relation $a = \lambda_0/(2\,\Delta z)$, remembering that a scan of one FSR of an etalon can be achieved by increasing the gap by $\lambda_0/2$. Comparing this dial range Δz with the values obtained from the actual gap (specified by manufacturer), we can get the scale in wavelength from the observed scale involving *z*.

Flux calibration in FPS observation is even more inaccurate and difficult than for the grating spectrograph for the simple reason that in FPS we observe one single emission line at a time. However, when we observe Hydrogen recombination lines in emission it is in principle possible to employ emission line stars that are considered spectroscopic standards as calibrators for which emission line fluxes are listed. Standard sources for emission line fluxes may include well-studied and well-documented bright gaseous nebulae, such as Orion HII region, but regions to be compared must be the same.

5.9.2.2 Single Interferogram Mode: For a single interferogram (see section 5.5.10) the data analysis is as follows.

Step 1: Obtain the object frame followed by a calibration frame using a spectral lamp source at appropriate wavelength. The exposure time of the object frame needs to be adequate to obtain a good *S/N*. If needed several frames may be taken at the same gap setting of the etalon. All object frames are dark & background-subtracted and flat-fielded.

Step 2: Find the centre of the fringe system from the calibration frame and confirm the same from outer fringes as well.

Step 3: Take radial scans at appropriate intervals of angles and arc lengths.

5.9.2.3 IFPS Data Cube Analysis: As already mentioned, this is the most accurate mode of operation of a FPS (see section 5.5.11). Let us enumerate sequential steps involved in this mode of operation (see for details Atherton et al 1982, Bland & Tully 1989).

Step 1: Obtain an adequate set of interferograms on the programme object at several gap positions of the etalon being used, covering a little more than 1 FSR – typically 1.2 times the FSR. Let us name this data set as Object Data Cube (2 space coordinates and one spectral coordinate – x, y, λ).

Step 2: Repeat Step 1 on a standard spectral source, for phase map generation. Let us call this as calibration data cube.

Step 3: Apply phase correction on the object data cube using the calibration data cube. This gives us phase-corrected object data cube.

Step 4: Using the phase-corrected object data cube extract line profiles in the entire field of view of interest at the spatial resolution limited by seeing (typically 2-3 arcsec taking Nyquist sampling criterion). In case the S/N is not adequate at this spatial resolution, we may bin the data so that a larger spatial element is sampled. This gives line profile data covering the entire field of view.

Step 5: Using this velocity profile map, generate iso-velocity contour maps – images at each velocity interval normalised to zero at the centre of the object. We may shift this centre to any reference point of our choice.

We may write down the peak transmission of FPS in terms of its Airy function and the object flux function as follows:

$$\Phi_{xy}(d) = \varphi_{xy}(\lambda)/[1 + \frac{4F^2}{\pi^2} sin^2\{ 2\pi d \frac{cos\theta}{\lambda}\}] \quad (5.101)$$

where d is now a function of "z" as defined in the earlier paragraphs and F is the effective finesse. The refractive index of the gap medium (air) is assumed 1. For central on-pixel $(x, y) = (0, 0)$, the wavelength λ_0 is transmitted maximally in the mth order for a value of z given by,

$$az_0 = m\lambda_0/2 - d(0)$$

For an off-axis pixel (x, y), maximal transmission for λ_0 occurs at a different z value given by,

$$az_{xy} = d(z_0)(\sec\theta - 1)$$

or,

$$z_{xy} = m\,\Delta z_0\,(\sec\theta - 1) \quad (5.102)$$

This then is called the phase map. We may notice that this function above is independent of the gap of the etalon.

5.9.2.4 FPS Fringes in Mono-Chromatic and Quasi-Monochromatic Light Beams: We may mention here that for a monochromatic light beam, the concentric rings that we obtain from an FP etalon represent orders of interference "m". Supposing that the central order is m_0 satisfying the equation, $m_0\lambda_0 = 2d$, then an off-axis fringe at angle θ at an order "m" satisfies the equation, $m\lambda_0 = 2d\cos\theta$. Therefore, we get,

$$\cos\theta = m/m_0 \quad (5.103)$$

For a quasi-monochromatic light beam (defined earlier as $\Delta\lambda/\lambda << 1$, where $\Delta\lambda$ is the range of wavelengths of the beam), however, the fringes of an etalon represent transmission at different wavelengths slightly different from the centrally transmitted wavelength. In this case the second equation for off-axis given above reads now as, $m_0\lambda = 2d\cos\theta$; it is the same order as for the central transmission but at a different wavelength. We can then show that the change in wavelength is given by,

$$\delta\lambda_{xy} = \lambda_0\,(1 - \cos\theta_{xy}). \quad (5.104)$$

Here the subscript "xy" indicates the position of an off-axis pixel (x, y).

CHAPTER 6

Data Quality Enhancement Techniques

Errors are not in the art but in the artificers – Isaac Newton

Astronomical data requires due processing to obtain meaningful results that can be later interpreted in terms of the physical processes governing a celestial object. The data obtained by using photometric and spectroscopic techniques, described in the foregoing chapters 4 and 5, is often fraught with noise from several sources: the environmental changes, sky variations, detector noise and errors associated with analysis or processing. It is therefore required to "clean" and enhance the quality of the data before starting interpretation.

Firstly, all the observation data obtained should have a reasonably good *S/N* ratio which may require co-addition of sets of observations and processing to remove all kinds of background and detector noise. Then the data requires calibration either photometric or spectral using standard sources. After this, especially spectroscopic data require deconvolution using instrumental function. The resulting data may then be subjected to further smoothing and other data enhancing techniques. We may refer to Starck & Murtagh (2017) for details.

We discuss here, some techniques of data processing in extracting photometric and spectral information from the observed data. We deal with the data quality enhancement techniques also. Convolution techniques to remove the instrumental effects are also presented with examples.

6.1 Accuracy and Precision

Used often as synonyms, the technical terms accuracy and precision actually mean quite differently. Suppose we make an observation of a physical parameter concerning an object using an instrument. Suppose that the expected value of the parameter is 'σ'. If the measured value $s \pm \Delta s$ matches with the expected value 'σ', within the limits of uncertainty Δs, then our observation is accurate within the specified uncertainty. If we repeat the observation arriving each time at an average value that matches with the expected value, then our observation has precision also within the domain or the limits of uncertainty. In some situations, it may so happen that our observations will have precision of repeatability in obtaining the value '*s*'; but do not have accuracy of reaching the expected value '*o*'. The reason for this discrepancy could be systemic. In many cases involving astronomical observations, we do not know the actual expected value, then in these cases precision (repeatability) is needed to rely upon the results. Our endeavour, therefore, should be to achieve precision. Accuracy may also be achieved if the systemic errors could be determined.

6.2 Noise Reduction

Noise is random and does not have a repeating pattern. It is usually quantified by root of mean square difference. Suppose we make *N* number of observations of same parameter *p*. We then find the mean of the observed quantity.

$$\bar{p} = \frac{1}{N}\sum_{1}^{N} p_i \tag{6.1}$$

The standard deviation (square root of variance σ^2) is then given by,

$$\sigma = \frac{1}{N}\left\{\sum_{1}^{N}(p_i - \bar{p})^2\right\}^{\frac{1}{2}} \tag{6.2}$$

This represents uncertainty in the observations.

6.2.1 SNR and Exposure Time: As mentioned earlier the *SNR* depends on the exposure time by its square root, $SNR \sim t^{1/2}$. Thus, by increasing the exposure time adequately we can achieve required *SNR* for any observation. But caution must be exercised to ensure that during the exposure time the extraneous factors such as atmospheric and environmental changes are insignificant. We must also avoid large zenith angle changes during the observation. Typically, a single exposure should not exceed 1 hour. During this period the object should be tracked meticulously so that it remains in the centre of the field of view. Many modern telescopes have excellent tracking accuracy over long observational times. For telescopes that do not have required tracking accuracy, we use a guide-star system with a dichroic/plane mirror in the optical arrangement of the focal-plane instrument. Typically, the dichroic/plane mirror reflects but only a small percentage of the light beam into the guide-star system. The corrections are applied to the telescope tracking system continually. In the case of digital array detectors, we do not need a separate guide-star arrangement because of their capability to image large areas of sky.

6.2.2 Co-adding Data: Signal and noise are uncorrelated. Noise being random, when we co-add several sets of observed data the signal-to-noise ratio improves. Instead of taking very long exposures that would be affected by physical changes dynamically occurring in the atmosphere, we can resort to making a number of shorter exposures and co-adding them while analysing the data. Each observation or exposure needs to be corrected for atmospheric effect especially the zenith angle effect involving air mass. As mentioned earlier, the *S/N* improves as square root of the exposure time. The final data set would simply be the average of several co-added individual sets of same exposure.

The *S/N* parameter can be improved best at the observation level itself by taking several sets of observations in similar conditions. However, there are techniques beyond the telescope floor that help improve the quality of data observed.

6.3 Data Quality Enhancement

Besides achieving improvement in the signal-to-noise ratio (*S/N*), we need other means to enhance the quality of the data obtained. This involves smoothing to get rid of unwanted spurious features – either systemic or environmental.

6.3.1 Moving Average Method for 1-D Data: In order to mitigate the noise effect and discern signal better, the simplest method adopted is to average neighbouring points and assign the value to the mid-point of the sample and move in forward direction each time doing the averaging and assigning. In this process of the averaging and assigning, we lose the first few data points and the last few. This requires the number of points chosen for averaging be an odd number. Suppose we average 3 data points in dependent variable coordinate (i.e., signal counts, for example), then we lose the first and the last points; if we average 5 points then we lose the first 2 and last two points, and so on. Care needs to be exercised while choosing the number of points since we lose information especially when the data has sharply peaked points that represent genuine information. Over-smoothing needs to be avoided. In general, we may formulate the moving average process as follows: suppose the original data (signal vs time or step size) is represented by s_i, s being the signal and i the time or step-size and $i = 1, 2, 3...n$. The number n is the total number of data points, considered here as an odd number. If the number of data points chosen for averaging is N (considered as an odd number), then the moving averaged data S_j (signal S

averaged over a set of N points and assigned to the middle point of the set of N) may be written as

$$S_j = \frac{1}{N}\sum_k s_k$$

$$with\ k = j - \frac{N-1}{2}\ to\ j + \frac{N-1}{2}$$

$$where,\ j = \frac{N+1}{2}\ to\ n - \frac{N-1}{2} \tag{6.3}$$

Figure 6.1 shows a raw spectrum obtained from Mt Abu 1.2 m telescope equipped with near-infrared camera and spectrograph (NICS) on the object V745Scorpii in H band along with the smoothed spectrum using 5-point moving-average. We see a little bit of smoothing without compromising with the positions of the real peaks that may actually represent faint emission lines. A 3-pt averaging (not shown here) almost coincides with the raw spectrum except at the peaks where the amplitude is a little bit reduced. But further smoothing beyond 5-point-average would certainly distort or partially remove genuine features and hence not advised.

6.3.2 Mean and Median Filtering for 2-D (image) Data: Very often when we make imaging observations using area detectors such as a CCD, we find abrupt intensity variations among neighbouring pixels. Usually, we remove inter-pixel sensitivity differences by flat-fielding as described earlier. But images still show fluctuations in intensity due to various factors, such as image wandering, seeing effects. To remove such effects, we resort to what is called a spatial filtering. The method involves choosing an appropriate kernel function and applying it in Fourier domain and taking an inverse transform, resulting in a cleaned or smoothed image. Kernel functions that are applied vary depending upon the purpose of enhancement. A simple **mean filter** does an averaging operation using a small square area consisting of 4x4 or 9x9 pixels (cf. moving average process in case of 1-D data). Median filtering, that involves choosing the median value of neighbouring pixels to represent the central pixel of a selected area, is often a better way to deal with smoothing than mean filtering, since the former's inherent nature of being more representative of the neighbouring pixels. Both filters may be considered as low-pass filters.

6.3.3 Filtering in Spatial Frequency Domain: The variation of signal with either time or space is smoother than the randomly varying noise. This allows us to use high-pass or low-pass (or even both) filtering in the Fourier domain (frequency for time series and spatial frequency for spatial signal). Care must be exercised in not to cut off signal frequencies. A trial-and-error exercise is often the best course; or for strong signals we can do power spectrum analysis to determine the range of signal frequencies. In case of very narrow signal frequency range, we can use the Gaussian type of filter that eliminates both low and high frequencies on either side of the signal frequency maximum. These methods may be applied for both 1-D as well as 2-D data sets.

The frequency filtering is done using both low- and high- pass filters – like a pass-band filter. Choosing of the correct pass band determines how much smoothing is done. Some trial-and-error needs to be done to arrive at an optimal band pass frequency. If the Nyquist critical spatial frequency f_c is defined as (with δx as interval between consecutive points in the data)

$$f_c = \frac{1}{2\,\delta x} \tag{6.4}$$

Usually a passband that allows +/- f_c may be adequate. A high frequency cut-off is usually given by (see Numerical Recipes 1999)

$$f_{hf} = Max\left[1 - \frac{6f^2}{f_c^2}\,,0\right] \tag{6.5}$$

Using smoothing or noise-reducing kernels such as mean (e.g., moving averaging for 1-D data as discussed above) or median filters for 2-D data would be equivalent to convoluting the data in the time or spatial coordinates. But in the Fourier space, it would be multiplicative. Both these processes are not the same but approximately give similar results. So, what we can do is to take FFT of the data series and multiply with a chosen filter function in Fourier space and then take inverse transform of the product. If *o(x)* is the observed function and *g(x)* is the filter function, then the filtered (observed) data function is

$$o'(x) = o(x) * g(x)$$

where the asterisk (*) denotes the convolution operation. In Fourier space, let the spatial frequency be denoted by X (the counterpart of x). Then the filtered data function O' is given by,

$$O'(X) = O(X) \times G(X) \tag{6.6}$$

The filter function *G(X)* above needs to be chosen judiciously in such a way as not to lose information. *G(X)* can have either a low-spatial-frequency cut-off or high-spatial-frequency cut-off or both (i.e., window) depending upon the data in question.

Alternatively, the filter function can be a Gaussian which will have both low and high frequency cut-offs. A Gaussian will not change its character in Fourier space – it would remain Gaussian with different width. By choosing an appropriate width we can apply a Gaussian function in spatial-frequency space. The inverse transform of the product of the FFT of the observed function and the Gaussian function in spatial frequency would give filtered data function $o'(x)$. The Gaussian filtering yields satisfactory results in both 1-D as well as 2-D data (imaging data).

Since a Gaussian remains a Gaussian in character in the frequency domain, the frequency filtering is very definitive in its limits in frequency (by choosing appropriate width of the Gaussian). This is not the case with the mean/median filtering as it shows in 2-D oscillations in frequency domain.

Wavelet transform is widely used for noise filtering due to its ability to separate signal from noisy data in the transformed domain (see Mallat 1999 for a comprehensive exposition of the subject). It allows the signal to be transformed into low-frequency and high-frequency components. The high-frequency components often represent noise and low-frequency components contain the primary signal information. In the methods using FFT, the transformed domain (time to frequency or space to spatial frequency) consists of base functions that are sine and cosine functions. In discrete wavelet transform base functions are not as simple and are called scaling functions and wavelets. Various types of wavelets exist with unique properties tailored for specific applications. Here we describe biorthogonal wavelets that are widely used to filter image noise. The advantage of the biorthogonal wavelets over the normal orthogonal wavelets is that it has more degrees of freedom and can have symmetrical wavelet functions compared to the latter. It consists of a set of primal and dual functions with scaling and wavelet functions in each for effective decomposition and reconstruction of the image. The scaling functions capture the low-frequency part of the signal while the wavelet functions are used for high-pass filtering. The primal scaling and wavelet functions are given respectively by

$$\begin{aligned} \phi(t) &= \sqrt{2}\ \textstyle\sum_k h_k\, \phi(2t-k) \\ \psi(t) &= \sqrt{2}\ \textstyle\sum_k g_k\, \phi(2t-k) \end{aligned} \tag{6.7}$$

where h_k, g_k are the low pass and high pass filter coefficients of the primal function. In the above equations, t represents the signal domain, 2 is the dyadic scaling factor used for scaling and k is an integer representing discrete shifts. The numerical factor is the normalisation factor between the primary and dual scaling function. The function φ allows multi-level decomposition.

Similarly, the dual scaling and wavelet functions are given by

$$\begin{aligned} \phi^{\sim}(t) &= \sqrt{2}\ \textstyle\sum_k h_k^{\sim}\ \phi^{\sim}(2t-k) \\ \psi^{\sim}(t) &= \sqrt{2}\ \textstyle\sum_k g_k^{\sim}\ \phi^{\sim}(2t-k) \end{aligned} \quad (6.8)$$

Here, $h^{\sim}_k$, $g^{\sim}_k$ are the low pass and high pass coefficients of the dual functions. The primal and dual functions satisfy orthogonality conditions; and decomposition, and reconstruction processes are invertible.

The low and high frequency decompositions are thus accomplished for the observed two-dimensional astronomical images using the primal and dual functions described above along their rows and columns. The high frequency noise that is manifested in the coefficients are removed by suitable thresholding. Inverse wavelet transform can be used to reconstruct the image after suppressing the coefficients that correspond to noise in the image.

The advantage of wavelet transform method over the Fourier techniques is that it can handle data that contains discontinuities more effectively than the latter. While the wavelet transform is localised in both the domains (e.g., time and frequency or space and spatial frequency), the Fourier transform is only localized in the frequency domain.

6.3.4 Examples of 1-D and 2-D Filtering: Figure 6.2 shows a raw spectrum obtained from Mt Abu 1.2 m telescope equipped with near-infrared camera and spectrograph (NICS) on the object V745Scorpii in H band and a frequency filtered spectrum using both low and high frequency cut-offs.

For Gaussian filtering a kernel in the shape of a Gaussian function is chosen and multiplied with the Fourier transform of the observed spectrum and the product in the spatial frequency domain is then inverse Fourier transformed to give filtered spectrum. One such example is shown in Figure 6.3. A Gaussian width of 256 elements was chosen to obtain the smoothing. Reducing the width would result in over-smoothing and may result in loss of information.

It must be said here that the image software that is supplied with imaging detectors do have routines to clean up data using various types of filtering techniques. Some of these techniques are also provided by internet astronomy sites. Therefore, the attempt here might seem audacious and superfluous. However, the purpose of including some of the methods here is to provide an overview of the basics involved in the methods. The filters, like instruments, should never be used as black boxes. We must understand the power and pitfalls of the tool in hand before using it.

As is clear from Figures 6.1, 6.2 and 6.3 all the three methods of smoothing yielded comparable results but the 5-point moving average method show rounded-off peaks (Figure 6.1) compared to the frequency filtering methods.

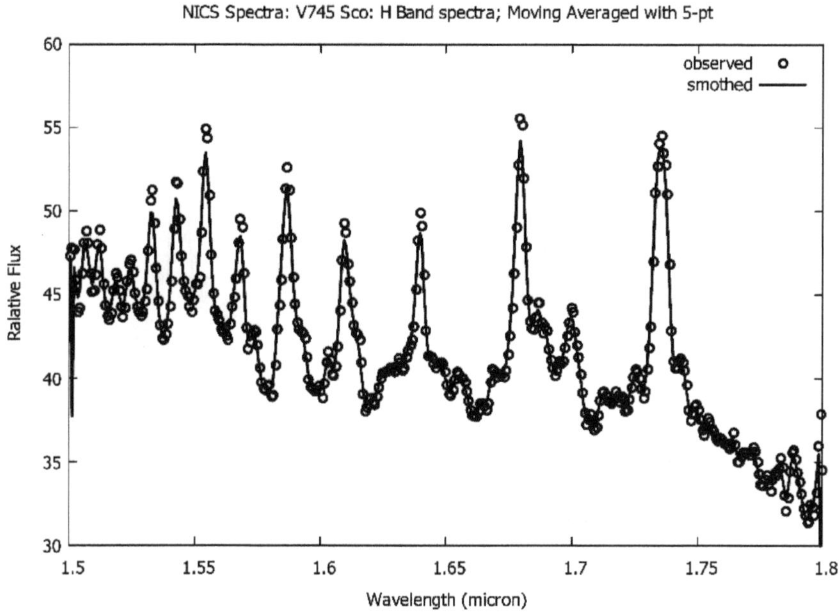

Figure 6.1: Moving averaged spectra of a Nova V745 Sco in H band taken at Mt Abu. A 5-point average was employed.

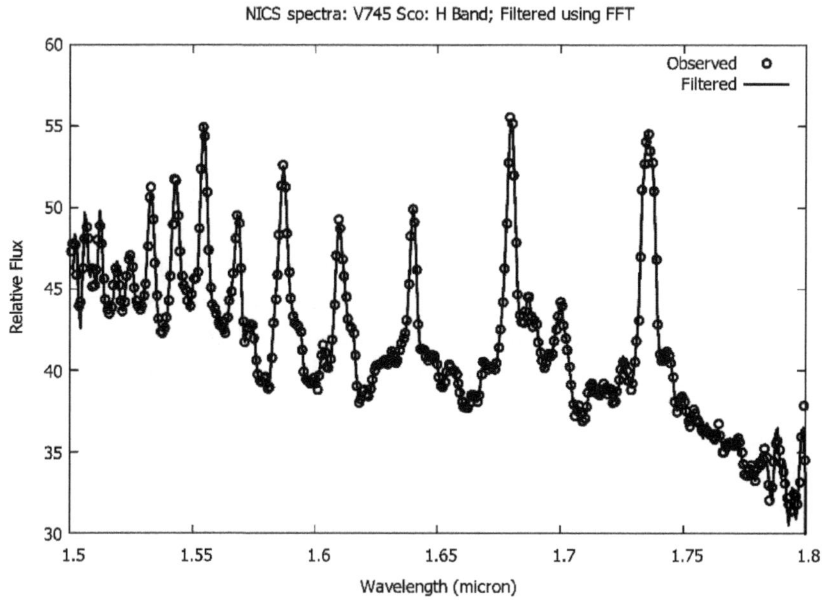

Figure 6.2: Frequency-filtered spectra of V745 Sco in H band: a combination of low- and high-pass filter was employed.

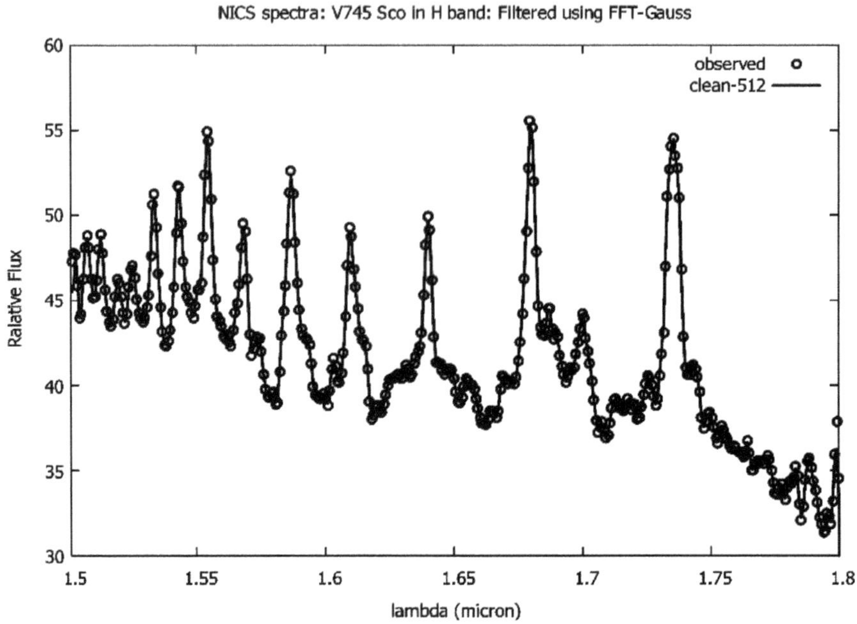

Figure 6.3: Gaussian-filtered spectra of V745 Sco in H band.

We showed a few smoothing methods to clean 1-D data in Figures 6.1, 6.2 and 6.3. Similar techniques are applied to the H band image of a Globular Cluster M53 (NGC 5024, in the Coma Berenice constellation) shown in Figure 6.4. The original image obtained from Mt Abu is of the size 1024x1024 pixels, each pixel corresponding to 0.5 arcsec in the sky. For this image, for various stars of different magnitudes, the *S/N* varied in the range of 20-35. The FWHM is typically 3 pixels. Figure 6.5 shows the image after using a Gaussian kernel filter. For this case, for the same set of stars, we obtained *S/N* in the range 40-75 with an FWHM of 4 pixels. Figure 6.6 shows the image after applying a high cut-off frequency filter. In this case the *S/N* varied in the range 40-65 with typical FWHM of 3 pixels. Finally, Figure 6.7 shows NICS H-band image of M53 cleaned using the wavelet transform method. For this case, for the same set of stars, we obtained *S/N* in the range 40-80 with an FWHM of 3 pixels. We find a substantial improvement in the *S/N*, by a factor of 2, for all the three image quality enhancement techniques described here. In the case of the Gaussian kernel method, the star image is a little rounded-off, as can be made out from the slight increase in the FWHM. Thus, the filtered images show smoothing of peripheral regions while preserving the genuine sources and improving the *S/N*.

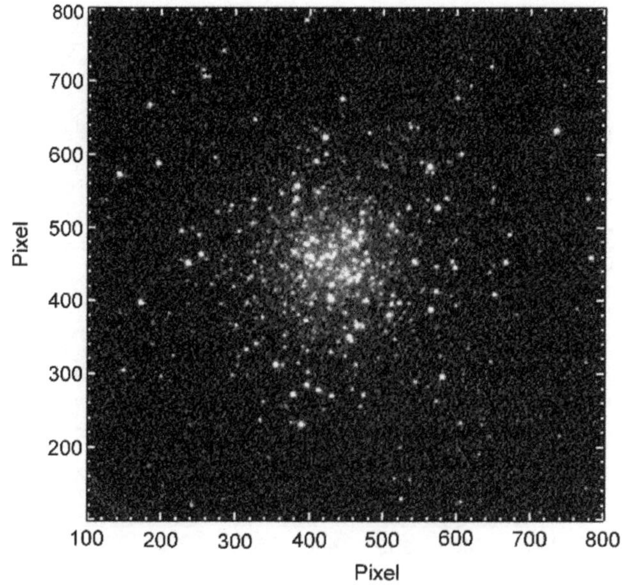

Figure 6.4: A sky-subtracted, un-smoothed image of the Globular Cluster M53 (NGC 5024) in the Coma Berenice constellation, imaged in H Band from Mt Abu using NICS; the original image is 1024x1024 (~ 8.5x8.5 arcmin^2 – each pixel being 0.5 arcsec).

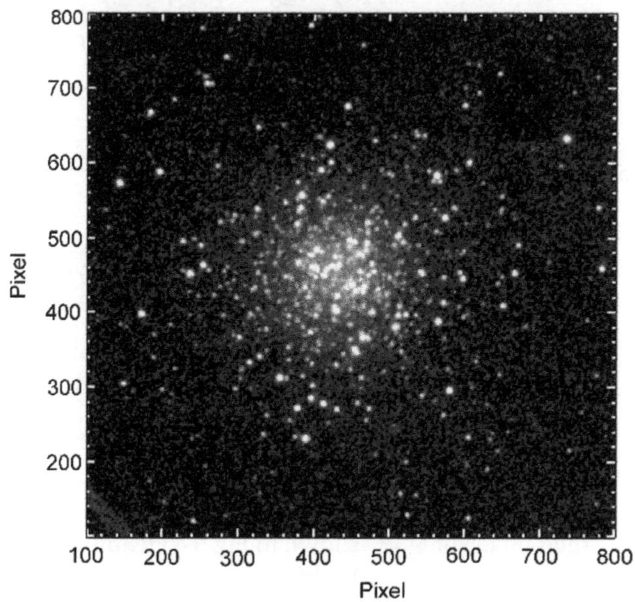

Figure 6.5: Smoothed image of M53 (shown in Fig 6.4) using a Gaussian kernel.

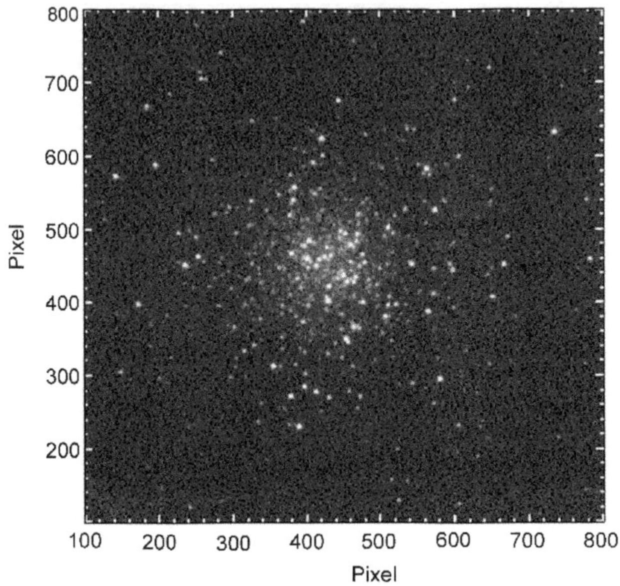

Figure 6.6: Smoothed image of M53 (shown in Fig 6.4) using high-pass filter.

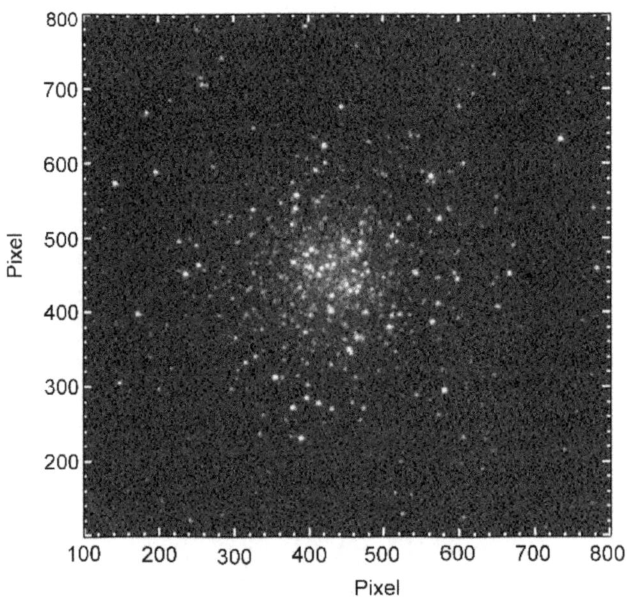

Figure 6.7: Smoothed image of M53 (shown in Fig 6.4) using a wavelet transform method.

In the following pages we describe two more techniques that are needed to complete the narrative on data analysis. The first one involves the removal of instrumental effects on the observed data. This requires a prior knowledge of the instrumental point spread function or the response of the instrument to a point source in imaging or spectroscopic devices. The second technique is to decompose an observed function into its constituent components. Many a time we encounter objects that emit complex photometric or spectroscopic signatures that need to be understood in terms of contributions from different components embedded in the source.

6.4 Deconvolution

When we make observations in photometry or spectroscopy, we measure a certain parameter function using a suitable instrument. The function that we measure or observe does not truly represent the actual or real function that originates from the object on which we make measurement or observation. It is always convoluted by the instrumental function. Instrumental function as defined in the earlier chapters is the response of an instrument to a point source in case of a photometer and to a narrow spectral line in case of a spectrometer. Let us suppose that the actual real function is $p(x)$ and the instrumental function, also sometimes called point spread function (PSF), is $a(x)$ then the observed function $s(x)$ (after removing noise by using methods discussed earlier) may be written as

$$s(x) = p(x) * a(x) \tag{6.9}$$

Here the asterisk (*) denotes convolution operation. This may be expressed as

$$s(t) = \int_{-\infty}^{+\infty} p(t')a(t - t')dt' \tag{6.10}$$

We can see that if $a(x)$ were to be a delta function $\delta(x)$, $s(x)$ would be the same as $p(x)$. But in real world, $a(x)$ has a finite width. We can prove this as follows: taking Fourier transform of the convolution equation, we have with upper case letter denoting the FT of the function and y is the independent variable in its Fourier space,

$$S(y) = P(y) \times A(y) \tag{6.11}$$

But we have $A(y) = 1$ for delta function. Hence by taking inverse FT we get back simply $P(y)$.

However, in real world the instrumental function has a finite width and cannot in general be approximated to a delta function. In such cases, how does one retrieve the object function $p(x)$ given the observed function $s(x)$

and the instrumental function *a(x)*? One way is to take FT of the convolution equation as above and write,

$$P(y) = \frac{S(y)}{A(y)} \tag{6.12}$$

And then we take inverse FT to get back *p(x)*. However, determining discrete FT (DFT) for the PSF could create numerous problems because for most of the *x*-space it is nearly zero except for the central narrow maximum. Further in the presence of noise in both *s(x)* and *a(x)* the inversion problem becomes unwieldy. We note that although we are discussing a 1-D case here, we may extrapolate these arguments for a 2-D imaging observation as well.

Alternatively, we can write the integral equation for convolution as a set of linear equations and solve for *p(x)* by matrix inversion. As mentioned already the near-zero values in *a(x)* matrix would render the entire exercise ill-posed.

In order to retrieve the object-function we need to resort to other methods. We present one such method here, called "regularization" (see Phillips 1962, Subrahmanya 1980, Anandarao and Rao 1986a, for details and cross references). We describe here the method that includes noise *n(x)* in the data (but added separately) and use the same to minimise it while obtaining a solution for *p(x)*.

The convolution integral with finite number of closely spaced terms is given by,

$$s(x_m) = \sum_{i=1}^{i=N} p(x_i)a(x_m - x_i)\, w_i \; + n(x_m) \tag{6.13}$$

Here $m = 1$ to N and the factor w_i are the weighting factors used for numerical integration given by, $w_i = k_i \Delta x$ where k_i are constants and Δx the step size of data. Now let us write down the variance of noise *n(x)* as basically the averaged noise, overall the terms,

$$\sigma^2 = \frac{1}{N}\sum_1^N n_m^2 \; = \; \frac{1}{N}\sum_{m=1}^N \left\{s(t_m) - \frac{1}{N}\sum_{i=1}^N p(x_i)a(x_m - x_i)w_i\right\}^2 \tag{6.14}$$

The condition for minimisation is written as

$$\frac{\partial \sigma^2}{\partial p_i} = 0, \qquad for\ i = 1,2, \dots N \tag{6.15}$$

Substituting the equations, we obtain

$$\sum_{j=1}^{N} A_{ij} p_j = B_i\,, \qquad for\ i = 1,2,..N \tag{6.16}$$

The parameters A_{ij} and B_i are given by,

$$A_{ij} = \frac{1}{N} \sum_{m=1}^{N} w_i a(x_m - x_i)\, w_j a(x_m - x_j) \tag{6.17}$$

$$B_i = \frac{1}{N} \sum_{m=1}^{N} w_i a(x_m - x_i) s_m \tag{6.18}$$

In principle, the object function could be obtained by solving the above equation but for the fact that the matrix A_{ij} is ill-conditioned due to PSF *a(x)* which produces spurious ripples in the solution thus obtained (Phillips 1962). We therefore need to constrain the smoothness of the solution in addition to the minimization of variance of noise. This is done by adding an additional term involving a parameter C as in $(\sigma^2 + \gamma\,(Cp)^2)$, where C is either 1, or first or second derivative operator. γ is a constant that needs to be chosen appropriately (see later paragraph). This function is then used for minimization. In this approach called regularisation (Phillips 1962), we choose C as the second derivative operator. We write the condition for minimisation as,

$$\frac{\partial(\sigma^2 + \epsilon^2)}{\partial p_i} = 0 \tag{6.19}$$

$$\epsilon^2 = \frac{\gamma}{N-1} \sum_{i=2}^{N} [p(x_{i+1}) - 2p(x_i) + p(x_{i-1})]^2 \tag{6.20}$$

$$\sum_{j=1}^{N} (A_{ij} + \gamma C_{ij}) p_j = B_i\ , \quad i = 1,2, \dots N \tag{6.21}$$

The above equation is then solved for *p(x)*. Using this method, Anandarao and Rao (1986a) deconvolved an air-borne photometric observation on the planet Venus in 100 μm band as shown in Figure 6.8. The 32.5 cm telescope on-board the French Caravelle air-borne observatory used a field aperture of 1.4 arcmin, nearly the same as the diffraction limit for a 32.5 m telescope at the wavelength of 100 μm. The angular size of the planet Venus varies between ~ 9 arcsec to ~ 60 arcsec, approximated to be an unresolved point source. Using a PSF of the photometer for a point source as 1.4 arcmin, the figure shows the deconvolved profile reasonably well-matched with the observed profile. The example is given only as an illustration of the method.

The constant γ in the above smoothing function needs to be chosen judiciously so as to keep smoothing at such a level as to not sacrifice the information. By iteration we can find out a critical value γ_c above which smoothing would start playing a dominating role. The observed signal-to-noise ratio (*S/N*) could be an indicator to a proper value for γ.

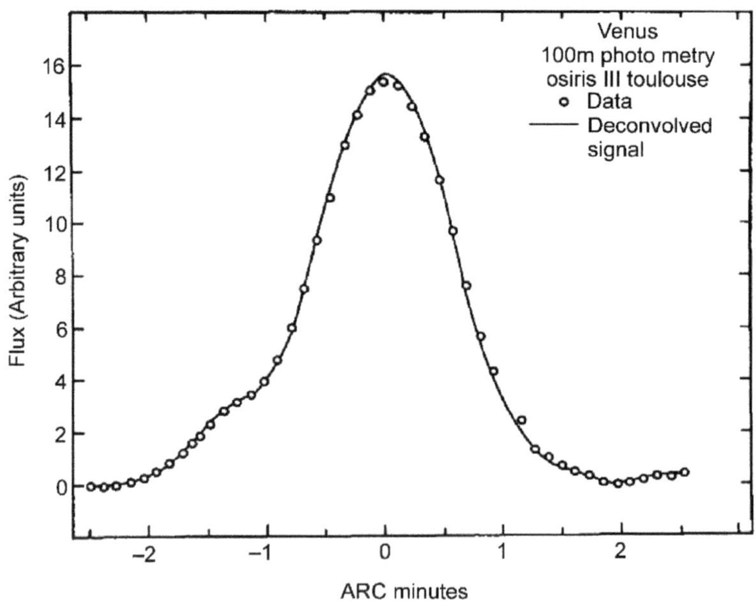

Figure 6.8: Deconvolution of observed Venus photometric profile: circles represent observed data (from van der Wal et al 1985) and the full curve shows the deconvolved profile (from Anandarao & Rao 1986a).

6.5 Decomposition of Composite Profiles

Sometimes a photometric or a spectroscopic observation is such that there could be more than one contributing source - for example, two or more overlapping point sources in photometric observations or two or more overlapping spectral lines or spatial variation in a single spectral line profile in spectroscopic observations. Further, this could happen in the case of photometry because of inadequate spatial resolution (for instance, an unresolved binary star) and in the case of spectrometry because of inadequate spectral resolution (for instance, unresolved components at different Doppler shifts with reference to line centre). Such an observation would have an asymmetric or broader than expected profile and hence could easily be surmised as to the presence of a secondary or tertiary source. Assuming that the basic profile can be approximated to a Gaussian, we describe here a method for resolving the observed composite data into constituent components. This process called decomposition is an important tool in astrophysical data analysis. We describe here one such approach adopted by Anandarao and Rao (1986b). Currently such algorithms are available online. But it is worth revisiting as to the methodology that goes into such techniques – only then we appreciate the advantages as well as the pitfalls.

Let us denote the observed data set $P(x_i)$ that needs to be decomposed. The model function $F(x_i)$ is supposed to be a composite of a set of functions of similar kind but with different parameters so that we can write,

$$F(x_i) = \sum_{k=1}^{M} f_k(x_i, \theta_j) \,, \qquad i = 1,2,..N \; and \; j = 1,2..L \tag{6.22}$$

Here M denotes the number of constituent functions f_k (of the same kind), L the number of parameters θ_j that define uniquely such a constituent function, and N the number of data points in Intensity – x-space, x could be either space coordinate or wavelength in case of a spectrum. For instance, if we assume the constituent function to be a Gaussian, then we have its height, full width at half maximum intensity (FWHM) and the position of the maximum in x-axis (say, wavelength) to completely define it. Thus, in this case, j takes 3 values; and our problem is to find a combination of these parameters so as to obtain a good fit of the data. Evidently there cannot be a unique solution given the non-linearity of the functions.

We can write the residual sum of squares of the deviation of the model from the observed data, i.e., $D(x_i) = F(x_i) - P(x_i)$ with $i=1$ to N, as

$$\phi(\theta_1, \theta_2, ..\theta_L) = \frac{1}{N}\sum_{i=1}^{N} D(x_i)^2 , \quad N \geq J \tag{6.23}$$

Here $J = L \times M$, the total number of parameters that need to be evaluated for a good fit.

The minimising condition for the deviation residual function φ can be written as

$$\frac{\partial \phi(\theta)}{\partial \theta_p} = 0, \quad for\ p = 1,2, \ldots J \tag{6.24}$$

The set of equations, that the above differentiability condition gives, needs to be solved in order to obtain all the parameters numbering J. The equations can be linear or non-linear depending up on the function $f(x,\theta)$. Employing methods required to solve non-linear least squares regularisation (see Anandarao and Rao 1986b, for details), we write the above equation as a Taylor series expansion,

$$\frac{\partial \phi}{\partial \theta_p} = \frac{\partial \phi}{\partial \theta_p^0} + \delta_p \sum_{q=1}^{L} \frac{\partial^2 \phi}{\partial \theta_p^0\, \partial \theta_q^0} + O(\delta_q^2) = 0; \quad p = 1\ to\ L \tag{6.25}$$

Here θ with superscript 0 represents approximate initial value of the parameter and the one without the superscript represents the next approximation with addition of an incremental value δ such that $\theta_p = \theta_p^0 + \delta_p$. Substituting for φ and neglecting second derivatives in $D(x)$ since they are approximately of the order of δ^2, we obtain, at $\theta = \theta^0$,

$$\sum_{i=1}^{N} \frac{\partial D_i}{\partial \theta_p} D_i = \delta_p \sum_{q=1}^{L}\sum_{i=1}^{N} \frac{\partial D_i}{\partial \theta_p}\frac{\partial D_i}{\partial \theta_q} = 0 \tag{6.26}$$

Defining the matrices in the first and second terms as

$$g = \sum_{i=1}^{N} \frac{\partial D_i}{\partial \theta_p} D_i\ ; \ and\ G = \sum_{i=1}^{N} \frac{\partial D_i}{\partial \theta_p}\frac{\partial D_i}{\partial \theta_q}; \ at\ \theta = \theta^0 \tag{6.27}$$

$$\delta_p = G^{-1} g \tag{6.28}$$

Solving the above equation, we get δ_p so that we can obtain the next approximation θ_p till the solutions converge to a desired pre-set accuracy parameter η so that,

$$\frac{\delta_p}{\theta_p^0} \leq \eta$$

Then we calculate the standard deviation ε (for difference *D(x)* between model and data) using the new values and check if

$$\varepsilon \leq \eta \tag{6.29}$$

If the above inequality is satisfied, then we accept the solution or else we iterate with the latest set of *D(x)* till we reach convergence to η.

Since the method searches in the neighbourhood of the initial guess values or priors of the parameters θ^0, caution should be exercised while choosing them lest the solutions diverge. A priori knowledge of the nature of the functions as well as the number of such functions would help in arriving at a physically meaningful solution.

Here we give an example of decomposition using Gaussian as the constituent function which is commonly assumed to represent emission (or absorption) profiles of atomic or molecular lines from cosmic sources. We can write the function as

$$f_{ki} = A_k \exp\left\{-\frac{(\lambda_{0k} - \lambda_i)^2}{2\sigma_k^2}\right\} \tag{6.30}$$

In the above equation for Gaussian profile, the parameters A_k, λ_{0k}, and σ_k denote the peak amplitude, central wavelength, and full width at half-maximum of the line profile component *k* respectively. We then substitute this into the equations involving derivatives and solve for the three parameters for each constituent function *k*. If one or two of the parameters is known to an acceptable degree of accuracy, then we hold those as constants and solve only for the unknown parameter.

Figure 6.9 shows an example of decomposition of emission line profiles into multi-Gaussian components. As already mentioned, the inversion problem being nonlinear the solutions are not unique. However, a priori knowledge of the source properties would help in choosing a viable or probable solution out of a bevy of possibilities. Thus, priors do play a critical part.

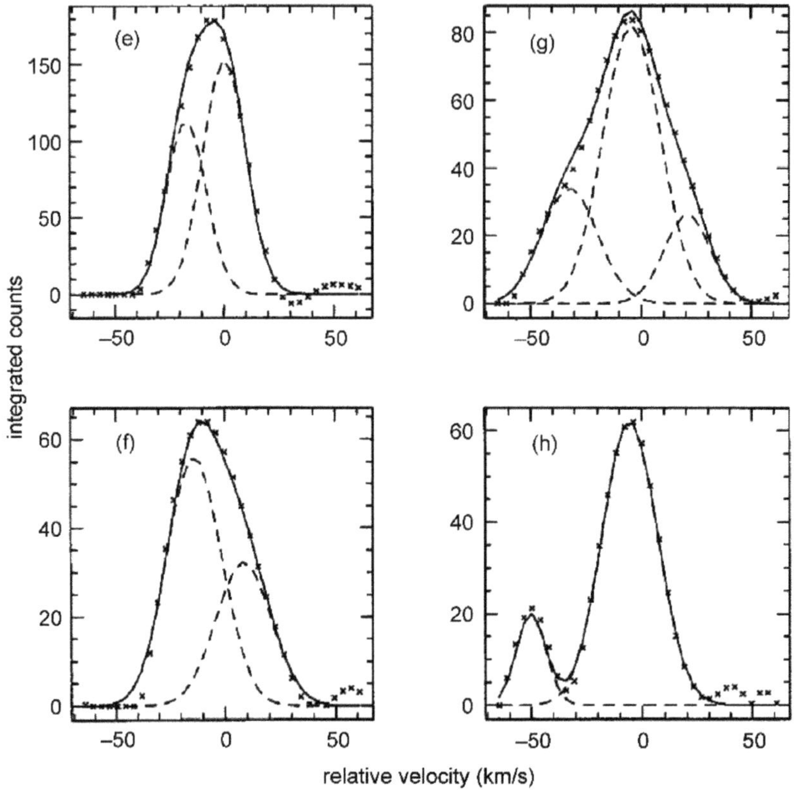

Figure 6.9: Decomposition of observed emission line profiles into multi-Gaussian components. The observations were made on the HII region M8 in Lagoon Nebula in [NII] line at 658.4 nm using a IFPS at Mt Abu 1.2 m telescope (from Chakraborty and Anandarao, 1997).

We can, in principle, solve for any functional form including Planck function while modelling spectral energy distribution involving two or more components contributing to the observed photometric distribution.

CHAPTER 7

Epilogue

O Telescope, instrument of knowledge, more precious than sceptre

– Johannes Kepler

In the foregoing chapters we discussed types of astronomical sources that need observations to unravel underlying physical processes governing their existence and evolution. In this endless endeavour for unravelling the Cosmos, we need instrumentation of high quality such as large telescopes and highly sensitive detectors and high-quality innovative optics to analyse the electromagnetic radiation (see e.g., Simons et al 2005). We need both accuracy and precision to confidently claim discovery of new processes and features as well as the existence of new objects. While efforts are being made to further improve the imaging quality and resolution in ground-based telescopes and instruments, the space-based telescopes would continue to play a key role even in the wavelength regions accessible from the ground. Already there have been several space-based explorations in wavelength regions that are partially or completely blocked by terrestrial atmosphere. The space-based explorations need supplementary and complementary support from ground-based telescopes with equally sophisticated back-end instrumentation. In the Appendix A5, a list of space observatories is given. We may access the data at the website https://irsa.ipac. caltech.edu. We may also access basic data and literature at SIMBAD Astronomical Database - CDS (Strasbourg). Useful data on atmospheric effects, photometric and spectroscopic standards, and optical and infrared spectral line data can be accessed from the web portals of ground-based observatories, such as Gemini (https://www.gemini.edu), CFHT(Canada France Hawai'i Telescope - Canadian Astronomy Data Centre).

The photometric observations of the future should have accuracies better than milli-magnitude approaching micro-magnitude. Spectroscopy needs both very high resolution and time-resolved multiplexing capabilities in both spatial and spectral domains. On large telescopes the integral field spectroscopic instruments would play a dominant role. In space-based telescopes, the slit-less spectroscopy is rapidly gaining ground for survey spectroscopy. We can find details on these in the web portals corresponding to the space agencies such as NASA, ESA and others.

In the past decade or two, the field of astronomy and astrophysics has seen tremendous advancements in terms of opening new spectral windows and improvements in the existing ones. Two new windows are being explored: (i) in the detection of neutrino detection (see Ahlers and Halzen 2017 and the references therein), and (ii) in the detection of gravitational waves in space-time (Agazie et al 2023 and the references therein). In the windows that are already being accessed, from high energy end to the radio region, the frontiers of spectral and spatial dimensions are being extended, exploiting emerging new technologies. Many new ground-based facilities are being created, and new space observatories are being planned. Some of the existing and future space missions are listed in Appendix A5. The future of astronomy lies in observations of very high spectral and spatial resolutions to unravel secrets of exoplanets, star formation, extragalactic astronomy at very high redshifts, mergers of highly compact stellar sources and observational cosmology that deals with origin of the Universe. What we have attempted to describe in the foregoing pages are some glimpses of cosmic phenomena and the measurement of light from such sources in the optical and near-infrared wavelength regions.

Very Large Telescopes: Due to technological limitations, monolithic telescope mirrors larger than ~ 8 m are untenable (the Japanese 8.2 m telescope in Mauna Kea, Hawaii is one of the largest monolithic telescopes in the world). The 10 m telescope in Hawaii, called the Keck telescope, is a mosaic of several smaller segments that make up the intended size. Based on such innovative technologies, New Age telescopes of apertures larger than the Keck 10 m telescope are being planned: (1) the 30 m segmented telescope (TMT International Observatory) at Mauna Kea, Hawaii is an international consortium including India. It will have 492 hexagonal segments, each measuring 1.44 m corner to corner; (2) the extremely large telescope or ELT by European Southern Observatory Consortium (https://elt.eso.org/) will be the world's largest telescope with an aperture of 39.3 m (a mosaic of 798 hexagonal segments each of which is 1.45 m in size) and will be situated at Atacama desert, Chile atop the Andes mountains; and, (3) the Giant Magellan Telescope (GMT) is being planned with 7 segments of 8.4 m sized mirrors arranged to give a 25.4 m effective size to be situated in Las Companas, Chile. All these giant telescopes will have photometric and spectroscopic capabilities in the visible and infrared regions with highly sophisticated instruments. Diffraction-limited observations will be achieved using the advanced Adaptive Optics (AO) technology.

Milli-arcsec Resolution and Large Base-line Interferometry: There are many astrophysical observations that require high angular resolutions, and the astronomers have been striving to achieve milli-arcsec (mas) or sub-milli-arcsec (sub-mas) resolutions. Notably, we need such high angular resolution, in both optical and infrared spectral regions, for direct imaging of exoplanets, for resolving the debris-disks in protostars, for resolving the central regions of the Active Galactic Nuclei, to give a few examples. We have shown in Chapter 2 that the angular resolution of a ground-based telescope is limited by the atmospheric effects, called the "seeing limit" that is typically 1 arcsec and significantly exceeds the diffraction limit for a telescope of aperture ≥ 2 m. Sophisticated techniques such as the AO can achieve nearly the diffraction-limited resolution for large ground-based telescopes. For example, the diffraction-limited angular resolution for Keck 10 m telescope at optical wavelengths (say, at $\lambda = 500$ nm) is 0.01 arcsec or 10 mas. The proposed/upcoming large telescopes give mas resolutions at their diffraction limits (with AO) – the TMT can resolve up to 3 mas while the ELT can reach 2.5 mas. Can we achieve such levels from multiple smaller apertures but separated by long baselines?

Radio astronomers have long mastered the technique of long-baseline interferometry or aperture synthesis to realise unprecedented angular resolution (since the radio wavelengths are much longer than those of optical region, we need unwieldly large apertures to achieve angular resolutions comparable to those with optical telescopes, even at the seeing-limited (see e.g., Thompson et al 2001). With these innovative technologies, high angular resolutions have been made possible in sub-mm, mm, cm, and meter wavelengths. However, to use the interferometric technique at the optical wavelengths, we need stability at higher precision than for the case of radio, for baselines, say of 100 m. Let us understand the basic principles of interferometry.

Optical interferometry is a technique that combines light from multiple smaller telescopes to create an effective aperture much larger than any individual telescope. It uses the principle of two-beam interference to achieve high angular resolution (see the section 5.4 on Michelson interferometry). For an illustration, let us consider two identical telescope apertures separated by a baseline of B, as shown in Figure 7.1. After suitably compensating for the path delay in the two beams (using "delay lines"), the optical radiation beams from the two telescopes are then combined at a beam combination laboratory where interference fringes are produced. These interference fringes are described in detail in section 5.4. The fringe visibility (see section 5.4) corresponds to the real part of the Fourier transform of the unresolved image, according to the van Cittert-Zernike theorem (see e.g., Born & Wolf 1999; van Belle 2014). Thus, we need to detect the visibility fringes that correspond to the path difference introduced by the two opposite extremes of a star being observed, over the zero-path difference. For telescopes not corrected for atmospheric turbulence effects, we need to keep exposures under the atmospheric coherence timescales of a few milli-seconds. This implies, therefore, that the technique of interferometry needs to be complemented by AO, for large apertures.

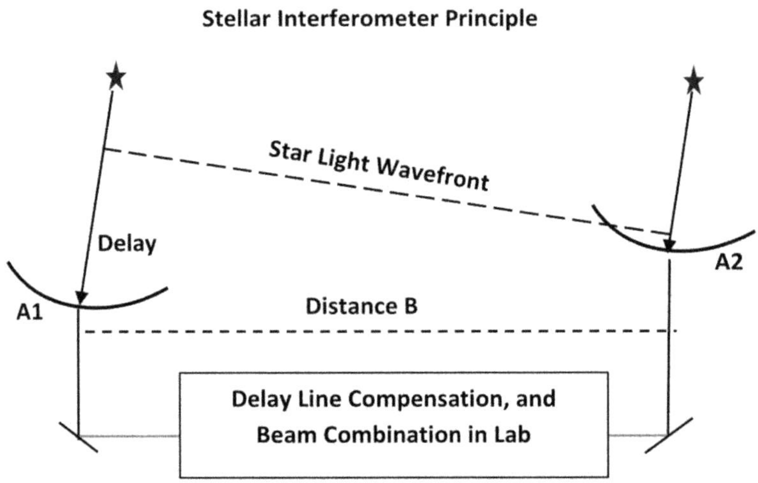

Figure 7.1: Schematic of a two-aperture Stellar Interferometer. The apertures A1 and A2 are separated by a distance *B*. Inside a Laboratory the beams are combined after delay line compensation. The wavefronts from star light are shown by arrows.

The angular resolution θ_i of an interferometer is characterized by

$$\theta_i = \frac{\lambda}{2B} \quad radians$$

where λ and B are the observing wavelength and baseline respectively.

The Very Large Telescope Interferometer (VLTI) and the Center for High Angular Resolution Astronomy (CHARA) array are the two prominent optical interferometers used in astronomy to achieve extremely high-resolution imaging. Both systems combine light from multiple telescopes to simulate a much larger aperture, enabling observations with fine angular resolution at milli-arcsec or sub-milli-arcsec levels. The VLTI is part of the European Southern Observatory's (ESO) Very Large Telescope (VLT) located in the Atacama Desert, Chile. It combines light from up to four 8.2-meter Unit Telescopes (UTs) or four 1.8-meter Auxiliary Telescopes (ATs) to achieve milliarcsecond resolution. Details on VLTI may be found in

Woillez et al (2024) and the references therein. For details on CHARA we may refer to Anugu et al (2020) and the references therein. Presently, the interferometric technique is successfully being used in the visible and infrared regions at these facilities.

Clearly, we can look forward to many more exciting events and discoveries in the coming years.

APPENDICES

A1. Table of Physical Constants and Astronomical Quantities

Parameter/Constant	Symbol	Value (MKS Units)
Velocity of light in vacuum	c	2.989×10^{8} m/s
Boltzmann constant	k	1.3806×10^{-23} J K^{-1}
Planck's constant	h	6.6261×10^{-34} J s^{-1}
Elementary charge	e	1.6022×10^{-19} C
Gravitational constant	G	6.6726×10^{-11} m^{3} kg^{-1} s^{-2}
Electron volt	eV	1.6022×10^{-19} J
Electron Mass	m_e	9.1094×10^{-31} kg
Proton Rest Mass	m_p	1.6726×10^{-27} kg
Stefan-Boltzmann constant	σ	5.6705×10^{-8} W m^{-2} K^{-4}
Photon energy	$h\nu$	$1.386 \times 10^{-15} / \lambda$ (Å) Joule
Avogadro number	N_A	6.0221×10^{23} $mole^{-1}$
Solar Mass	$M_\odot$	1.989×10^{30} kg
Solar Radius	$R_\odot$	6.960×10^{8} m
Solar Luminosity	$L_\odot$	3.8300×10^{26} W
Sun's effective temperature	$T_\odot$	5770 K
Sun's absolute magnitude	$M_{V\odot}$	+4.83
Sun's apparent magnitude	$m_{V\odot}$	- 26.74
Angular size of Sun	$\theta_\odot$	0.5 degree
Solar constant (irradiance)	$W_\odot$	1.368×10^{3} W m^{-2}
Earth's Mass	$M_\oplus$	5.9736×10^{24} kg
Earth's Radius	$R_\oplus$	6.3781×10^{6} m
Astronomical Unit	AU	1.4960×10^{11} m
Light Year	Ly	9.4607×10^{15} m
Parsec	pc	9.4607×10^{15} m
Sidereal Day	-	23h 56m 4.09s
Sidereal Year	-	365.2563 days
Rayleigh	R	10^{10} photons m^{-2} s^{-1}

A2. Probable Errors of a Function of Many Variables

Let us define y as a function of n variables,

$$y = f(x_1, x_2, x_3, \dots \dots x_n) \tag{A2.1}$$

If the probable errors in the measured quantities x are p then the probable error in y is given by

$$p_y = \left[p_1^2 \left(\frac{\partial y}{\partial x_1} \right)^2 + p_2^2 \left(\frac{\partial y}{\partial x_2} \right)^2 + \dots \dots \dots \right]^{1/2} \tag{A2.2}$$

where p_1, p_2 ...... are probable errors in x_1, x_2, ... respectively.

A3. Table of Vega (α Lyrae: A0V star) Standard Fluxes (from Glass 2000; Tokunaga 2000)

Band	λ (μm)	Δλ (μm)	F_λ (W/m²/μm)	F_ν (Jy) (W/m²/Hz)	N_p (p/s/m²/μm)
U	0.366	0.065	4.175 (-8)	1790	7.69 (10)
B	0.438	0.089	6.320 (-8)	4063	1.39 (10)
V	0.545	0.084	3.631 (-8)	3636	9.61 (10)
R_C	0.641	0.158	2.177 (-8)	3064	7.02 (10)
I_C	0.798	0.154	1.126 (-8)	2416	4.48 (10)
J	1.22	0.26	3.150 (-9)	1590	1.93 (10)
H	1.654	0.29	1.140 (-9)	1020	9.35 (10)
K	2.19	0.41	3.960 (-10)	640	4.36 (9)
L	3.45	0.57	7.10 (-11)	290	1.23 (9)
L'	3.76	0.65	5.26(-11)	248	9.94 (8)
M	4.8	0.45	2.20 (-11)	170	5.27 (8)
N	10.6	5.19	9.60 (-13)	36	5.17 (7)
Q	21	7.8	6.40 (-14)	9.4	6.75 (6)

A4. Table of Colours for Luminosity Classes I, III and V (Tokunaga 2000)

Sp.Ty.	Log T_{eff}	L	M_{bol}	U - B	B - V	V - K	J - H	H - K
O3 I	4.675	2.2 (6)	- 11.2	- 1.175	- 0.330	--	--	--
B0 I	4.415	2.6 (5)	- 8.9	- 1.070	- 0.301	- 0.69	-0.04	-0.10
A0 I	3.988	3.5 (4)	- 6.7	- 0.005	0.00	0.19	0.09	-0.02
F0 I	3.886	3.2 (4)	- 6.6	0.053	0.295	0.64	0.15	0.04
G0 I	3.744	3.0 (4)	- 6.6	0.057	0.595	1.44	0.33	0.08
K0 I	3.645	2.9 (4)	- 6.5	0.436	0.816	2.15	0.46	0.12
M0 I	3.562	4.1 (4)	- 6.9	1.204	1.420	3.80	0.73	0.18
O3 III	4.618	2.1 (6)	- 11.2	-1.22	--	--	--	--
B0 III	4.463	1.1 (5)	- 8.0	-1.08	--	--	--	--
A0III	4.005	106	0.30	--	-0.02	--	--	--
F0 III	3.854	43	1.4	--	0.30	--	--	--
G0 III	3.767	34	0.8	--	0.58	1.75	0.37	0.07
K0 III	3.676	60	0.2	--	0.81	2.31	0.54	0.10
M0 III	3.580	330	- 1.6	--	--	3.85	0.83	0.19
O3 V	4.720	1.4 (6)	- 10.7	-1.175	-0.330	--	--	--
B0 V	4.486	5.2 (4)	- 7.1	-1.070	-0.301	- 0.83	-0.159	-0.067
A0 V	3.978	54	0.3	-0.005	0.000	0.0	-0.032	0.028
F0 V	3.857	6.5	2.6	0.053	0.295	0.70	0.096	0.045
G0 V	3.780	1.5	4.2	0.057	0.595	1.41	0.262	0.067
K0 V	3.720	0.42	5.6	0.436	0.816	1.96	0.387	0.091
M0 V	3.585	7.7 (-2)	7.4	1.184	1.445	3.65	0.627	0.190

A5. Space Probes Past/Present/Future by NASA/ESA/Japan/India

Space probes	Space agency	Launch year	Photometry	Spectroscopy
International Ultraviolet Explorer (IUE)	ESA, NASA, SERC	1978	115-197 nm, 175-330 nm, 175-330 nm	Echelle (115 nm to 198 nm and 180 nm to 320 nm) Resolving power: 10^4
Infrared Astronomical Satellite (IRAS)	NASA, SERC, NIVR	1983	12, 25, 60 and 100 μm bands	7.5 to 23 μm Resolving power: 20
Hipparcos	ESA	1989	Broad band, and Astrometry to milli-arcsec precision	---
Hubble Space Telescope	NASA & ESA	1990	Ultraviolet, Optical, and Near-Infrared channels with Broad, Medium, and Narrowband filters	UV, Visible and near-IR 115–27000 nm
Infrared Space Observatory (ISO)	ESA	1995	2.5–240 μm	2.4–45 μm 43–197 μm
Midcourse Space Experiment (MSX)	Ballistic Missile Defence Organization	1996	4 bands in 8.28 -21.3 μm region	Fourier Transform Spectrometer (FTS): 2.5 to 28 μm
Spitzer Space Telescope	NASA	2003	3.6, 4.5, 5.8 and 8.0 μm 24, 70 and 160 μm	Infrared Spectrograph (IRS): 5.2–38 μm
Galaxy Evolution Explorer (GALEX)	NASA	2003	FUV (155 nm) NUV (220 nm)	---
Herschel Space Observatory	ESA & NASA	2009	70 - 250, and 250 - 500 μm	57 - 210 μm 194 - 672 μm

Contd....

Space probes	Space agency	Launch year	Photometry	Spectroscopy
Wide-field Infrared Survey Explorer (WISE)	NASA	2009	3.4, 4.6, 12 and 22 μm bands	--
Kepler	NASA	2009	420 to 900 nm Exoplanet transit detection	--
Akari (Astro-F)	Japan Aerospace Exploration Agency (JAXA)	2011	2.8 - 18 μm 65 - 160 μm)	2.5 to 18 μm 50 to 180 μm
GAIA	ESA	2013	Blue/Red band Photometry, Astrometry	Radial velocity spectrometer
AstroSat	Indian Space Research Organisation (ISRO)	2015	UVIT - FUV (130-180 nm), NUV (180-300 nm) & VIS (320-530 nm) Soft X-ray (0.3-8 keV), Hard X-ray (10 - 150 keV)	Grism spectrograph
Transiting Exoplanet Survey Satellite (TESS)	NASA	2018	Broadband Wide-angle survey with 4 CCD cameras	----
James Webb Space Telescope (JWST)	NASA/ESA/Canadian Space Agency (CSA)	2021	Near-Infrared 0.9 - 4.44 μm Mid-Infrared 5.6 - 21 μm	0.6 μm – 5.0 μm 5.0 μm – 28 μm R ~ 100 to 2000 Near Infrared Imager and Slitless Spectrograph 0.8 μm – 5.0 μm at R ~ 1000

Contd....

Space probes	Space agency	Launch year	Photometry	Spectroscopy
Aditya-L1	ISRO	2023	Visible Emission Line Coronagraph (VELC) Imaging solar corona at 500 nm Solar Ultraviolet Imaging Telescope (SUIT) in 214 – 340 nm Solar wind Particle Experiment (ASPEX) Plasma Analyser (PAPA) Advanced Tri-axial High Resolution Digital Magnetometers (MAG)	VELC Multi-slit spectroscopy at 530.3 nm [Fe XIV], 789.2 nm [Fe XI], and 1074.7 nm [Fe XIII] Solar Low Energy X-ray Spectrometer (SoLEXS) 1 to 30 keV High Energy L1 Orbiting X-ray Spectrometer (HEL1OS) 1 to 150 keV
PROBA-3	ESA	2024	Polarimetric and Imaging of the Corona of the Sun 400 to 700 nm 700 to 1000 nm	----
Spectro-Photometer for History of Universe at the Epoch of Reionization and Ices Exploration (SPHEREx)	NASA	2025	90 spectral bands in visible and near-IR region	0.75 – 5 μm with resolving power 35-130

ESA – European Space Agency

NASA – National Aeronautics and Space Administration

SERC – Science and Engineering Research Council, UK

NIVR – Netherlands Agency for Aerospace Programmes

A6. Examples

1. A star has an apparent visual magnitude of $m_V = 12.5$ and an absolute magnitude of $M_V = 7.2$. Calculate the distance to the star in parsecs.
2. A star has the H-alpha line in its spectrum shifted by 0.2 nm. What is the velocity of the star along the line of sight?
3. What would be the velocity of a nebula if the [OIII] line at 500.7 nm it emits is blue shifted by 0.5 nm?
4. A post-AGB star has a temperature of 20,000 K and a luminosity of 0.4 $L_\odot$. Calculate its radius.
5. A star is classified as a spectral type B2V. Using the mass-luminosity relation for main-sequence stars, estimate its mass.
6. Given magnitudes/fluxes for a known star (m_0) and an unknown star (m_1), find exposure times to achieve S/N ratios of 5 and 10.
7. Estimate the central core temperature of the Sun from equations of stellar interiors.
8. Explain how elements heavier than helium are formed in stars and identify the process responsible for forming elements beyond iron.
9. A star's luminosity on the main sequence scales as $L \alpha M^{3.5}$. If a star is of mass 3 $M_\odot$, what is its luminosity compared to the Sun's?
10. Analyse the difference between a galaxy and a globular cluster.
11. How does one demarcate regions of different densities from collisional-excited molecular transitions in gaseous clouds?
12. Discuss evolution of binary stars vs evolution of single star.
13. A spectral line from a binary star has a splitting of 1.5×10^{-4} of its wavelength. If the period of the binary is 15 days, find the distance between the stars if their masses are equal.
14. Planet detection may be done by transit photometry and precision high-resolution spectroscopy. Discuss the physical parameters that one can obtain from each method?
15. If an exoplanet has a radius $R_p = 1.2$ times the Jupiter's radius and is transiting its parent star that has a radius $R* = 2\ R_\odot$, calculate the relative transit depth.
16. How much would be the velocity fluctuation experienced by the Sun when it rotates around the centre of mass of the Sun-Earth system?
17. A star cluster contains stars with masses ranging from 0.2 $M_\odot$ to 50 $M_\odot$. Assuming the most massive stars have completed the main sequence phase, estimate the age of the cluster.

18. What is an inverse P Cygnii line?
19. Compute plate-scales for a telescope of diameter (aperture) of 3 m at prime and Cassegrain foci of *f*/8 and *f*/15 respectively. Find diffraction limit at 500 nm for the telescope.
20. A telescope has an effective focal length of 15600 mm. Calculate its plate scale in arcseconds per millimeter.
21. A telescope observes a star through the atmosphere, and the wavefront distortion has an RMS error of 200 nm. If the observing wavelength is 550 nm, calculate the Strehl ratio.
22. Analyse the difference between the active and the adaptive optics for a telescope.
23. Compare the merits and demerits of an imaging photon-counting detector, a CCD and a CMOS array.
24. What are principal differences in a CCD and an infrared array detector?
25. A gaseous nebula is at an equilibrium temperature of 10000 K. Find the temperature broadening for a given transition of Hα and [OIII] at 656.3 nm and 500.7 nm respectively.
26. Assume that the Sun's magnetic field strength in sunspots is approximately $B = 1000$ Gauss. Calculate the wavelength separation between the split components for the Na D line at 589 nm due to the presence of the magnetic field.
27. Find critical densities for atomic/ionic [OIII] forbidden transitions in the visible region.
28. Find the magnitude of magnetic field from the Zeeman splitting separation of 0.01 nm in a HI line of 656.3 nm.
29. The Hydrogen Balmer alpha line from a star shows a difference in the wavelength of 10 pm at the two opposite edges of the equatorial diameter. What is the rotational velocity of the star?
30. Draw and distinguish typical shapes of curves of for temperature and rotational broadening.
31. Find turbulent broadening given line widths at two wavelengths and the gas temperature.
32. Given magnitude of a star at the zenith, calculate the photometric corrections at different zenith angles. Find optimum exposure times for this (note that longer observations would average out the corrections).
33. What do we learn from Colour-Colour and Colour-Magnitude diagrams?

34. Design a grating spectrograph at a given spectral resolution of 1 nm/px, and spatial resolution of 1 arcsec using an area detector. For visible and infrared wavelengths.
35. Discuss how to find an unknown wavelength using a Michelson Spectrometer and an FPS.
36. Discuss advantages and issues with a Fourier Transform Spectrometer.
37. What is apodization?
38. Find effective finesse in FPS for given plate defects and reflectivity of 95%.
39. Find the minimum number of lines/grooves of diffraction grating required to resolve spectral lines separated by delta lambda in 1st order.
40. How much is the linear separation on an imaging detector of two lines 650 nm and 655 nm resolved by a diffraction grating of 2500 groves/cm, in the 2nd order? The focal length of the imaging optics is 120 cm.
41. A diffraction grating has 3000 lines/mm. Two wavelengths, 300 nm and 320 nm are incident normally. Calculate the angular separation ($\Delta\theta$) between the two wavelengths in the first-order spectrum.
42. Calculate the free spectral range (FSR) at the wavelength of HI Balmer α line of an airgap Fabry-Pérot etalon with a plate separation of $d = 1$ mm.
43. A Fabry-Pérot etalon used in a spectrometer has its plates slightly misaligned. Describe the effect of this misalignment on a spectral line.
44. If the temperature of the FP etalon increases resulting in thermal expansion of the plates leading to change in spacing. How will this change the FSR and the position of the transmission peaks?
45. Define and compare the product of Luminosity and Resolving Power for various spectrometers.
46. Discuss the advantages of various filtering techniques for image enhancement.
47. Discuss the need for deconvolution in spectroscopy.

REFERENCES

Abe, H., Abe, K., Abe, S., Aguasca-Cabot, S., et al, 2023, Astroph. J., 956, 80A

Agazie, G., et al 2023, Astrophys. J., 951, L8

Ahlers, M., & Halzen, F., 2017, Prog. Theor. Exp. Phys., 12A105

Aitken, D.K., 1988, Vistas in Astronomy, 31, 755

Akhlaghi, M., 2021, GNU Astronomy Utilities, v.0.16, 2021

Akhlaghi, M., & Ichikawa, T., 2015, Astroph. J. Suppl. Ser., 220:1

Albert L., et al, 2023, Publ. Astron. Soc. Pacific, 135, 075001

Allamandola, L.J., Tielens, A.G.G.M., & Barker, J.R., 1987, Astrophys. J. Suppl. Series, 71, 733

Allington-Smith, J., 2006, New Astron. Rev., 50, 244

Allington-Smith, J., 2000, Astronomical Spectroscopy with a diffraction grating, Unpublished document accessible on internet; See also the document "Grating Physics", available in Newport web pages (Spectroscopy Instruments).

Anandarao, B.G., & Banerjee, D.P.K., 1988, Astron. & Astroph., 202, 215

Anandarao, B.G., & Rao, S.R., 1986a, Bul. Astron. Soc. India, 14, 34

Anandarao, B.G., & Rao, S.R., 1986b, Bul. Astron. Soc. India, 14, 25

Anandarao, B.G., Nanda Kumar, M.S., Jog, N.S., & Patel, R.T., Bul. Astron. Soc. India, 28, 687, 2000

Anandarao, B.G., Richardson, E.H., Chakraborty, A., & Epps, H., 2008, Proc. SPIE, 70142Y, 7014

Anandarao, B.G., Venkataraman, V., Ghosh, S.K., Ojha, D.K., & Nanda Kumar, M.S., 2008, MNRAS, 390, 1185

Anugu et al (2020), AJ, 160, 158

Arsenault R., & Roy J.-R., 1984, PASP, 96, 496

Aschwanden, M., 2019, New Millennium Solar Physics, Astrophysics and Space Science Library, vol. 458, Springer.

Atherton, P.D., Reay, N.K., Ring, J., & Hicks, T.R., 1981, Opt. Engg., 20, 806

Atherton, P.D., 1995, 3D Optical Spectroscopic Methods in Astronomy, Eds. G. Compte, M.Marcelin, ASP Conf. Ser., Vol.71, p. 50.

Atherton, P.D., Taylor, K., Pike, C.D., Harmer, C.F.W., Parker, R.N., & Hooke, R.N., 1982, Mon. Not. Roy. Astron. Soc., 201, 661

Bacon, R., et al 1995, Astron. & Astroph. Suppl. Ser., 113, 347

Bacon, R., & Monnet, G., Optical 3D-Spectroscopy for Astronomy, Wiley-VCH Verlag GmbH, 2017

Bahcall, J., et al 1991, as cited in Rieke, 2007

Balick, B., & Frank, A., 2002, Annu. Rev. Astron. Astrophys., 40,439

Banerjee, D.P.K., & Anandarao, B.G., 1991, Astron. & Astroph., 250, 165

Banerjee, D.P.K., Anandarao, B.G., Desai, J.N., et al, 1987, Astroph. Space. Sci., 139,327

Banerjee, D.P.K., Anandarao, B.G., Jain, S.K., & Mallik, D.C.V., 1990, Astron. & Astroph., 240, 137

Banerjee, D.P.K., et al 1999, Bull. Astr. Soc. India, 27, 425

Barbieri, C., & Bertini, I., Fundamentals of Astronomy, CRC Press, 2020

Beckers, J.M., 1993, Annu. Rev. Astron. Astrophys., 31, 13-62

Berger, E., 2014, Annu. Rev. Astron. Astrophys., 52, 43

Bershady, M.A., Andersen, D.R., Harker, J., Ramsey, L.W., & Verheijen, M.A.W., 2004, Publ. Astron. Soc. Pacific, 116, 565

Bershady, M.A., et al, 2005, Astroph. J. Suppl. Ser., 156, 311

Bessell, M., 2005, Annu. Rev. Astron. Astrophys., 43, 293

Bessell, M., & Brett, J.M., 1988, Publ. Astron. Soc. Pacific, 100, 1134

Binney, J., & Tremaine, S., Galactic Dynamics, Princeton University Press, Princeton NJ, 2008

Black, J.H., & van Dishoeck, E.F., 1987, Astroph. J., 322, 412

Bland, J., & Tully, B., 1989, Astron. J., 98,723

Blasberger, A., et al., 2017, Astroph. J., 836, 173

Bohlin, R.C., Harrington, J.P., Stecher, T.P., 1978, Astroph. J., 219, 575

Bohren, C.F., & Huffman, D.R., Absorption and Scattering of Light by Small Particles, Wiley Science Series, New York, 1983

Boksenberg, A., & Burgess, D.E., Astronomical Observations with TV type Sensors, Univ. British Columbia, Vancouver, 1973

Born M., and Wolf, E., Principles of Optics, Cambridge, 1999

Boyd T.J.M., & Sanderson, J.J., 2003, in The Physics of Plasmas, Cambridge University Press

Bradt, H., Astrophysics – Processes, Cambridge Univ. Press, 2008

Bransden B.H., & Joachain, C.J., 2003, Physics of Atoms and Molecules, Prentice-Hall

Carroll, B.W., & Ostlie, D.A., An Introduction to Modern Astrophysics, Addison & Wesley, 2007; 2nd Edition, Cambridge Univ. Press, 2017

Carruthers, G., 1970, Astroph. J. Lett., 161, L81

Carter A.L., et al., 2023, Astroph. J. Lett., 951, L20

Chakraborty, A., & Anandarao, B.G., 1997, Astron. J., 114, 1576

Chakraborty, A., et al, 2014, Publ. Astron. Soc. Pacific, 126, 133

Chakraborty, P., Das, H.K., & Tandon, S.K., 2005, Bull. Astron. Soc. India, 33, 513

Chen, G-Y., & Wang, Y-F., 2017, IOP Conf. Ser., Earth & Environmental Science, 69, 012099

Chomiuk, L., Metzger, B.D., & Shen, K.J., 2021, Annu. Rev. Astron. Astrophys., 59, 391-444

Churchwell, E., 2002, Annu. Rev. Astron. Astrophys., 45, 481-563

Connes, P., & Michel, G., 1975, Appl. Optics, 14, 2067

Desai, J.N., 1998, Bull. Astron. Soc. India, 26, 241

Dewangan, L.K., & Anandarao, B.G., 2011, Mon. Not. Roy. Astron. Soc., 414, 1526

Draine, B.T., & Bertoldi, F., 1996, Astroph. J., 468, 269

Draine BT and Flatau PJ, 1994, J. Opt. Soc. Amer. A, 11, 1491;

Draine BT and Flatau PJ, 2003, User Guide, DDSCAT 6.0, http://arxiv.org/abs/astro-ph/0309069

Draine, B.T., Roberge, W.G., & Dalgarno, A., 1983, Astroph. J., 264, 485

Dyson, J.E., & Williams, D.A., Physics of the Interstellar Medium, Manchester Univ. Press, 1980

Elmegreen, B.G., 1997, Astroph. J, 477, 196

Emerson, D., Interpreting Astronomical Spectra, John Wiley & Sons, 1996

Esakkiappan, A., Rai, A., Ganesh, S., & Vadawale, S., 2019, J. Astron. Tel. Instr. Syst., 5 (3), id. 035006

Fabry, C., & Perot, A., Ann. Chim. Phys., (7), 16 (1899), 115

Fellget, P., 1951, Thesis, Cambridge University (as cited in Roesler 1974)

Fitzpatrick, E.L., 1999, PASP, vol. 111, 63-75

Fitzpatrick, E.L., 2004, Astrophysics of Dust, ASP Conf. Ser., Vol. 309, Eds. AN Witt, GC Clayton & BT Draine.

Glass, I.S., Handbook of Infrared Astronomy, Cambridge, 1999

Gray, D. F., The Observation and Analysis of Stellar Photospheres, Cambridge University Press, Cambridge, 2021

Greene et al, 2017, J. Astron. Telesc. Instrum. Syst., 3(3), 035001

Grundahl, F., Stetson, P.B., & Anderson, M.I., 2002, Astron. Astroph., 395, 481

Guyon, O., 2018, Annu. Rev. Astron. Astrophys., 56, 315-355

Hann, James, 1866, The Elements of Spherical Trigonometry, pres2014-0682-02.pdf (uchicago.edu)

Harlow, S.B., 1992, Astron. Soc. Pacific Conference Series, vol. 23, 105

Hartley, K.E., & Wilson, R.W., 2023, Mon. Not. Roy. Astron. Soc., 526, 3482

Heidt, J., Astronomy in the Near-IR: Observing Strategies and Data Reduction Techniques, Springer, Berlin/Heidelberg, 2022

Hernandez, O., et al., 2008, Publ. Astron. Soc. Pacific, 120, 665

Herwig, F., 2005, Annu. Rev. Astron. Astrophys., 43, 435

Hoare, M.G., 2007, Protostars and Planets V, B. Reipurth, D. Jewitt, and K. Keil (eds.), University of Arizona Press, Tucson, 951 pp., 2007., p.181-196

Howell, S. B., Sobeck, C., Haas, M., Still, M., Barclay, T., Mullally, F., et al., 2014, Publ. Astron. Soc. Pac. 126, 398

Hoffman, A.W., Love, P.J., & Rosbeck, J.P., 2004, Proc. SPIE, 5167, 194

Hubble, E.P., 1929, Proc. Nat. Acad. Sci., 15, 168

Hunt, L.K., et al., 1998, Astron. J., 115, 2594

Iben, I., 1974, Annu. Rev. Astron. & Astrophys., 12, 215

Iben, I., 1985, Qtr. J. Roy. Astron. Soc., 26, 1

Ivezic, Z. & Elitzur, M., 1997, MNRAS, 287, 799

Ivezic, Z. Nenkova, M. & Elitzur, M., 1999, User Manual for Dusty, accessible at http://www.pa.uky.edu/moshe/dusty

Jacquinot, P. J., 1954, Opt. Soc. Amer., 44, 761

Jacquinot, P.J., 1960, Rep. Prog. Phys., 23, 267

Janka, H-T., 2012, Annu. Rev. Nucl. Part. Sci., 62, 407

Jenkins, C. R., 1987, Mon. Not. Roy. Astron. Soc., 226, 341

Jenkins, F.A., and White, H.E., Fundamentals of Optics, McGraw-Hill, 1981

Johnson, H.L., 1966, Annu. Rev. Astron. Astrophys., 4, 193

Jorden, A.R., & Fordham, J.L.A., 1986, Qtr. J. Roy. Astron. Soc., 27, 166

Kaltenegger, L., 2017, Annu. Rev. Astron. Astrophys., 55, 433-485

Kirkpatrick, J. Davy, Annu. Rev. Astron. Astrophys., 43, 195, 2005

Kitchin, C.R., Astrophysical Techniques, IOP Publishing, 2003

Krisciunas, K., et al., 1987, P.A.S.P., 99, 887

Kumar A., et al, Ultraviolet to Gamma Ray, Proc. SPIE, vol 8443, art. Id. 84431N, 2012

Kulshrestha, A., Deshpande, M.R., & Joshi, U.C., 1987, Astron. Astrophys. Suppl. Ser., 71, 565

Kurucz, R.L., 1979, Astroph. J. Suppl. Ser., 40, 1 (Stellar atmosphere models)

Kwok, S., The Origin and Evolution of Planetary Nebulae, Cambridge University Press, 2000

Kwok, S., 2004, Nature, 430, 985

Kwok, S., Star Dust, Springer, Berlin/Heidelberg, 2013

Lancon, A., & Rocca-Volmerange, B., 1992, Astron. Astroph. Suppl. Ser., 146, 217

Lang, K.R., Astrophysical Quantities, Springer-Verlag, 1974

Lang, K.R., NASA Cosmos 2010

Lang, K.R., Essential Astrophysics, Springer, Berlin/Heidelberg, 2013

Langer, N., 2012, Annu. Rev. Astron. & Astrophys., 50, 107

Labeyrie, A., 1970, Astron. & Astroph., 6, 85

Labeyrie, A., 1995, Astron. & Astroph. 298, 544

Lamers, H.J.G.L.M., & Cassinelli, J.P., Introduction to Stellar Winds, Cambridge Univ. Press, 1999

Latham, D.W., Detective Performance of Photographic Plates, in Methods of Experimental Physics, Astrophysics, Part A: Optical and Infrared, Ed. N. Carlton, Pergamon Press, 1974

Leger, A., & Puget, J.-L., 1984, Astron. & Astroph., 137, L5

Lèna, P., Observational Astrophysics, Springer-Verlag, 1979

Lesser, M., 2015, Publ. Astron. Soc. Pacific, 127, 1097

Longair, M.S., High Energy Astrophysics, Cambridge Univ. Press, 2011

Luhman, K.L., Engelbracht, C.W., & Luhman, M.L.,1998, Astroph. J., 499, 799

Maillard, J.-P., & Michel, G., 1982, in Instrumentation in Astronomy with Large Optical Telescopes, Astrophysics & Space Science Library, vol 92, Springer, Dordrecht

Majewski, S.R., Lecture Notes, ASTR 3130 on Photometry, U. Virginia, 2015

Mallat, S.G., in A Wavelet Tour of Signal Processing, Academic Press, 1999

Marchant, P., & Bodensteiner, J., 2024, Annu. Rev. Astron. Astrophys., 62, 21

Massey, P., Gronwall, C., & Pilachowski, C.A., 1990, Publ. Astron. Soc. Pacific, 102, 1046

Massey, P., & Gronwall, C., 1990, Astroph. J., 358, 344

Massey, P., et al, 1988, Astroph. J., 328, 315

Mathis J.S., 2000, in Allen's Astrophysical Quantities, 4th edition, edited by AN Cox

Mathis, J.S., Rumpl, W., Nordseick, K.H., 1977, Astroph. J., 217, 475

Mayor, M., & Queloz, D., 1995, Nature, 378, 355

McLean, I.S., Electronic Imaging in Astronomy, Springer, Berlin/Heidelberg, 2008

Meaburn, J., 1970, Astroph. Space. Sci., 9, 206, 197

Michelson, A.A., Phil. Mag., 31, 256, 1891; ibid, 34, 280, 1892

Muthu, C., & Anandarao, B.G., 2001, Astron. J., 121, 2106

Muthu, C., & Anandarao, B.G., 2003, Astron. J., 126, 2963

Muthu, C., Anandarao, B.G., & Pottasch, S.R., 2000, Astron. & Astroph., 355, 1098

Nanda Kumar, M.S., Anandarao, B.G., & Davis, C.J., 1999, Astron. Astrophys. Lett., 344, L9

Naylor, D.A. & Tahic, M.K., 2007, J. Opt. Soc. Amer. A, 24, 3644

Netzer, H., 2015, Annu. Rev. Astron. Astrophys., 53, 365

Oke, J.B., 1990, Astron. J., 99, 1621

Van Paradijs, J., Kouveliotou, C., & Wijers, R. A. M. J., 2000, Annu. Rev. Astron. Astrophys., 38, 379

Pedrotti, SJ, F.L., Pedrotti, L.M., & Pedrotti, L.S., Introduction to Optics, Cambridge University Press, 2017

Pendleton, Y.J., & Allamandola, L.J., 2002, Astroph. J. Suppl. Ser., 138, 75

Percy, J.R., Understanding Variable Stars, Cambridge Univ. Press, 2007

Petitdidier, M., 1983, Appl. Optics, 22, 1797

Phillips, D.L., 1962, J. Assoc. Comp. Math., 9, 84

Pecaut, M.J., & Mamajek, E.E., 2013, Astroph. J. Suppl. Ser., 208, 9

Pottasch, S.R., Planetary Nebulae, Planetary Nebulae, Astrophysics & Space Science Library, Vol. 107, D. Reidel, Dordrecht, 1983

Pottasch, S.R., 1993, in Infrared Astronomy, Mampaso, A., Prieto, M., & Sanchez F. (Eds), Cambridge U. Press, Cambridge, p.63

Prialnik, D., An Introduction to the Theory of Stellar Structure and Evolution, Cambridge Univ. Press, 2000

Puget, J.L., Leger, A., Boulanger, F., 1985, Astron. Astroph., 142, L19

Puget, J.-L., & Leger, A., 1989, Annu. Rev. Astron. Astrophys., 27, 161

Reipurth, B., & Bally, J., 2001, Annu. Rev. Astron. Astrophys., 39, 403

Ricker, G. R., Vanderspek, R., Winn, J., Seager, S., Berta-Thompson, Z., Levine, A., et al. (2016). The transiting exoplanet survey satellite. *Proc. SPIE* 9904:99042B. doi: 10.1117/12.2232071

Rieke, G.H., Annu. Rev. Astron. Astrophys., 2007, 45, 77

Roddier, F., Adaptive Optics in Astronomy, Cambridge Univ. Press, Cambridge UK, 1999

Roesler, F.L., Fabry-Perot Instruments in Astronomy, in Methods of Experimental Physics. Vol. 12 – Astrophysics, Part A Optical and Infrared Astronomy. Ed. N. Carleton, Academic Press, 1974

Rubin, V.C., Burstein, D., Ford, W.K. Jr., & Thonnard, N., Astroph. J., 289, 81, 1985

Ryter, C., as cited in Smith 2004

Sahu, K.C., Desai, J.N., & Jog, N.S., 1984, Proc. SPIE (Instrumentation in Astronomy V), 445, 33

Sanchez, S.F., Aceituno, J., Thiele, U., Perez-Ramirez, D., & Alves, J., 2007a, Publ. Astron. Soc. Pacific, 119, 1186

Sanchez, S.F., Cardiel, N., Verheijen, M.A.W., Pedraz, S., & Covone, G., 2007b, Mon. Not. Roy. Astron. Soc., 376,125

Schady, P., 2017, R. Soc. Open Sci., 4:170304; see also Schady, P., 2015, J. High Energy Astroph., 7, 56-63

Schaller, G., Schaerer, D., Meynet, G., & Maeder, A., 1992, Astron. & Astroph. Suppl. Ser., 96, 269

Schnopper, H.W., and Thompson, R.I., Ch. 11: Fourier Transform Spectroscopy, in Methods of Experimental Physics, Vol. 12 – Astrophysics, Part A Optical and Infrared Astronomy. Ed. N. Carleton, Academic Press, 1974

Schroeder, D.J., Astronomical Optics, Elsevier Science, 2000

Schulz, N.S., From Dust to Stars, Springer, 2005

Schuster, W.J., et al., 2004, Astron. & Astroph., 422, 527

Seema, P., 1996, Publ. Astron. Soc. Pacific, 108, 461

Seema, P., Anandarao, B.G., Banerjee, D.P.K., Jog, N.S., & Patel, R.T., 1992, Publ. Astron. Soc. Pacific, 104, 1091

Sellgren, K., Werner, M.W., & Dinerstein, H.L., 1983, Astroph. J. Lett., 271, L13

Serkowski, K., 1974, in Methods of Experimental Physics, Ed. N. Carlton, Vol. 12, Astrophysics, Acad. Press

Shu, F., The Physical Universe, University Science Books, 1982

Shull, J.M., & Beckwith, S., 1982, Annu. Rev. Astron. & Astrophys., 20, 163

Shure, M.A., Toomey, D.W., Rayner, J.T., Onaka, P.M., & Denault, A.J., 1994, Proc. SPIE, 2198, 614

Simons, D.A., Tokunaga, A.T., 2002, Publ. Astron. Soc. Pacific, 114, 169

Simons, D.A., Amico, P., Baade, D., et al, 2005, Exp. Astron., 19, 15

Smartt, S. J., 2009, Annu. Rev. Astron. & Astrophys., 47, 63

Smith, M.D., The Origin of Stars, Imperial College Press, 2004

Smith, R.C., Contemporary Physics, 2007

Spitzer, L., Physical Processes in Interstellar Medium, Wiley-Interscience Publ., Wiley, New York, 1978

Stahler, S.W., & Palla, F., The Formation of Stars, Wiley-VCH, 2004

Starck, J.-L., & Murtagh, F., Astronomical Image and Data Analysis, Springer, Heidelberg, 2006

Sternberg, A., & Dalgarno, A., 1989, Astroph. J., 338, 197

Stone, R.P.S., 1996, Astroph. J. Suppl. Ser., 107, 423

Stromgren, B., 1966, Annu. Rev. Astron. Astrophys., 4, 433

Subrahmanya, C., 1980, Astron. & Astroph., 89, 132

Sutton, E.C., Observational Astronomy, Cambridge Univ. Press, 2012

Taam, R. E., & Sandquist, E. L., 2000, Annu. Rev. Astron. Astrophys., 38, 113

Taylor, G.B., Lecture Notes Astro 421, Univ. New Mexico, Albuquerque, 2022

Tennyson, J., Astronomical Spectroscopy, Imperial College Press, 2005

Thompson, A.R., Moran, J. M., & Swenson, G.W., Interferometry and Synthesis in Radio Astronomy, 2nd Edition, Wiley-Interscience, New York, 2001

Tielens, A.G.G.M., The Physics and Chemistry of Interstellar Medium, Cambridge Univ. Press, 2005

Tielens, A.G.G.M., 2008, Interstellar Polycyclic Aromatic Hydrocarbon Molecules, Annu. Rev. Astron. Astrophys., 46, 289

Tielens, A.G.G.M., & Allamandola, L.J., in Physical Processes in Interstellar Clouds, Proc. NATO Adv. Study Inst., Reidel, p. 333, 1987

Tohline J. E., 2002, Annu. Rev. Astron. Astrophys., 40, 349

Tokunaga, A.T., Infrared Astronomy, Chapter 7, in Allen's Astrophysical Quantities, 4th Edition, Ed. A.N. Cox, Springer-Verlag, 2000

Tokunaga, A.T., Simons, D.A., & Vacca, W.D., 2002, Publ. Astron. Soc. Pacific, 114, 180

de Vaucouleurs, G., & Pence, W.D., 1980, Astroph. J., 242, 18

de Vaucouleurs, G., Pence, W.D., & Davoust, E., 1983, Astroph. J. Suppl. Ser., 53, 17

van Belle, G., in "New windows on massive stars: asteroseismology, interferometry, and spectro-polarimetry", Proceedings IAU Symposium No. 307, 2014, G. Meynet, C. Georgy, J. H. Groh & Ph. Stee, eds.

van der Veen, W. E. C. J., & Habing, H. J., 1988, Astron. & Astroph., 194, 125

Venkataraman, V., & Anandarao, B.G., 2008, Mon. Not. Roy. Astron. Soc., 385, 1076

Venkataraman, V., Anandarao, B.G., Janardhan, P., & Pandey, R., 2017, Mon. Not. Roy. Astron. Soc., 470, 1593

Vink, J. S., 2022, Annu. Rev. Astron. Astrophys., 60, 203

van der Wal, P.B., Anandarao, B.G., Rouan, D., Lèna, P., de Muizon, M., Wijnbergen, J., 1985, Astroph. Space. Sci., 117, 209

Walmsley, C.M., Natta, A., Oliva, E., & Testi, L., 2000, Astron. Astroph., 364, 301

Warner, B., Cataclysmic Variable Stars, Cambridge University Press, 1995

Weiler, K. W., & Sramek, R. A., 1988, Annu. Rev. Astron. Astrophys., 26, 295

Whittet, D.C.B., Dust in the Galactic Environment, Series in Astronomy & Astrophysics, Eds., M. Birkinshaw, M. Elvis & J. Silk, IOP Publishing Ltd., 2003

Willott, C.J., et al, 2022, Publ. Astron. Soc. Pacific, 134, 025002

Wing, R.F., 1969, PhD thesis, U. California, Berkeley (as cited in Bessell 2005)

van Winkel, H., 2003, Annu. Rev. Astron. Astrophys., 41, 391

Wiza, J.D., 1979, in Nuclear Instruments and Methods, vol. 162, pp. 587-601

Woan, G., The Cambridge Handbook of Physics Formulas, Cambridge Univ. Press, 2000

Woillez et al., (2024), A&A, 688, A190

Young, A.T., Photomultipliers, in Methods of Experimental Physics, Vol. 12: Part A: Optical and Infrared Astronomy, Ed. N. Carlton, Academic Press, p.1, 1974

Zhang, C.Y., & Kwok, S., 1991, Astron. & Astroph., 250, 179

Zinnecker, H., & Yorke, H.W., 2007, Annu. Rev. Astron. Astrophys., 45, 481

Zuckerman, B., 2001, Annu. Rev. Astron. Astrophys., 39, 549

INDEX

L

M

N

O

P

V

W

Z